Marie Odile Bristeau
Roland Glowinski
Jacques Periaux
Henri Viviand (Eds.)

Numerical Simulation of Compressible Navier-Stokes Flows

Notes on Numerical Fluid Mechanics
Volume 18

Marie Odile Bristeau
Roland Glowinski
Jacques Periaux
Henri Viviand (Eds.)

Numerical Simulation of Compressible Navier-Stokes Flows

A GAMM-Workshop

Friedr. Vieweg & Sohn Braunschweig / Wiesbaden

CIP-Kurztitelaufnahme der Deutschen Bibliothek

Numerical simulation of compressible Navier-Stokes flows: a GAMM workshop / Marie Odile Bristeau ... (eds.). — Braunschweig; Wiesbaden: Vieweg, 1987.
(Notes on numerical fluid mechanics; Vol. 18)

NE: Bristeau, Marie Odile [Hrsg.]; Gesellschaft für Angewandte Mathematik und Mechanik; GT

Manuscripts should have well over 100 pages. As they will be reproduced photomechanically they should be typed with utmost care on special stationary which will be supplied on request. In print, the size will be reduced linearly to approximately 75 %. Figures and diagrams should be lettered accordingly so as to produce letters not smaller than 2 mm in print. The same is valid for handwritten formulae. Manuscripts (in English) or proposals should be sent to the general editor Prof. Dr. E. H. Hirschel, Herzog-Heinrich-Weg 6, D-8011 Zorneding.

The addresses of the editors of the series are given on the last page.

Produced by W. Langelüddecke, Braunschweig

ISBN-13: 978-3-528-08092-1 e-ISBN-13: 978-3-322-87873-1
DOI: 10.1007/978-3-322-87873-1

PREFACE

With the advent of super computers during the last ten years, the numerical simulation of viscous fluid flows modeled by the Navier-Stokes equations is becoming a most useful tool in Aircraft and Engine Design. In fact, compressible Navier-Stokes solvers tend to constitute the basic tools for many industrial applications occuring in the simulation of very complex turbulent and combustion phenomena. In Aerospace Engineering, as an exemple, their mathematical modelization requires reliable and robust methods for solving very stiff non linear partial differential equations.

For the above reasons, it was clear that a workshop on this topic would be of interest for the CFD community in order to compare accuracy and efficiency of Navier-Stokes solvers on selected external and internal flow problems using different numerical approaches.

The workshop was held on 4-6 December 1985 at Nice, France and organized by INRIA with the sponsorship of the GAMM Committee on Numerical Methods in Fluid Mechanics.

The organizers wish to express their thanks to :

- Professor U. Schumann, Chairman of the GAMM Committee for giving them the possibility to hold this workshop,
- Professor R. Peyret for presenting a survey paper and for fruitful discussions during the preparation of the workshop,
- Professor J. Allegre, Drs M. Raffin and J.C. Lengrand for providing experimental results which give an added interest to the workshop,
- chaipersons for their active parts in directing open discussions,
- all the attendees for their effort in complying with the requirements on the presentation of results,
- Mrs O. Labbé who efficiently contributed to the comparison of the results of Problem (B) by carrying out the plotting of the numerical data,
- the "Service des Relations Extérieures" at INRIA whose help contributed for a large part to the successof the meeting,
- Mrs C. Barny and C. Dubois for their careful typing during the preparation of the workshop and of the synthesis.

M.O. BRISTEAU, INRIA

R. GLOWINSKI, Univ. of HOUSTON/INRIA

J. PERIAUX, AMD/BA

H. VIVIAND, ONERA.

CONTENTS

PRESENTATION OF PROBLEMS AND DISCUSSION OF RESULTS

M.O. BRISTEAU
INRIA, B.P. 105, 78153 LE CHESNAY CEDEX, FRANCE

R. GLOWINSKI
UNIVERSITY OF HOUSTON, TEXAS, U.S.A. and INRIA

J. PERIAUX
AMD/BA, B.P. 300, 92214 SAINT CLOUD, FRANCE

H. VIVIAND
ONERA, 92320 CHATILLON, FRANCE

1. INTRODUCTION

The workshop was organized with the specific purpose to bring together a small number of scientists highly concerned with the numerical solution of the compressible Navier-Stokes equations.

Two test problems namely external flows past an airfoil (A) and internal flows in a double throat nozzle (B) were defined with the following features :

i) they should be on simple analytical geometries, suited for a wide range of methods (finite differences, finite volumes, finite elements, spectral, etc...).

ii) they should be 2-D laminar viscous flows with moderate gradients in order to avoid the introduction of turbulent models and also large instabilities due to strong shock-boundary layer interactions, too much costly in CPU time for researchers with limited computer facilities.

In practice the proposed test cases correspond to transonic or supersonic steady flows at low or moderate Reynolds numbers. Two tests cases for problem (A) were chosen because of existing experimental results.

For both problems (A) and (B) the main challenge was the computation of the pressure and skin friction coefficients on the wall.

2. THE PROBLEMS FOR ANALYSIS

2.1. Fluid characteristics

For both problems (A) and (B), we assume that

- the fluid is a perfect gas with constant specific heats C_p, C_v, of ratio :

$$\gamma = C_p / C_v = 1.4$$

- the viscosity coefficient λ, μ verify the Stokes relation :

$$3\lambda + 2\mu = 0$$

- the viscosity coefficients and the thermal conductivity coefficient k are constant and the Prandtl number is :

$$Pr = \mu C_p / k = 0.72$$

2.2. Problem (A) : External 2D flow around a NACA0012 airfoil

Geometry

We consider the NACA0012 airfoil extended to a closed trailing-edge (with zero thickness). It is a symmetrical airfoil and the equation of the upper surface is :

$$\tilde{y}(\tilde{x}) = 5\, t(0.2969\, \tilde{x}^{1/2} - 0.126\, \tilde{x} - 0.3516\, \tilde{x}^2 + 0.2843\, \tilde{x}^3 - 0.1015\, \tilde{x}^4),$$

where

$\tilde{x}$ = the chordwise distance from the leading edge at $\tilde{x} = 0$ referred to c_1,

$\tilde{y}$ = the upper-surface coordinate referred to c_1,

$t = 0.12$ (value of thickness parameter for NACA0012).

For the closed profile

$$0 \leq \tilde{x} \leq 1.\ 008930411365$$

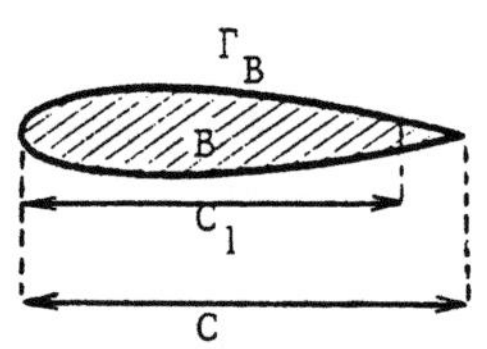

$$c = 1.0089\ c_1$$

Taking the chord c of the closed profile as reference length, the dimensionless coordinates to be used are :

$$x = \frac{\tilde{x}}{1.0089} \quad , \quad y = \frac{\tilde{y}}{1.0089}$$

Note : The extended profile has a relative thickness of 0.12/1.0089.

Boundary Conditions

The following boundary conditions are prescribed :

at Γ_B,

$\underset{\sim}{u}_B = \underset{\sim}{0}$, adherence condition

$T_B = T_0 = T_\infty \left(1 + \frac{\gamma-1}{2} M_\infty^2\right)$, free stream total temperature ;

at infinity, a uniform flow defined by the following parameters :

M_∞ = Mach number at infinity,

α = angle of attack,

Re = Reynolds number, $Re = \frac{|\underset{\sim}{u}_\infty| \rho_\infty c}{\mu}$.

Test Cases

Seven test cases are proposed :

A1 : M_∞ = .8	α = 10°	Re = 73	(mandatory)
A2 : M_∞ = .8	α = 10°	Re = 500	(mandatory)
A3 : M_∞ = 2.	α = 10°	Re = 106	(mandatory)
A4 : M_∞ = 2.	α = 10°	Re = 1000	(optional)
A5 : M_∞ = .85	α = 0°	Re = 500	(mandatory)
A6 : M_∞ = .85	α = 0°	Re = 2000	(optional)
A7 : M_∞ = .85	α = 0°	Re = 10000	(optional)

The six cases A_1-A_6 give steady flows, only the optional case A_7 is associated with an unsteady flow.

The cases A_1 and A_3 have been prescribed because experimental results were available for these data of the problem. It would have been interesting to associate slip conditions to these cases because they are closer to the conditions of the experiments.

The test case A_2 leads to a separated flow. The main feature of the supersonic cases A_3 and A_4 is a detached bow shock.

The Reynolds numbers of case A_5, A_6, A_7 associated with the same Mach number are chosen in order to allow the comparison of the thickness of the boundary layers and to check that they vary as $1/\sqrt{Re}$.

Numerical results

The authors have selected the most representative results among the following numerical outputs of the proposed test cases :

- Plots of the mesh,
- Wall distributions :

  a) pressure coefficient, $C_p = \dfrac{p-p_\infty}{\frac{1}{2}\rho_\infty|\underset{\sim}{u}_\infty|^2}$,

  b) skin friction coefficient, $C_f = \dfrac{\tau_w}{\frac{1}{2}\rho_\infty|\underset{\sim}{u}_\infty|^2}$,
  (τ_w = wall shear stress)

  c) heat flux coefficient, $C_h = \dfrac{q}{\frac{1}{2}\rho_\infty|\underset{\sim}{u}_\infty|^3}$

  (q = heat flux, positive if directed from the wall to the fluid)

- Contour maps

  a) Mach number,

  b) pressure,

  c) density.

- Convergence history plots of the following residuals :

$$\frac{1}{\Delta t}\;\frac{\|\rho^{n+1}-\rho^n\|_{L^2}}{\|\rho_\infty\|_{L^2}},$$

$$\frac{1}{\Delta t}\;\frac{\|\rho\underset{\sim}{u}^{n+1}-\rho\underset{\sim}{u}^n\|_{L^2}}{\|\rho_\infty\underset{\sim}{u}_\infty\|_{L^2}},$$

$$\frac{1}{\Delta t}\;\frac{\|e^{n+1}-e^n\|_{L^2}}{\|\rho_\infty|\underset{\sim}{u}_\infty|^2\|_{L^2}}.$$

2.3. Problem (B) : Internal flow in a double throat nozzle

The double throat nozzle of test problem (B) was designed with the aim of generating strong viscous interaction phenomena in steady, laminar, compressible flows in a well-bounded domain. Supersonic flow conditions are first obtained by means of a simple converging-diverging nozzle. Then the wall is turned concave and a converging channel region is thus formed. It is in this diverging-converging part of the nozzle, with partly supersonic flow conditions, that we expect compression waves, shock waves and separation phenomena to occur. Then the flow goes through a second throat and is allowed to expand rapidly in a second diverging channel.

Geometry

This plane symmetrical nozzle is shown on the figure below. The wall is made up of 5 polynomial arcs, with continuity of slope and of curvature, except for the points $x = x_2$ and $x = x_5$ where the curvature is discontinuous.

The origin of the abscissa, $x = 0$, is taken at the first throat. The half-height at $x = 0$ is chosen as reference length in the following equations.

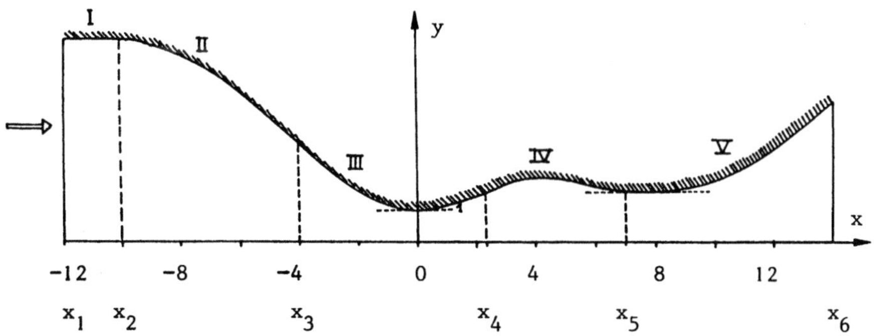

Nozzle geometry

Arc III (first throat region)

$x_3 \leq x \leq x_4$, $\boxed{x_3 = -4, \; x_4 = 2.3}$

$$y = 1 + \frac{a}{2} x^2 \left[\frac{x^2}{6} - \frac{x_3+x_4}{3} x + x_3x_4 \right], \quad \boxed{a = -0.03}$$

We deduce :

$$y(x_3) = 1 + \frac{a}{6} x_3^3 (2x_4 - \frac{1}{2} x_3)$$

$$y'(x_3) = \frac{a}{2} x_3^2 (x_4 - \frac{1}{3} x_3)$$

$$y(x_4) = 1 + \frac{a}{3} x_4^3 (x_3 - \frac{1}{4} x_4)$$

$$y'(x_4) = \frac{a}{2} x_4^2 (x_3 - \frac{1}{3} x_4)$$

The points $x = x_3$ and $x = x_4$ are inflexion points ($y'' = 0$). The throat radius of curvature is $R = \frac{1}{a\, x_3\, x_4} = \frac{1}{0.276}$.

<u>Arc II (convergent)</u>

$$x_2 \le x \le x_3 \quad , \quad \boxed{x_2 = -10}$$

$$y = y(x_3) + y'(x_3)(x - x_3) \left[1 - \frac{1}{3} \left(\frac{x - x_3}{x_2 - x_3}\right)^2\right]$$

We deduce :

$$y(x_2) = y(x_3) + \frac{2}{3} (x_2 - x_3)\, y'(x_3)$$

At $x = x_2$, $y' = 0$.

<u>Arc I (constant section inlet)</u>

$$x_1 \le x \le x_2 \; , \quad \boxed{x_1 = -12.}$$

$$y = y(x_2).$$

<u>Arc IV (divergent-convergent)</u>

$$x_4 \le x \le x_5 \; , \quad \boxed{x_5 = 7.}$$

The second throat is at $x = x_5$, $y'(x_5) = 0$, with $y(x_5)$ given :

$$\boxed{y(x_5) = 1.6}$$

$$y = y(x_4) + (x-x_4)\ [y'(x_4) + A_5\ X^2 + B_5\ X^3]$$
$$X = (x-x_4)\ /\ (x_5-x_4)$$

$$A_5 = 4\ C_5 - 3\ y'(x_4)$$
$$B_5 = -3\ C_5 + 2\ y'(x_4)$$
$$C_5 = (y(x_5) - y(x_4))\ /\ (x_5-x_4)\ .$$

We deduce

$$y''(x_5 - 0) = \frac{6}{x_5 - x_4}\ [-2\ C_5 + y'(x_4)] > 0$$

Arc V (divergent)

$$x_5 \le x \le x_6$$

$x_6 = 14.$

$y(x_6) = 5.85$

$$y = y(x_5) + [y(x_6) - y(x_5)]\ z^3\ [2-z]$$

$$z = (x-x_5)\ /\ (x_6-x_5)$$

Note that $y'' = 0$ at $x = x_5 + 0$ and at $x = x_6$.

Discussion of problem (B)

Uniqueness of steady solutions of the compressible Navier-Stokes equations for this problem remains an open question from a mathematical point of view, and special attention must be given to the geometry and to some of the flow boundary conditions in order to come as close as possible to a well-posed problem. Since we do not have a complete mathematical proof that this will be the case, we must rely upon the physical understanding of the flow problem. In this respect, it can be noted that an internal flow bounded by inflow and outflow section appears to raise more questions than the unbounded flow past a finite obstacle since, in this latter case, uniform flow conditions apply at infinity (of course practical difficulties may arise because the outer boundary is located at a finite distance).

In order to define unambiguous reference quantities, it is best to assume that the incoming flow at the entrance section possesses an inviscid isentropic core, so that the corresponding reservoir conditions provide the needed reference values (total enthalpy or temperature, sound speed, pressure...). Furthermore, to avoid the problem of the choice of the flow conditions to be specified across a viscous layer at the inflow boundary, we take zero boundary layer thickness at this boundary, which amounts to assuming inviscid isentropic flow across the entire entrance section, except at the wall itself where to no-slip condition is imposed. In fact, whether or not the no-slip condition is imposed at the entrance section does not make any appreciable difference if this section is large enough (compared to throat section) : for example, when the Mach number changes from $M_1 = 0$ to $M_2 \neq 0$ at constant total enthalpy and constant pressure, the relative density variation is $(\rho_2-\rho_1)/\rho_1 = \frac{\gamma-1}{2} M_2^2$, which gives 2×10^{-3} at $M_2 = 0.1$. Moreover this alternative is more or less equivalent to a variation in the location of the nozzle entrance section equal to one mesh size Δx, which is negligible.

Thus entropy and total enthalpy are known over the entire entrance section. If we assume that inviscid flow theory can be applied at this section, then an additional condition is needed in subsonic flow ; the upstream part of the nozzle is made with parallel walls, so that it is reasonable to assume parallel flow ($v = 0$) at the entrance section.

At the exit section, the flow is in a state of rapid expansion, with a large supersonic part, which makes it reasonable to assume that upstream influence (i.e. the elliptic nature of the equations), can be neglected i.e. that the flow conditions in this section are entirely determined from the flow properties upstream without the need to impose some boundary conditions.

Finally, taking into account all the above conditions, we assume that, for given reservoir conditions and given physical reference length (hence given Reynolds number), as in inviscid flow there is a maximum value Q_M of the mass flow rate Q such that no solution exists for $Q > Q_M$ and the solution exists and is unique for $Q = Q_M$. Problem (B) is concerned with this solution corresponding to the maximum mass flow rate which is unknown.

The second throat has been chosen large enough compared to the first one so that the maximum mass flow rate corresponds to a change from subsonic to supersonic flow across the first throat at least in the core around the nozzle axis. In other words the solution for maximum mass flow rate is determined by the first throat.

Boundary conditions

At the wall

- no-slip condition $\underset{\sim}{u}_W = \underset{\sim}{0}$
- temperature equal to the reservoir temperature : $T_W = T_0$

Upstream boundary (inlet section, $x = x_1 = -12.$)

- total enthalpy equal to reservoir enthalpy, i.e. :

$$\frac{\gamma}{\gamma-1}\frac{p}{\rho} + \frac{1}{2}\underset{\sim}{u}^2 = \frac{\gamma}{\gamma-1}\frac{p_0}{\rho_0} = \frac{a_0^2}{\gamma-1} \qquad (1)$$

- entropy equal to reservoir entropy, i.e. :

$$p\,\rho^{-\gamma} = p_0\,\rho_0^{-\gamma} \qquad (2)$$

An additional condition may be specified, but all the flow properties cannot be known in advance since the mass flow rate is not arbitrary. A natural condition is that of parallel flow, i.e. : $v = 0$, where v is the y-component of velocity.

Test Cases

The only remaining parameter is the Reynolds number defined as :

$$Re_0 = a_0\, L\, \rho_0 / \mu$$

where the subscript 0 refers to reservoir conditions, and L is the nozzle half-height at the first throat $x = 0$. (we recall that the viscosity is a constant).

The flow in this nozzle should be computed for the following three values of the Reynolds number :

test case B_1 : $Re_0 = 100$
test case B_2 : $Re_0 = 400$
test case B_3 : $Re_0 = 1600$

Numerical results

The suggested numerical outputs were the following :

- Plots of the mesh,
- Wall distributions :
  a) pressure , p/p_0
  b) skin friction coefficient, $C_f = \dfrac{\tau_w}{\frac{1}{2}\,\rho_0\, a_0^2}$,
  (τ_w = wall shear stress)

c) heat flux coefficient, $C_h = \frac{q}{\frac{1}{2} \rho_0 a_0^3}$

(q = heat flux, positive if directed from the wall to the gas)

- Distributions on symmetry axis :
  a) pressure p/p_0,
  b) Mach number,
  c) quantity related to entropy, $S_1 = \frac{p}{p_0} (\frac{\rho_0}{\rho})^\gamma$
- Transverse profiles at x = -10, x = 4 and x = 8 of :
  . u/a_0 (u = x-velocity component)
  . v/a_0 (v = y-velocity component)
  . Mach number M
  . pressure p/p_0
  . $S_1 = \frac{p}{p_0} (\frac{\rho_0}{\rho})^\gamma$
- Mass flow rate
  Plot $Q(x)/(\rho_* a_* L)$ where Q(x) is the mass flow rate through the transverse mesh line starting at abscissa x on the symmetry axis, and ρ_* and a_* are the sonic values of density and sound speed in isentropic flow.
- Contour maps of
  a) Mach number
  b) pressure, p/p_0
  c) entropy, S_1.

3. RESULTS

3.1. Contributors

Participation to the workshop was based on invitation to people who were known to have softwares for solving the compressible Navier-Stokes equations. After two years, sixteen scientists (or teams) contributed to this event and we thank all of them for their interest to the workshop (cf. Table 1.). Very few people run optional test cases of problem (A) which require, at higher Reynolds number, a great number of nodes mainly distributed in the boundary layer and a control of the numerical viscosity of the code to be able to compute an unsteady solution. However some qualitive comparisons exist for these cases.

Different approaches for solving the Navier-Stokes equations including finite differences (4), finite elements (4) and finite volumes (8) have been used by authors as shown on Tables 2(a), 2(b). Many of them used an unsteady formulation in order to capture steady solutions.

3.2. Comparison of results of problem (A)

At first sight the results appear quite similar on the pressure coefficients on the airfoil. However significant differences appear in the computed skin friction and heat flux coefficients and also integral values of lift, drag and moment coefficients. Computed density contours can be superposed with the measurements provided by Allegre et al for test cases A_1 and A_3.

To compare the required outputs of the different contributors, one should pay particular attention to the following features :

i) stability of the solution at different Reynolds numbers via the smoothness of the isolines which depends strongly on the number of nodes and on the numerical viscosity.

ii) symmetry of the solution, in particular of the heat flux and skin friction coefficients on the airfoil at angle of attack $\alpha = 0°$.

iii) possibility to handle mainly supersonic flows.

iv) quality of the mesh : density and regularity in the vicinity of the leading and trailing edges.

v) maximum and minimum values of the skin friction and heat flux coefficients which depend highly on the discretization in the vicinity of the leading edge.

vi) influence of the Reynolds number on the thickness of the boundary layer via the closeness of the Mach lines near the wall.

vii) comparison of the position of the separation point for the flow of test case A3 via the change of sign of the skin friction coefficient (cf. Table 7.).

viii) comparison of integral values of lift, drag and moment coefficients (see Tables 4, 5, 6).

Unsteady solution for test case A7 can be obtained only with a large number of nodes (≈15000) on a super computer. Transient interaction of the shock with large eddies of the wake can then be observed for different time cycles via isodensity lines in the flow field and pressure coefficient on the body.

Finally very substantial differences on CPU time can be observed depending on the number of nodes (cf. Table 3) and on the computer used (cf Table 8). These comparisons can provide a useful guideline for choosing a method for 3-D viscous flow simulations.

3.3. Comparison of results of problem (B)

Global comparison

Cross-plots of results pertaining to five numerical solutions to test-problem (B) published in this volume are presented in Figs. 1 to 22. On these plots, the solutions (when distinguishable) are referred to by numbers, the correspondence with the authors being given in Table 2(b).

The results shown are those provided by the authors on magnetic tapes after the workshop. The results of Y. Marx were not available at the time the comparison was made.

Details on these solutions and on the numerical methods can be found in the separate papers in this volume, and it is sufficient here, in order to characterize the solutions, to specify the mesh size :

Solution Nr.	mesh size ($N_x \times N_y$)	
1....................	152 × 32	
2....................	208 × 29	
3....................	121 × 81	for Re_0 = 100
	246 × 81	for Re_0 = 400 and 1600
4....................	218 × 61	
5....................	201 × 32	

We recall the main features of the flow field at the three values 100, 400 and 1600 of the Reynolds number Re_0. All the five solutions agree quite well about these features which are revealed by contour maps of Mach number, pressure and entropy to be found in the separate papers.

The flow deceleration brought about by the concave diverging-converging portion of the nozzle gives rise to a compression zone normal to the axis at Re_0 = 100 and 400 (and looking very much like a normal shock at Re_0 = 400 but not satisfying the Rankine Hugoniot relations), and to two oblique shock waves crossing each other on the axis at Re_0 = 1600. A separated flow region exists in this concave wall region at all three Reynolds numbers, and a second much smaller separated flow region appears downstream of the second throat at Re_0 = 1600 ; this second separation is caused by the impingement of the oblique shock on the viscous layer, but the effect is small because of the expansion of the flow in the diverging channel downstream (in fact the wall curvature has been made discontinuous at x = 7, equal to zero at x = 7 + 0, in order to lessen the expansion effect).

The abscissae x_S and x_R of the separation and reattachment points are given in Table 9. The agreement between the solutions is very good, the maximum variation in the abscissa x_S or x_R being less than 0.06 for the first separated region and less than 0.1 for the second one.

The general conclusion which can be drawn from the cross-plots presented in Figs. 1 to 21, taking into account the scales used in these plots, is that there is a very good agreement between all five solutions despite the differences in grids and in numerical methods. It is clear that these numerical solutions relate to the same flow field, and this fact supports the initial assumption that the problem is well-posed at least in a global manner (it cannot be excluded that small differences observed in particular near the outflow boundary reflect the need for some additional boundary condition).

Discrepancies and particular features

Here we point out at particular features and small discrepancies revealed by the cross-plots.

a) The plots of C_f and C_h along the wall (Figs. 2, 3) for Re_0 = 100 and 400 show appreciable differences over a short distance upstream of the outflow boundary. These differences denote a local sensitivity of the solution to the numerical treatment at this boundary, especially in the highly viscous part, and as already mentioned above, it may signify that an additional boundary condition is necessary to have a unique solution. For Re_0 = 1600 this effect is quite small, but still visible.

In general, the agreement on C_h (Fig. 3) is not as good as that on C_f, but it should be noted that C_h is about ten times smaller than C_f. At Re_0 = 1600, the solution Nr. 4 gives higher values of C_h in the first throat region.

At $Re_0 = 100$, all the solutions show the same small spike in C_f and C_h at the second throat ($x = 7$) where the curvature is discontinuous, and the wall pressure curve (Fig. 1) has an inflexion point.

b) The distributions of pressure, Mach number and entropy along the symmetry axis (Figs. 4, 5, 6) generally show very good agreement between all five solutions.

At $Re_0 = 400$, some differences can be seen around the sharp maximum of pressure (Fig. 4) or around the sharp minimum of Mach number (Fig. 5) which occur near $x = 5.5$ at the downstream end of the compression region, and also in the weak recompression which occurs around $x = 8$.

At $Re_0 = 1600$, a weak recompression occurs between $x = 9$ and $x = 9.5$ (Figs. 4 and 5) ; solution Nr. 1 does not capture this phenomenon as accurately as the other solutions probably because of the coarser mesh.

The quantity S1 (Fig. 6), which is related to entropy, shows appreciable differences at $Re_0 = 400$ and 1600 (Fig. 6.b) in the compression region and downstream.

It is striking to see that, at all three Reynolds numbers, S1 on the axis (Fig. 6) remains practically constant and equal to one from the inflow boundary, as long as the pressure decreases on the axis, and then increases rapidly with the pressure. This statement must be mended on two points : at $Re_0 = 100$, S1 starts increasing somewhat upstream of the point of minimum pressure, and at $Re_0 = 400$, there is a curious decrease of S1 below unity for $0 < x < 2$ which must result from a heat loss through a transversal conduction effect.

c) Concerning the profiles of velocity, Mach number, pressure and entropy at the station $x = -10$ (Figs. 7 to 11), we observe a rather good agreement which tends to indicate that the upstream boundary conditions are correctly defined. The velocity is small, corresponding to a Mach number about 0.1 on the axis, and a viscous layer is already well established as can be seen in particular on the S1 profiles (Fig. 11).

d) At $x = +4$ (Figs. 12 to 16), small discrepancies can be noted on the v-profiles and on the p-profiles at $Re_0 = 100$ and 400.

There is better agreement at $Re_0 = 1600$, for all the profiles at $x = +4$.

Looking at the S1-profiles on Fig. 16, we find a very good agreement at all three Reynolds numbers.

e) The profiles shown at $x = +8$ (Figs. 17 to 21) do not include solution Nr.1 (data not available).

At $Re_0 = 100$, the agreement is very good for all profiles except the pressure profile with a slightly higher value given by solution Nr. 4.

At $Re_0 = 400$, some differences can be seen on the v-profiles and on the p-profiles, whereas at $Re_0 = 1600$ all the profiles except that of S1 show some discrepancies, particularly near $y = 1.1$ which is the region where the oblique shock impinges on the viscous layer.

Of course the variations observed for S1 on the x-axis downstream of the compression region at $Re_0 = 400$ and 1600 (Fig. 6-b) are no more visible on Fig. 21 because of the scale used. This draws attention to the fact that the agreement between the solution is relative to the scale which is appropriate for the particular plot considered.

f) The non-dimensional mass flow rate $\bar{Q} = Q/(\rho_* a_* L)$ is plotted versus x on Fig. 22 for solutions Nr. 1, 2 and 4 only.

At all three Reynolds numbers, sudden variations in $\bar{Q}$ are observed near the outflow boundary, suggesting again the need for some boundary condition. At $Re_0 = 100$, if we exclude the region $x > 13$, we find that for these three solutions $\bar{Q}$ remains within bounds equal to 0.811 and 0.817, i.e. a total relative variation of 0.74 %.

At $Re_0 = 400$, large oscillations in $\bar{Q}$ occur for all three solutions near $x = 5$ in the region of the strong normal compression. These oscillations determine the extremal values of $\bar{Q}$, namely 0.880 and 0.920, corresponding to a total relative variation of 4.4 %. If we exclude the region of these oscillations $(4.5. < x < 5.5)$, and also the region $x > 14$ for solution Nr. 1, then $\bar{S}$ is found to vary between 0.902 and 0.906, giving a total relative variation of 0.44 %.

At $Re_0 = 1600$, we also observe strong oscillations near $x = 8$, i.e. in the region where the oblique shock impinges on the viscous layer. The extremal values of $\bar{Q}$ are 0.944 and 0.956, corresponding to a total relative variation of 1.3 %.

Conclusion for problem (B)

The comparison of five numerical solutions to problem (B) shows a very good quantitative agreement at the three values of the Reynolds number.

Small differences observed close to the outflow section may indicate that some outflow boundary condition is needed especially in the highly viscous part of the flow. Of course such a boundary condition will comprise some arbitrariness from a physical point of view, but the point here is to see whether the agreement between different numerical solutions can be improved locally by imposing these solutions to verify the same constraint at the exit section.

4. CONCLUSIONS

From a first comparison of the results, it appears that the scatter of computed solutions is much larger for Problem (A) than for Problem (B). Since the methods used are the same or very similar, a possible source of this difference can be that problem (A) deals with external flows on non bounded domains while Problem (B) is defined on bounded ones.

Concerning Problem (A) even if the results look quite similar on the pressure and Mach number in the fluid and also on the pressure distribution on the wall, significant differences appear in the computed skin friction and heat flux coefficients. This discrepancy can be explained by the fact that these outputs involve first derivatives of the primitive variables (velocities and temperature) making of importance the choice of the formulation and also the number and the location of nodes in the spatial discretization.

The good agreement of results for Problem (B) seems to indicate that the chosen boundary conditions are appropriate with the small restriction mentioned concerning the outflow condition.

Problem (A) appears to be more sensitive to numerical features (mesh, location of outer boundary) than Problem (B) and the trailing-edge singularity may play a role in this behaviour.

For many codes the stability of steady solutions obtained at moderate Reynolds number is remarkable and the numerical viscosity is an important feature of many Navier-Stokes solvers to get converged solutions.

At the present time the difficulties encountered by several existing methods can be explained by the following remarks :

i) existence and uniqueness of the solution of the discretized problem are not yet proved. At high Reynolds number, steady solutions of the Navier-Stokes equations may not exist.

ii) the dispersion of results issued from problem (A) can be partly explained by the sensitive influence (more or less reflecting) of the boundary conditions at infinity, therefore a special attention has to be devoted to the use of absorbing ones.

iii) accurate prediction of separated external flows on simple geometries at moderate Reynolds numbers is not yet achieved. The computation of accurate solutions of 3-D complex flows on industrial complicated geometries with adapted meshes, remains a difficult objective in a near future.

To approach this ambitious objective, some lines of research are suggested thereafter :

i) improvement of the accuracy of the solution by local mesh refinement in preferred directions (boundary layers, wakes, shocks, vortex sheets),

ii) acceleration of the convergence by very fast solvers such as multigrids or iterative methods with appropriate preconditionners,

iii) systematic use of vectorizable (size of regular arrays) and parallelizable (block structure or domain decomposition) algorithms on supercomputers.

For instance, it is noticeable that the unsteady solutions of test case A7 were provided by Navier-Stokes solvers exhibiting a part of the above properties.

It is clear that much effort remains to be done to achieve good accuracy at acceptable computing cost for practical 3D applications. This volume should provide valuable basic reference cases for the assessment of the progress in numerical methods and techniques for viscous fluid flow simulation.

TABLE 1. List of Contributors

Authors	Country	Problem solved
J. Allègre, M. Raffin, J.C. Lengrand	France	A, experiments
F. Angrand	France	A
F. Bassi, F. Grasso, A. Jameson, L. Martinelli, M. Savini	Italy / U.S.A	B
M.O. Bristeau, R. Glowinski, B. Mantel, J. Periaux, C. Pouletty	France	A
L. Cambier	France	A
F. Grasso, A. Jameson, L. Martinelli	Italy / U.S.A	A
W. Haase	F.R.G	A
D. Kalfon, D. Volpert, O. Brocard	France	A
W. Kordulla	F.R.G	A
O. Labbé	France	B
Y. Marx	France	B
B. Müller, T. Berglind, A. Rizzi	Sweden	A
N. Satofuka, X. Morinishi, Y. Nishida	Japan	A
D. Schwamborn	F.R.G	B
J.N. Scott, M.R. Visbal	U.S.A	B
Y. Secretan, G. Dhatt, D. Nguyen	Canada	A
J.L. Thomas, R.W. Walters, B. Van Leer, C.L. Rumsey	U.S.A / Netherlands	B

TABLE 2(a) : List of contributors to Problem (A)

Number	Authors	Affiliation	Space discretization	Time discretization
1	J. Allègre, M. Raffin, J.C. Lengrand	Sessia/Lab. d'Aerotherm. du CNRS, Meudon, France	Experiments	
2	F. Angrand	INRIA, France	F.E.	Expl., Pred/Corrector
3	M.O.Bristeau, R.Glowinski, B. Mantel, J. Périaux, C. Pouletty	AMD/BA, INRIA, France	F.E.	Implicit
4	L. Cambier	ONERA, France	F.D.	Expl., Pred/Corrector
5	F. Grasso, A. Jameson, L. Martinelli	Ist.di Gasdinamica, Italy, Univ. of Princeton, USA.	F.V.	Expl. Runge-Kutta Impl. Smoothing
6	W. Haase	Dornier GmbH, FRG	F.V.	Expl. Runge-Kutta
7	D. Kalfon, D. Volpert, O. Brocard	ONERA-CERT, France	F.E.	Stationary method
8	W. Kordulla	DFVLR, FRG	F.V.	Impl., Pred/Corrector
9	B. Müller, T. Berglind, A. Rizzi	FFA, Aero. Res. Inst. of Sweden, Sweden.	F.D.	Implicit
10	N. Satofuka, K. Morinishi, Y. Nishida	Kyoto Inst. of Technology, Japan	F.D.	Expl., Rational Runge-Kutta Impl. averaging
11	Y. Secretan, G. Dhatt, D. Nguyen	Univ. Laval, Canada	F.E.	Implicit, Newton

F.D. : Finite Difference, F.V. : Finite Volume, F.E. : Finite Element.

TABLE 2(b) : List of contributors to Problem (B)

Number	Authors	Affiliation	Space discretization	Time discretization
1	F. Bassi, F. Grasso, A. Jameson, L. Martinelli, M. Savini	Milan, Naples, Italy, Princeton, U.S.A	F.V.	Expl. Runge Kutta
2	O. Labbé,	ONERA, France	F.V.	Impl. ADI
3	J.L. Thomas, R.W. Walters, B. Van Leer, C.L. Rumsey	Nasa Langley, Virginia Pol.Inst., U.S.A. Delft Univ. of Techn., Neth.	F.V.	Impl.
4	J.N. Scott, M.R. Visbal,	Univer. of Dayton, AFWAL, U.S.A.	F.D.	Expl. Mac Cormack
5	D. Schwamborn,	DFVLR, R.F.A.	F.V.	Expl. Runge Kutta
6	Y. Marx,	E.N.S.M., Nantes, France	F.V.	Expl.

TABLE 3 : Mesh sizes

Case / Number	A1	A2	A3	A4	A5	A6	A7
2	2354	Id°	Id°		Id° 3330		
3	3114	3060	2646		2354	4154	
4	193×45	265×55	193×56	193×72	193×47	193×47	265×58
5	265×64	Id°	Id°		Id°	Id°	128×32 Id°
6		Id°	Id°		Id°		
7	X	X			X	X	
8	126×32	Id°			Id°		Id°
9	163×65	Id°	Id°	Id°	Id°	Id°	Id°
10	257×65	Id°	Id°	257×257	257×65	257×65	Id°
11	5152	Id°	Id°		Id°		

TABLE 4 : Lift Coefficients

Case / Number	A1	A2	A3	A4	A5	A6
2	0.5916	0.517	0.367		10^{-5}	
3	0.551	0.461	0.331		10^{-5}	10^{-5}
4	0.5465	0.4342	0.3289	0.3427	$7\,10^{-4}$	$21\ 10^{-4}$
5	0.6657	0.4797	0.3716		0	
6	0.5959	0.4468	0.3063		$<10^{-5}$	
7	0.5584	0.4831			0.	0.
8	0.5380	0.4261			$4\,10^{-4}$	
9	0.5242	0.4199	0.3261	0.3388	$4\,10^{-5}$	$6\,10^{-6}$
10	0.5268	0.4145	0.3173		0.	
11	0.6127	0.4438	0.4059		$<10^{-4}$	

TABLE 5 : Drag Coefficients

Case / Number	A1	A2	A3	A4	A5	A6
2	0.58	0.243	0.412		0.2184	
3	0.608	0.257	0.443		0.179	0.101
4	0.6280	0.2656	0.4826	0.2535	0.2221	0.1164
5	0.2522	0.1597	0.1851		0.0964	
6	0.6377	0.2721	0.4772		0.2176	
7	0.6568	0.2868			0.23	0.136
8	0.7047	0.2845			0.2420	
9	0.6276	0.2632	0.4771	0.2515	0.2199	0.1160
10	0.6447	0.2644	0.4910		0.2261	
11	0.6377	0.2627	0.4791		0.2181	

TABLE 6 : Moment coefficients

Case / Number	A1	A2	A3	A4	A5	A6
2	0.233	0.166	0.1817		10^{-6}	
3	0.216	0.133	0.171		10^{-5}	10^{-4}
4	0.2118	0.119	0.1698	0.1560	-1.10^{-4}	$6\ 10^{-4}$
5						
6	-0.05119	-0.00324	-0.05061		$<10^{-5}$	
7	0.1912	0.1312			0.	0.
8	-0.2377	0.1624			$4 10^{-4}$	
9	0.2019	0.1145	0.1652	0.1549	-10^{-5}	$-4 10^{-6}$
10	0.2050	0.1124	0.1640		0.	
11	0.2453	0.1238	0.2044		$<10^{-4}$	

TABLE 7 : Case A2. Abscissa of separation point

Number	Separation point
2	0.5
3	0.45
4	0.36
5	
6	0.36
7	0.39
8	0.362
9	0.371
10	0.345
11	0.37

TABLE 8 : CPU times in seconds

N^r	Case / Machine	A_1	A_2	A_3	A_4	A_5	A_6
2	CRAY 1S	1500	1500	1500		620	
3	IBM 3090	7000	9500	11500		3200	16800
4	CRAY 1S	3280	3580	1570	1950	2230	2230
5	IBM 3081	9000	Id°	Id°		Id°	Id°
6	CRAY 1S	4237	1002	2808		1599	
7	CRAY 1S	700	840			1000	1200
8	-	-				-	
9	VAX 11	83640	59340	32820	42180	57480	69600
10	FACOM VP 200	186	Id	Id	Id	Id	Id
11	VAX 785	-	-	-	-		

TABLE 9 : Problem (B). Abscissae of separation and reattachment points on the nozzle wall.

	Re_o = 100	Re_o = 400	Re_o = 1600	
Sol. Nr.	x_S/x_R	x_S/x_R	x_S/x_R	x_S/x_R
2	3.000/4.8125	3.000/6.188	3.438/4.688	8.063/8.188
3	3.004/4.820	2.986/6.221	3.432/4.709	8.077/8.216
4	2.951/4.811	3.005/6.241	3.432/4.728	8.043/8.234
5	2.970/4.848	2.962/6.230	3.424/4.739	8.006/8.280

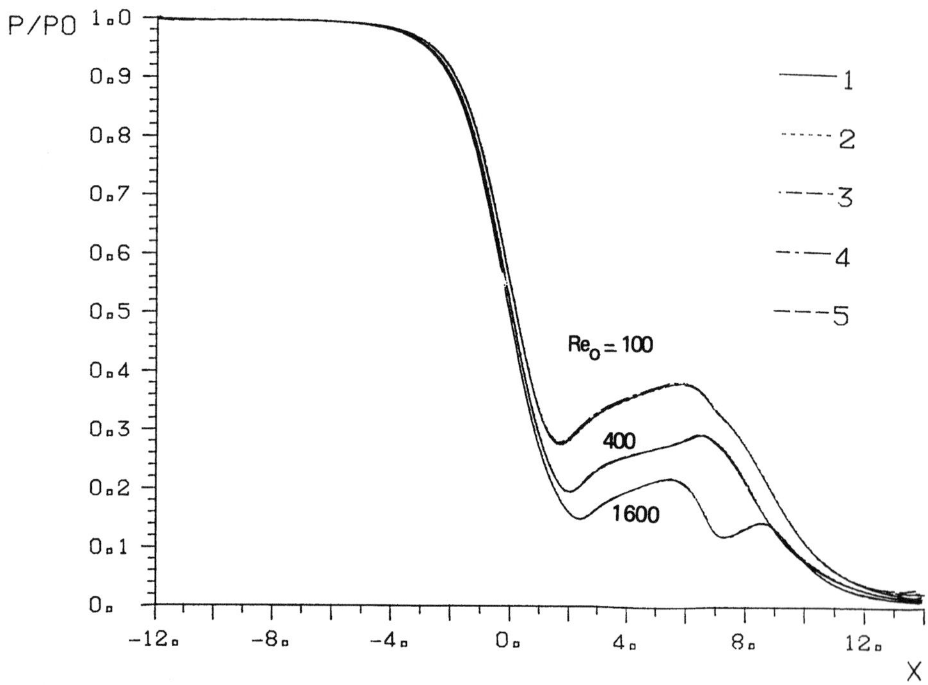

Fig. 1 Pressure distributions on the wall.

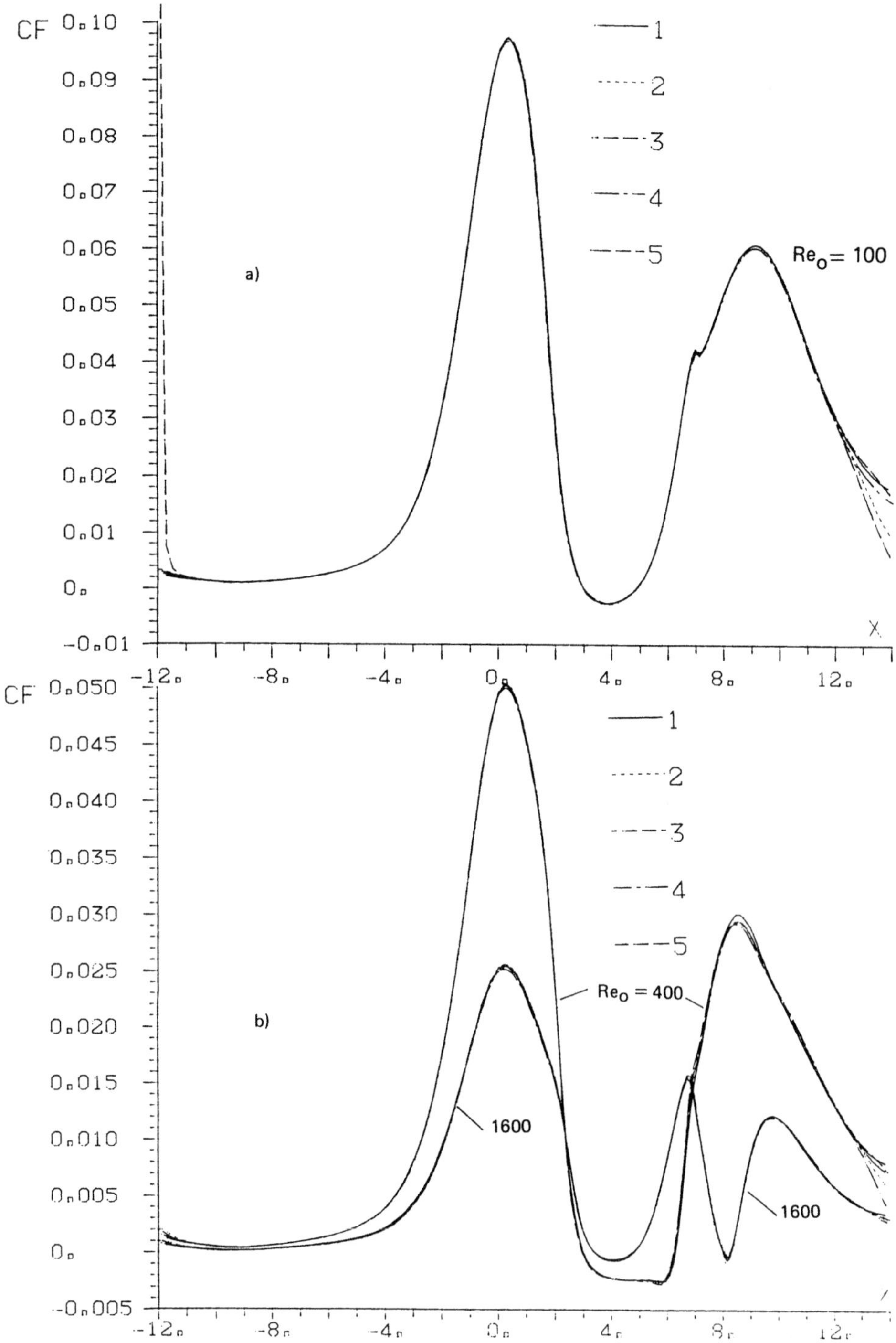

Fig. 2 Skin friction coefficient distributions. a) $Re_o = 100$, b) $Re_o = 400$ and 1600.

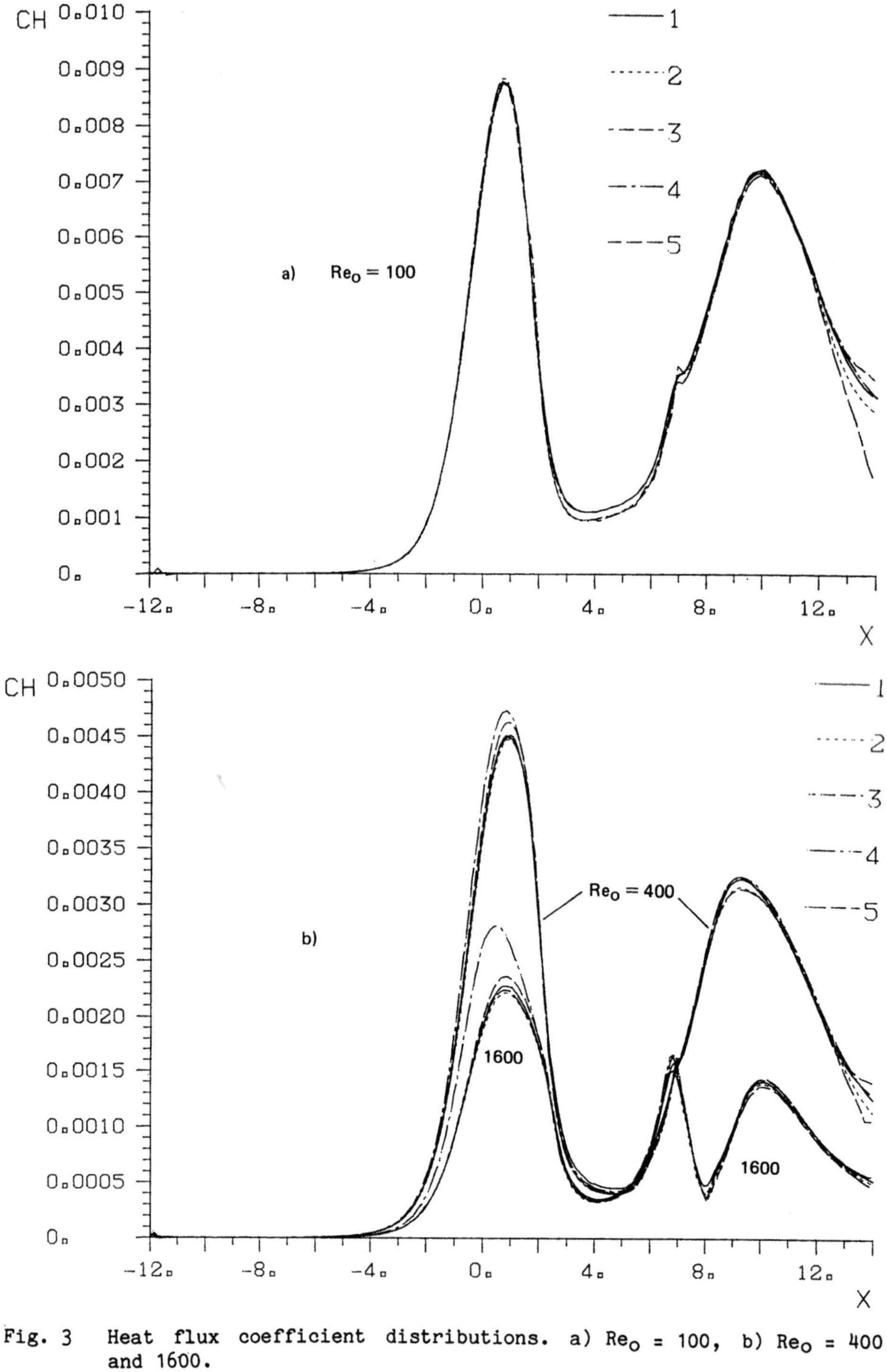

Fig. 3 Heat flux coefficient distributions. a) Re_o = 100, b) Re_o = 400 and 1600.

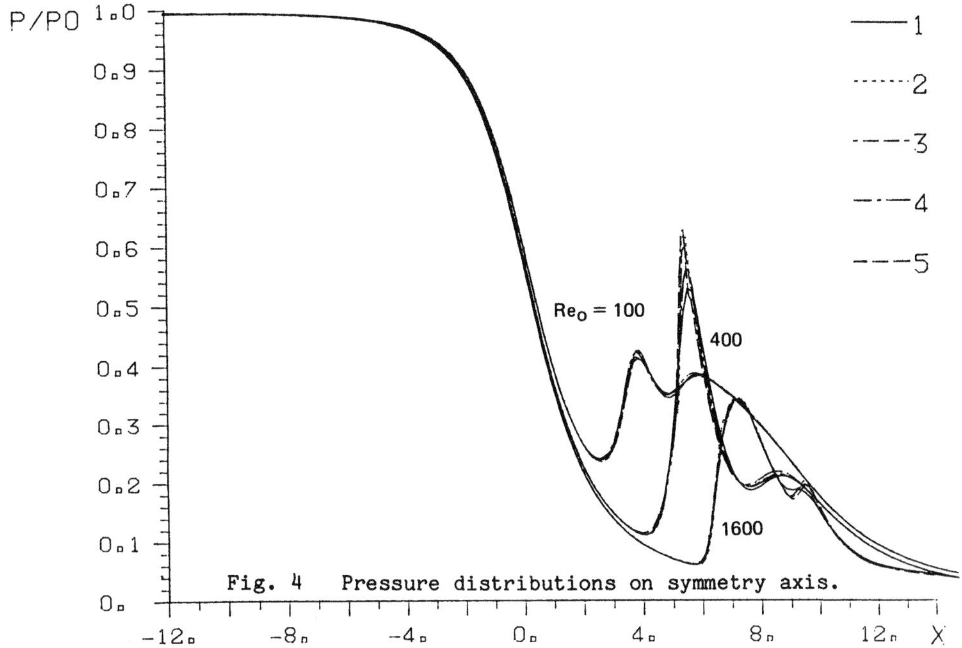

Fig. 4 Pressure distributions on symmetry axis.

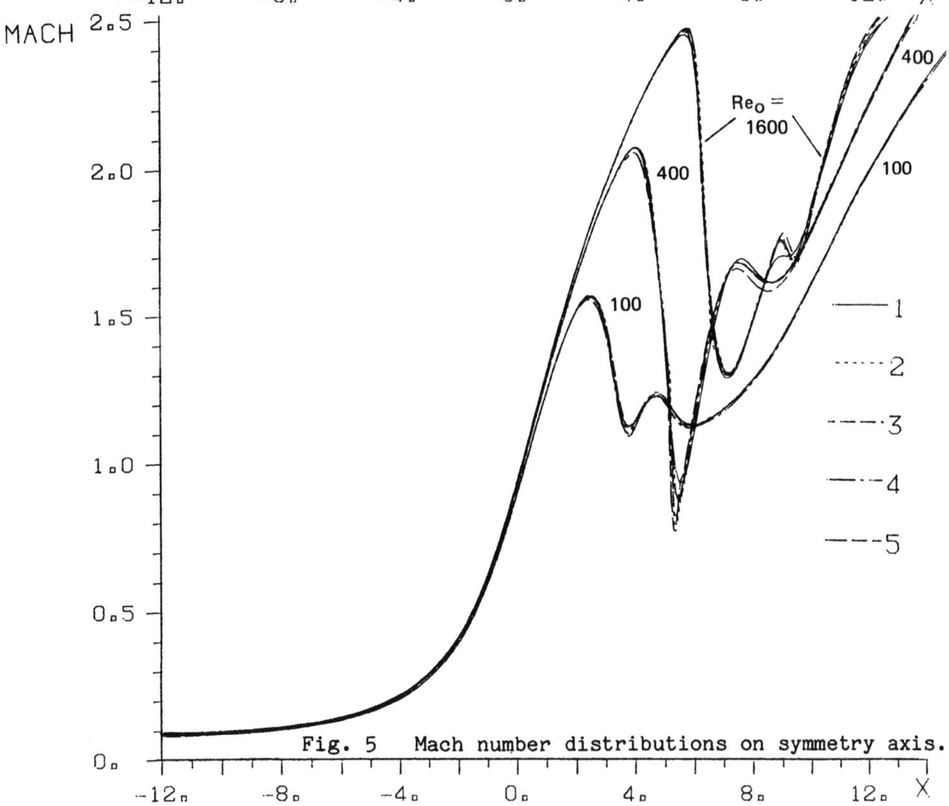

Fig. 5 Mach number distributions on symmetry axis.

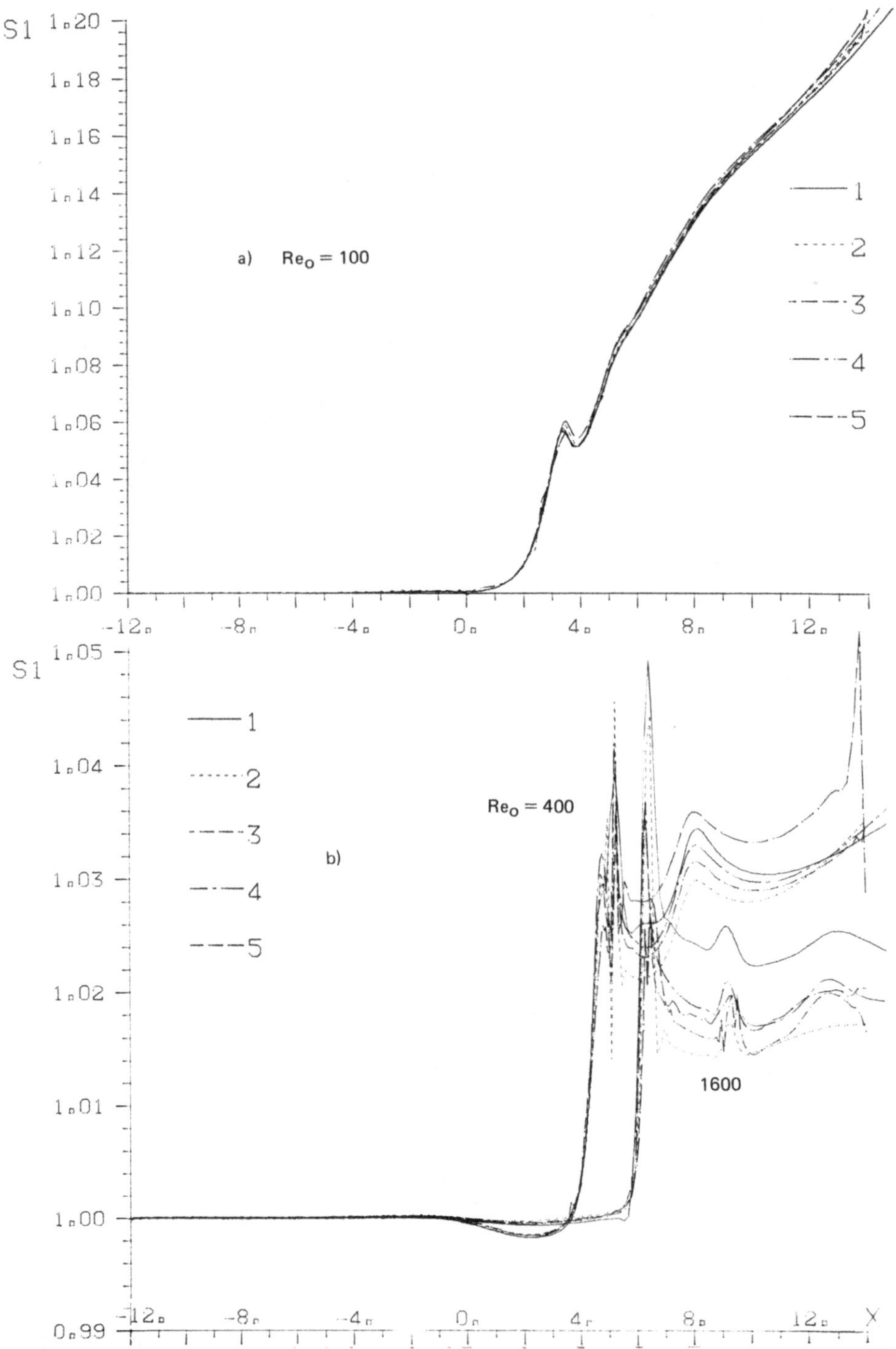

Fig. 6 S1 distributions on symmetry axis $(S1 = (p/p_o).(\rho_o/\rho)^{\gamma})$.
a) Re_o = 100, b) Re_o = 400 and 1600.

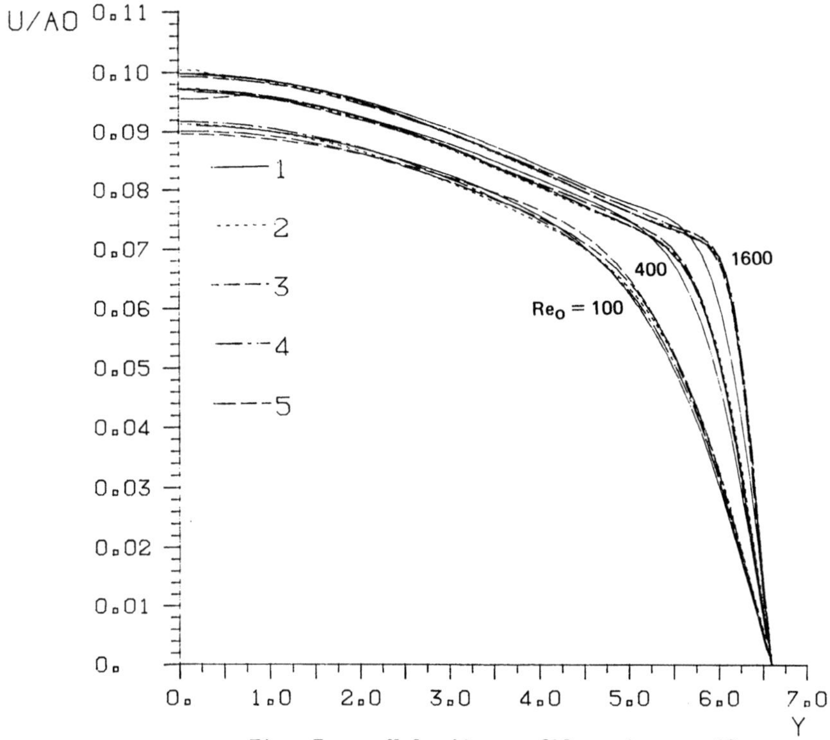

Fig. 7 x-Velocity profiles at x = -10.

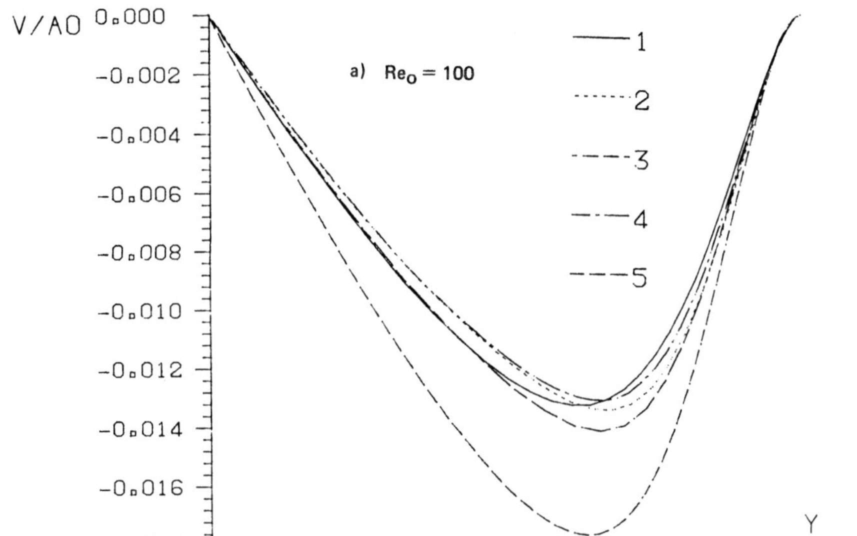

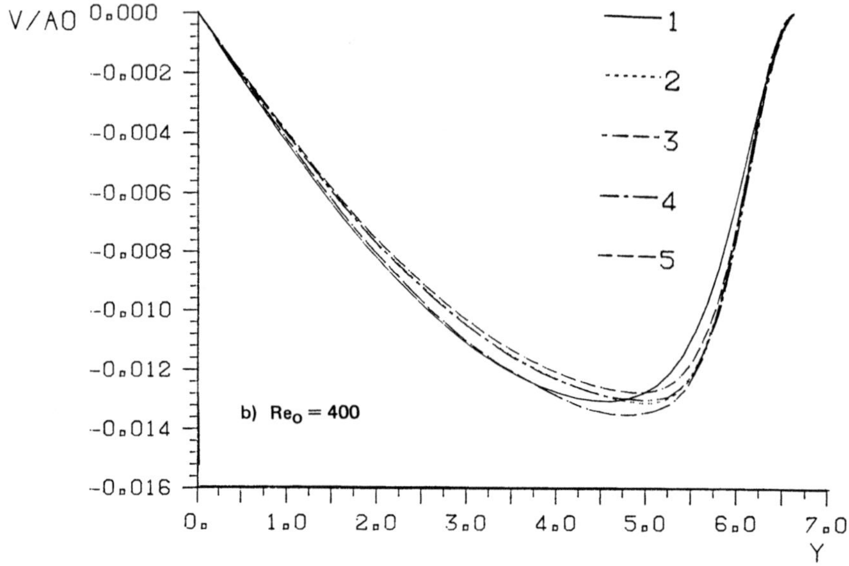

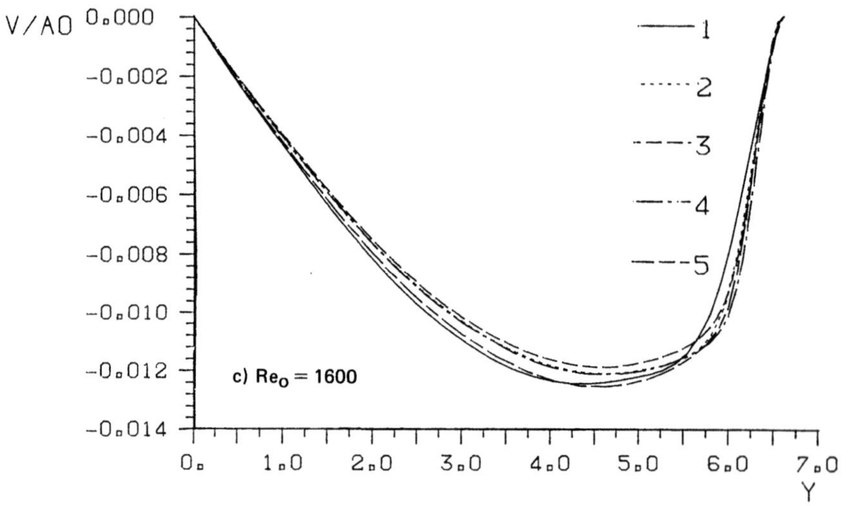

Fig. 8 y-Velocity profiles at x = -10. a) Re_o = 100, b) Re_o = 400, c) Re_o = 1600.

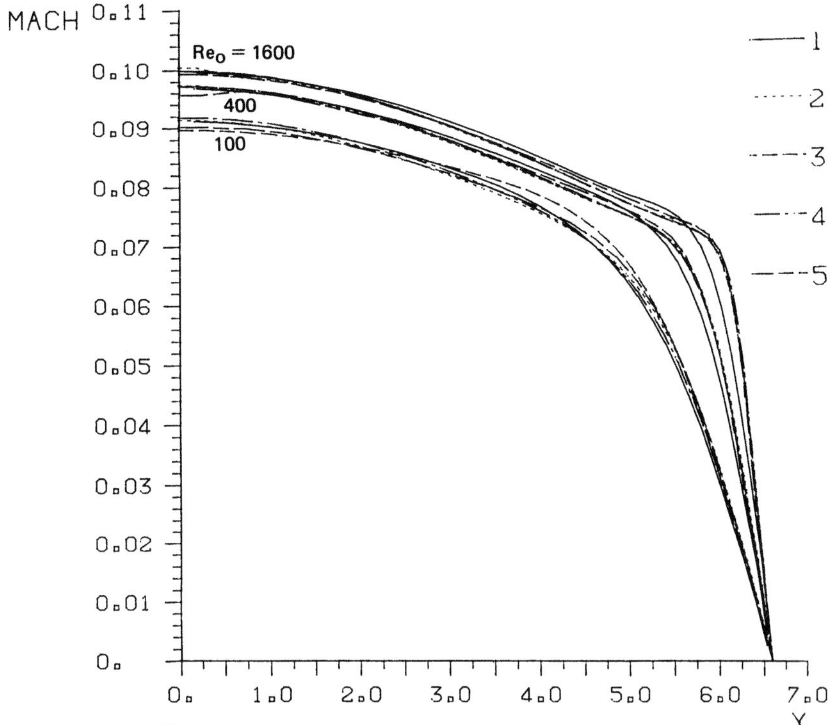

Fig. 9 Mach number profiles at x = -10.

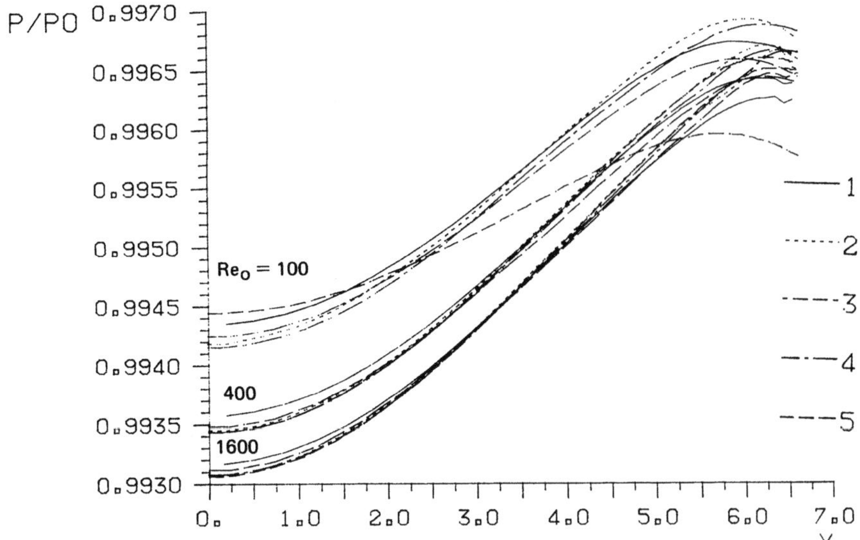

Fig. 10 Pressure profiles at x = -10.

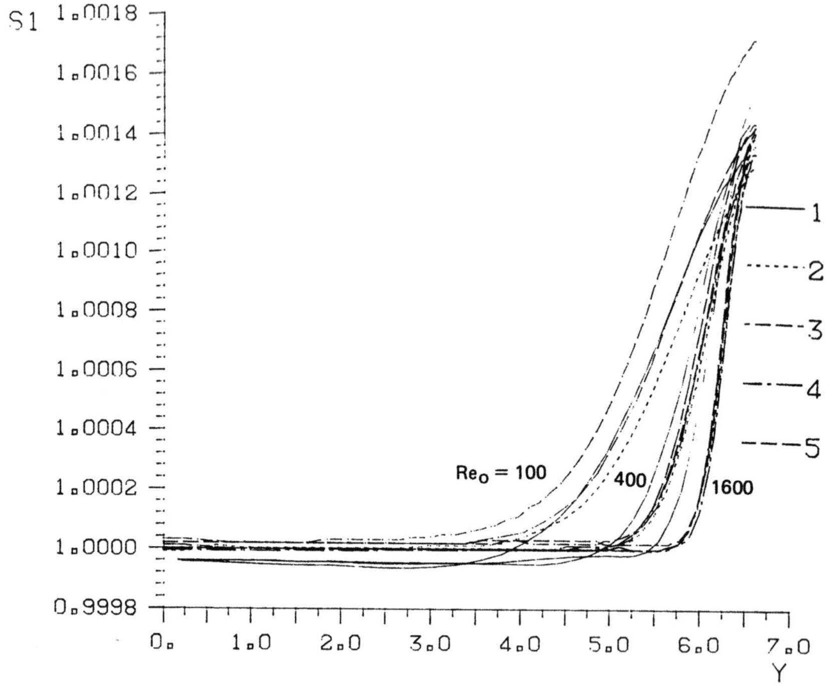

Fig. 11 S1 profiles at x = -10 ($S1 = (p/p_o) \cdot (\rho_o/\rho)^\gamma$).

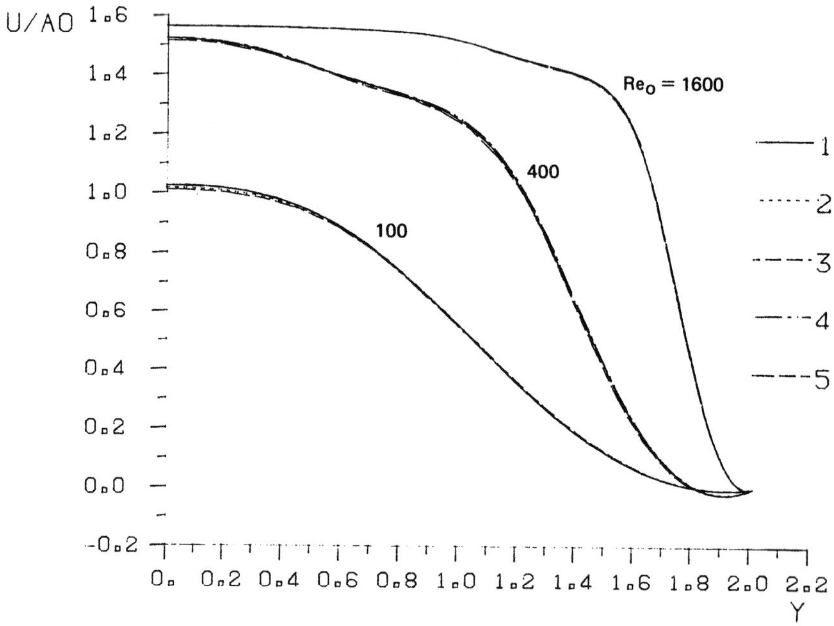

Fig. 12 x-Velocity profiles at x = 4.

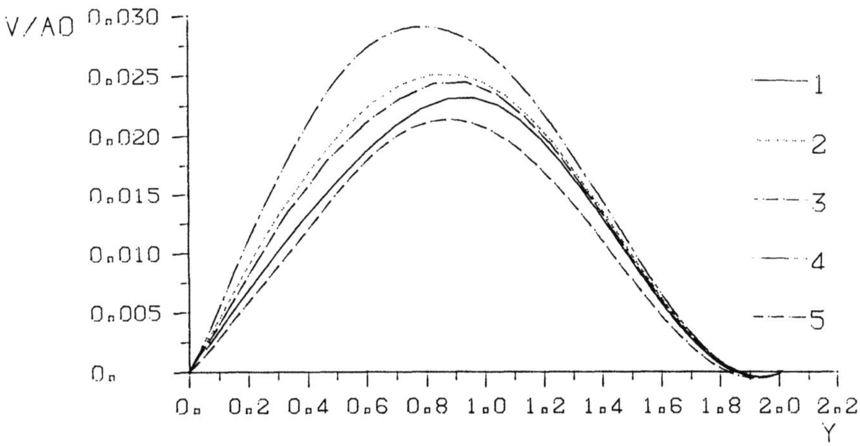

a) $Re_O = 100$

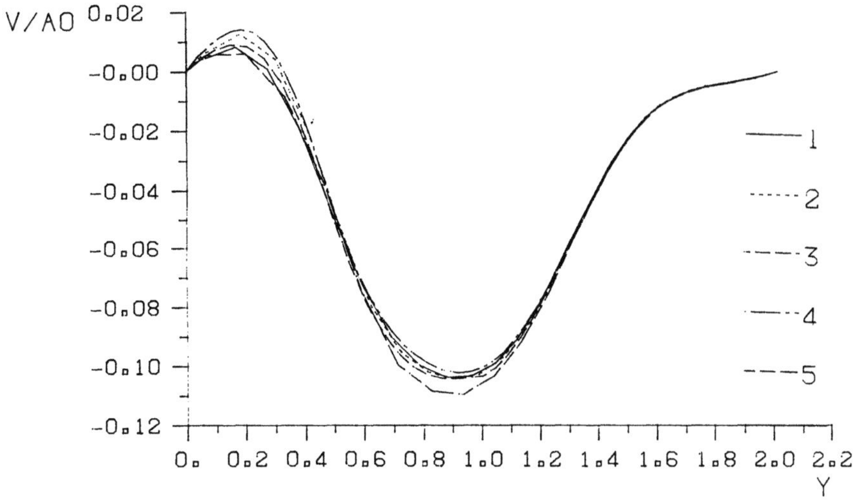

b) $Re_O = 400$

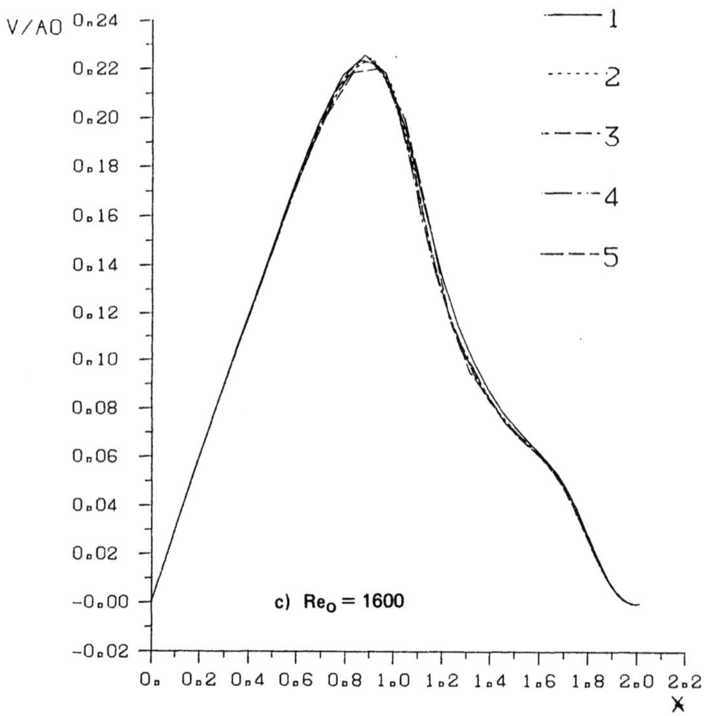

Fig. 13 y-Velocity profiles at x = 4. a) Re_o = 100, b) Re_o = 400, c) Re_o = 1600.

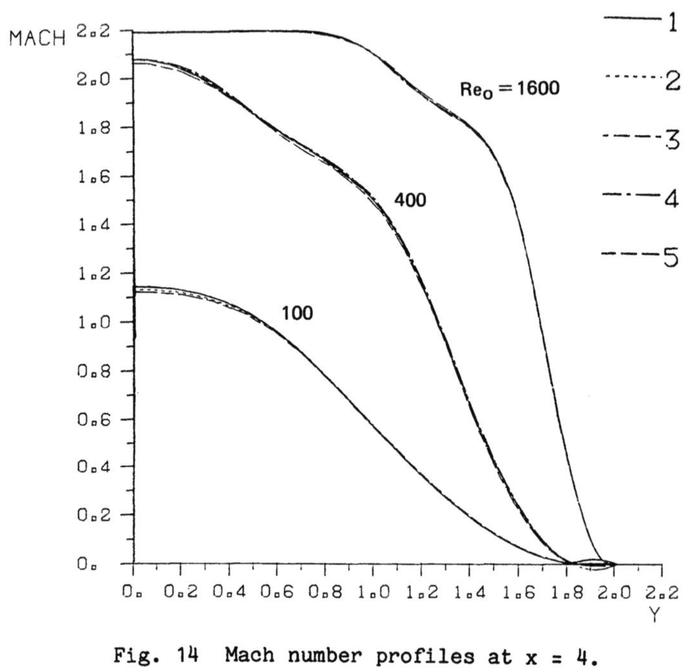

Fig. 14 Mach number profiles at x = 4.

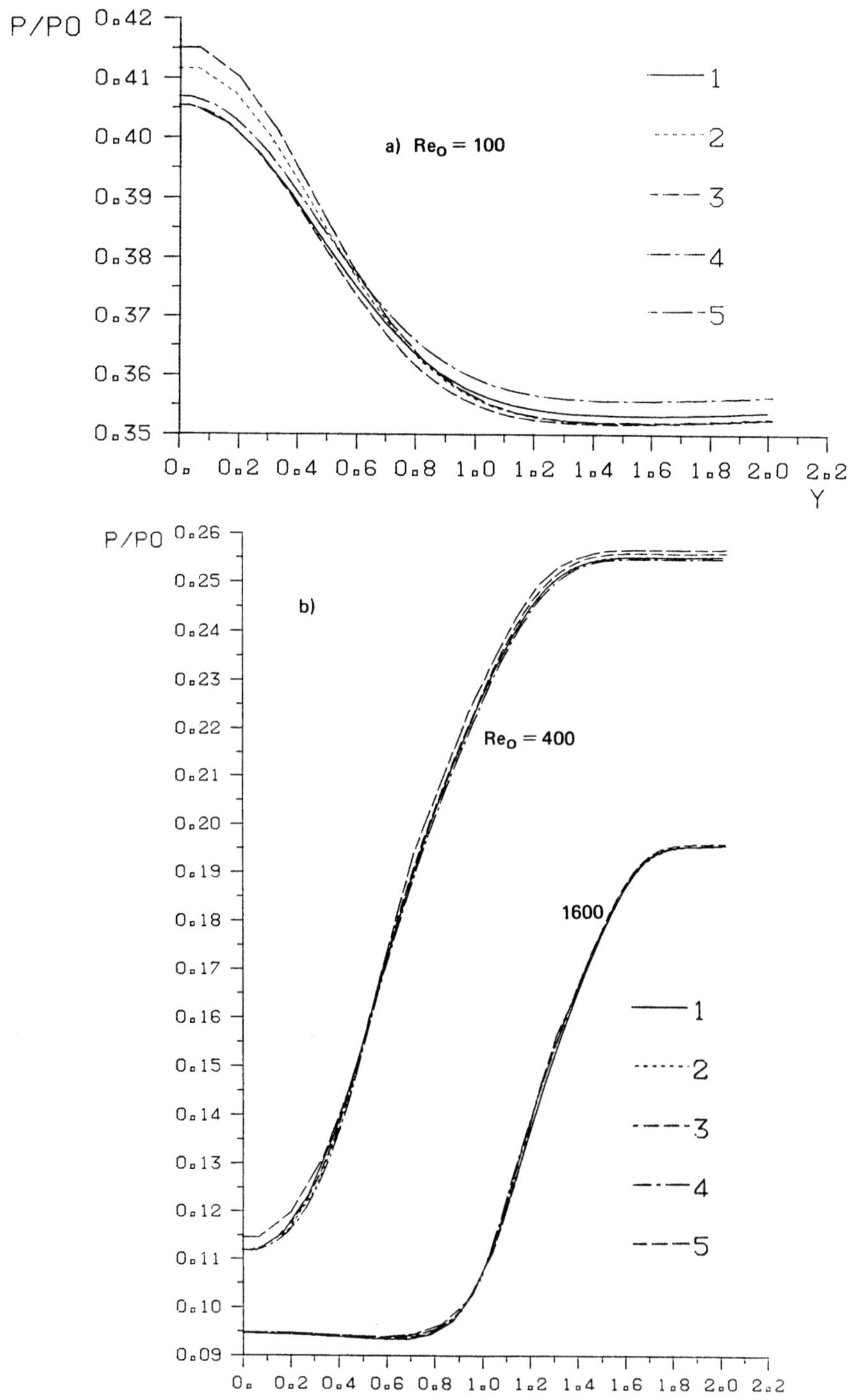

Fig. 15 Pressure profiles at x = 4. a) Re_O = 100. b) Re_O = 400 and 1600.

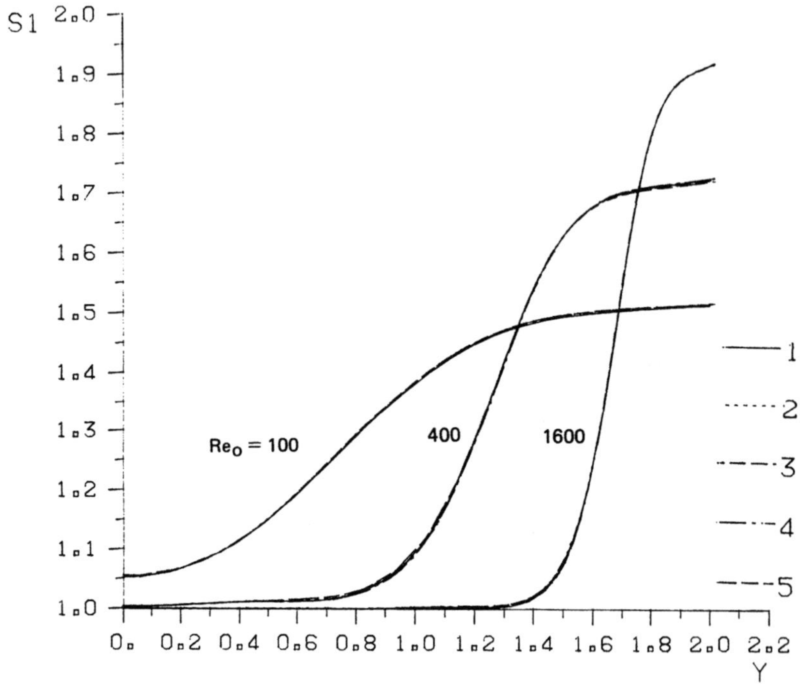

Fig. 16 S1 Profiles at x = 4 ($S1 = (p/p_o)(\rho_o/\rho)^{\gamma}$).

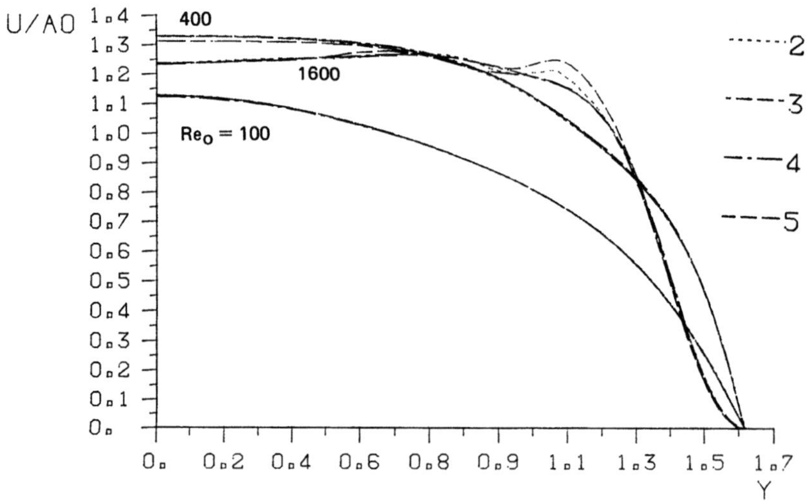

Fig. 17 x-Velocity profiles at x = 8.

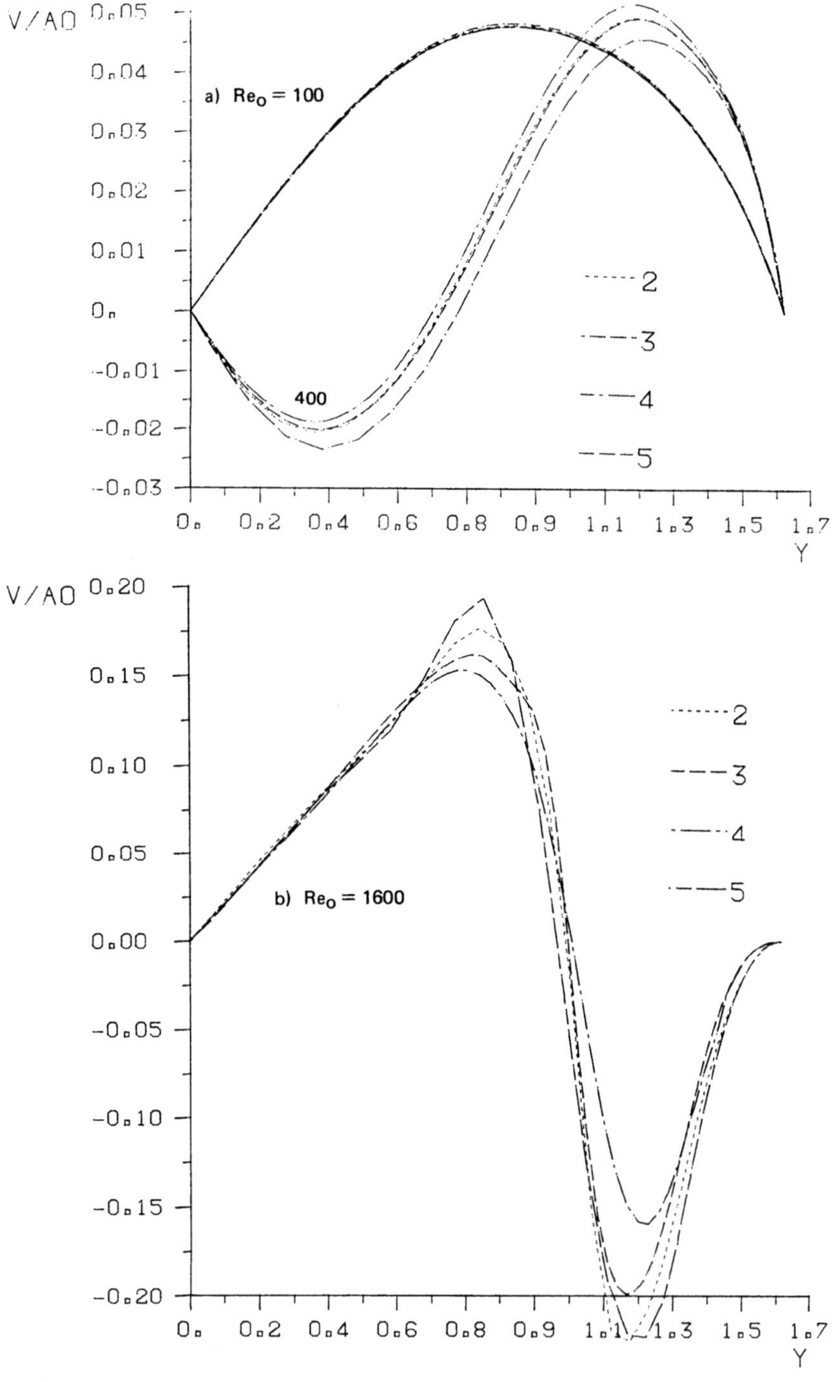

Fig. 18 y-Velocity profiles at x = 8. a) Re_0 = 100 and 400, b) Re_0 = 1600.

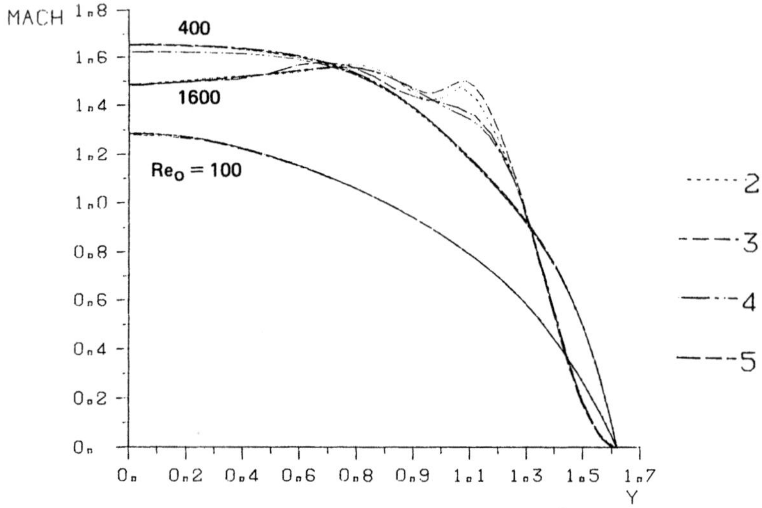

Fig. 19 Mach number profiles at x = 8.

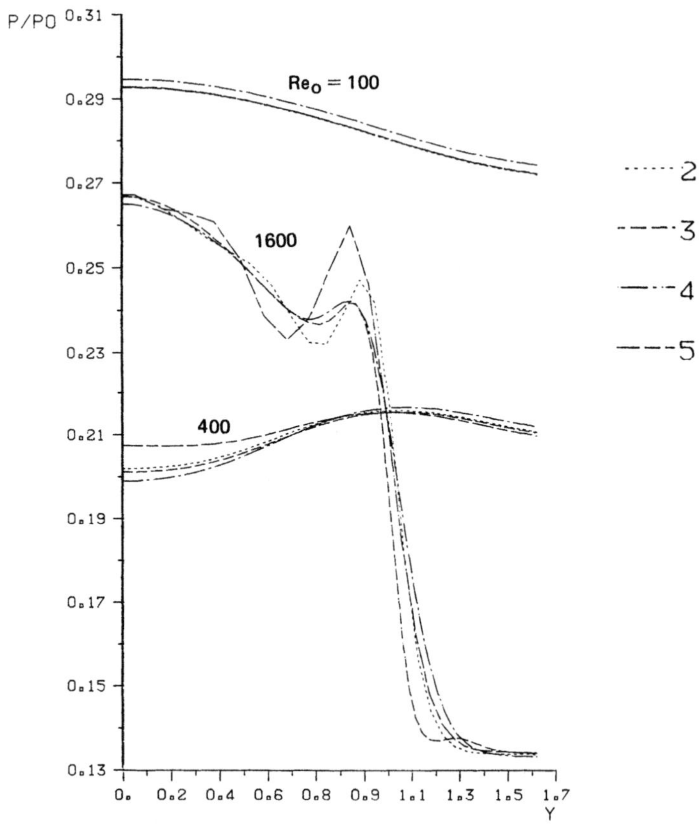

Fig. 20 Pressure profiles at x = 8.

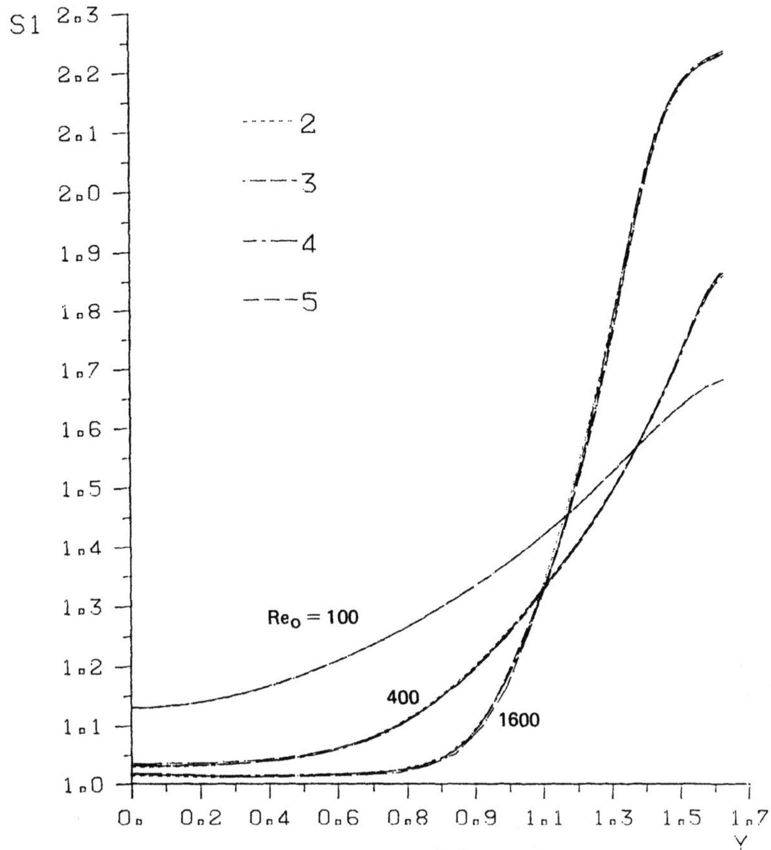

Fig. 21 S1 profiles at x = 8 ($S1 = (p/p_o)\,(\rho_o/\rho)^\gamma$).

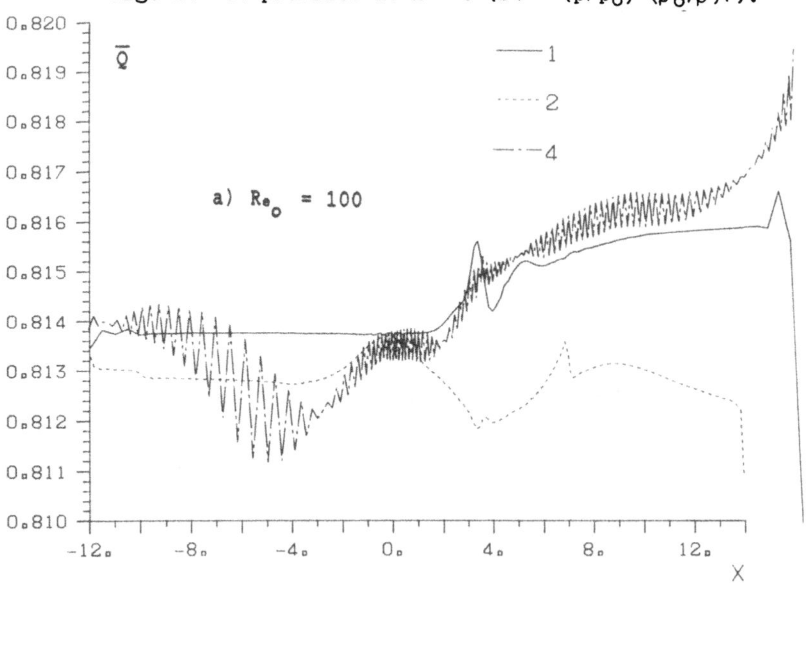

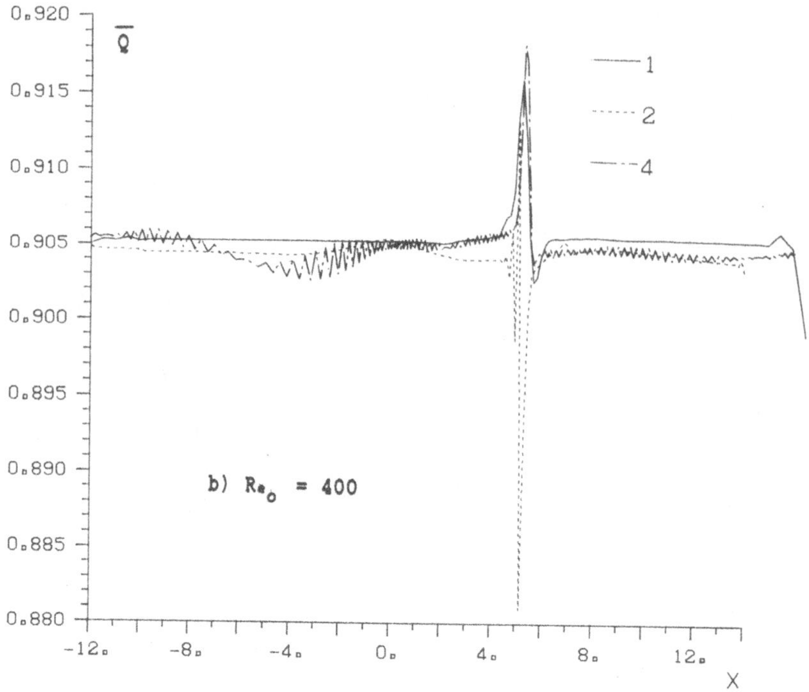

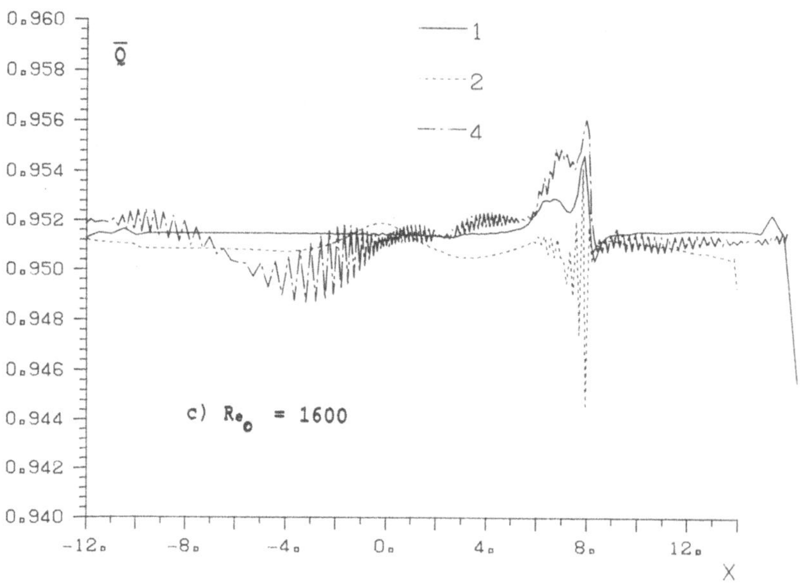

Fig. 22 Mass flow rate distributions. a) Re_o = 100, b) Re_o = 400, c) Re_o = 1600.

NUMERICAL SIMULATION OF COMPRESSIBLE VISCOUS FLOWS
- A SURVEY -

by

Roger Peyret
Département de Mathématiques
Université de Nice, Parc Valrose
06000 - NICE
and
INRIA, Sophia-Antipolis
06560 VALBONNE

1. INTRODUCTION

With the advances in theoretical and applied numerical analysis and in computer technology, the calculation of viscous compressible flows becomes a useful tool for industrial applications.

During the present GAMM Workshop, the reliability of methods for solving the Navier-Stokes equations has received an indisputable demonstration in case of two-dimensional (external and internal) flows at small or moderate Reynolds numbers. Obviously, it is necessary to verify also this reliability on more complicated problems involving higher Reynolds number, unsteady and three-dimensional flows in both laminar and turbulent cases. In case of turbulent flow, the limitation of the existing computers forbids a direct simulation through the exact nonaveraged Navier-Stokes equations, and it is necessary to have recourse to turbulence modelling. Hence, the success of turbulent flow calculations depends more on the validity of the turbulence model than on the ability of the numerical analysis.

Even if progress in the construction of the methods and in the construction of algorithms adapted to super-computers will be made in the future, the present state of the numerical solution of the equations governing compressible viscous flows allows us to apply it to a large variety of real flows.

This lecture is devoted to a presentation of the basic principles of numerical methods for solving the Navier-Stokes equations and some approximations of them. The numerical difficulties associated with this solution are discussed and some ways to surmount them are described. Only the laminar case will be considered here. We refer to [1] - [5] for general reviews on methods and applications to laminar and turbulent flows.

2. THE NAVIER-STOKES EQUATIONS

The motion of a compressible viscous fluid is governed by the Navier-Stokes equations (hereafter designated by the abbreviation "N.S" equations) derived from the application of the principles of classical mechanics and thermodynamics for the conservation of mass, momentum and energy. The general N.S. equations, written in integral form, are

$$\frac{d}{dt}\int_{\mathcal{V}} W\,dv + \int_{\Sigma} \vec{H}.\vec{N}\,d\sigma = 0 \qquad (2.1)$$

where Σ is the boundary of the fixed domain $\mathcal{V}$, $\vec{N}$ is the outward normal unit vector to Σ ; the unknown vector W is

$$W = (\rho, \rho u, \rho v, \rho w, \rho \mathcal{E}) \qquad (2.2)$$

where ρ is the density , u,v,w are the components of the velocity vector $\vec{V}$, and $\mathcal{E}$ is the total specific energy. The flux term $\vec{H} = (E,F,G)$ depends on W and its derivatives with respect to spatial coordinates. The flux term $\vec{H}$ can be divided in an inviscid part $\vec{H}_I = (E_I, F_I, G_I)$ and a viscous one $\vec{H}_V = (E_V, F_V, G_V)$ such that $\vec{H} = \vec{H}_I - \vec{H}_V$. The expression of $\vec{H}$ may be found, for example, in [1], [3] . Finally, the equation of state must be added to (2.1).

Now, by using the divergence theorem in (2.1), we obtain the differential form of the N.S equations :

$$\frac{\partial W}{\partial t} + \frac{\partial E}{\partial x} + \frac{\partial F}{\partial y} + \frac{\partial G}{\partial z} = 0 \qquad (2.3)$$

in a cartesian coordinate system (x,y,z). In case of complex geometry problems, the use of body-fitted curvilinear coordinate systems may be of interest. Let us denote by (ξ, η, ζ) the curvilinear system which may depend on time $t = \tau$; then, the equations (2.3) can be transformed in the conservative form [6]:

$$\frac{\partial \bar{W}}{\partial \tau} + \frac{\partial \bar{E}}{\partial \xi} + \frac{\partial \bar{F}}{\partial \eta} + \frac{\partial \bar{G}}{\partial \zeta} = 0 \qquad (2.4)$$

where $\bar{W} = W/J$, J being the Jacobian of the transformation. In the following we shall write occasionally Eq. (2.3) or (2.4) under the general symbolic form

$$\frac{\partial W}{\partial t} = A(W) \qquad (2.5)$$

where W is put for $\bar{W}$ in case of Eq. (2.4).

According to the method of approximation either the integral or the differential form has to be used. So, finite-difference (FD) methods make use of the differential form (2.3) or (2.4), while that finite-volume (FV) methods are based on the integral form (2.1). The finite-element (FE) methods make use of the differential form (2.3) associated with a weighted residuals formulation [7] . The spectral methods (not discussed here) can also be described in terms of weighted residuals methods.

3. BOUNDARY CONDITIONS

The unsteady N.S. equations for a compressible fluid form a hybrid hyperbolic-parabolic system called "incompletely parabolic" system. The parabolic nature is related to the presence of second-order derivatives in the momentum and energy equations which characterize the dissipative effects. On the other hand, the absence of such derivatives in the continuity equation makes the system different since itself it would be a hyperbolic equation assuming the velocity given. In the time-independent case, the N.S.equations form a hybrid hyperbolic-elliptic system.

Problems to be solved in practice are boundary-value problems, either time-dependent with given initial value of W or time-independent. Only a few special theoretical results are known about the well-posedness of N.S. problems. Heuristic arguments [1] - [3], justified by mathematical statements in the linearized case [8], allow us to define the number of boundary conditions needed at a (physical or artificial) boundary of the computational domain.

Let us consider the problem of the two-dimensional external flow around a fixed body with uniform flow conditions given at infinity. The domain of computation is bounded by the body surface Γ_b and by an artificial external boundary Γ_e.

At an inflow boundary Γ_e^i, the number of boundary conditions needed to solve the problem is equal to the number of equations, that is to say 4 in the present two-dimensional case. On an outflow boundary Γ_e^o, only 3 conditions have to be prescribed.

However, when the Reynolds number Re is large, so that the flow may be inviscid near the external boundary, it is preferable to consider boundary conditions compatible with the Euler equations and taking into account of the subsonic or supersonic nature of the flow near a boundary. In these respects, conditions that remain valid when Re $\to \infty$ are proposed in [8], [9]. Also, truly inviscid boundary conditions may be used [10] - [12] complemented with compatibility relations associated with the outgoing characteristics. The details of this technique are given in [13]. In some cases, as in [11], a simple extrapolation at the outflow boundary gives good results. Also, we point out that the viscous terms can be neglected near an inflow boundary when the flow in this region is practically inviscid.

On the body Γ_b assumed to be an impermeable wall and in cases where rarefaction effects are not present, the usual no-slip condition is

$$\vec{V} = 0. \tag{3.1}$$

For the temperature, two types of condition are usually considered, either

$$T = T_W \tag{3.2}$$

where T_W is the given wall temperature, or the adiabatic condition

$$\frac{\partial T}{\partial N} = 0. \tag{3.3}$$

In case of a perfect gas with constant specific heats, the above condition (3.2) or (3.3) gives a condition for the internal energy e. There is no boundary condition for the density ρ, and it therefore must be calculated. This can be done efficiently by using the normal derivative of the pressure $\partial p/\partial N$ at the wall deduced from the momentum equation projected normally to the body. In case of a high Reynolds number boundary-layer type flow, the condition $\partial p/\partial N = 0$ can be used.

4. SOME APPROXIMATIONS OF THE N.S. EQUATIONS

Because of the complexity of the N.S. problems (geometry and equations) particularly in case of three-dimensional flows, the use of approximated equations can be of interest. These approximations being based on specific assumptions, they may be used either in the whole computational domain if these hypotheses are satisfied everywhere or only in some regions of the domain according to the local nature of the flow. This second possibility leads to the multi-model/multi-zonal approach consisting in a division of the computation domain into a number of regions in which various systems of equations are employed, generally Euler equations, thin-layer N.S. equations, parabolized N.S. equations and N.S. equations themselves. This approach considered for example in [13] where the matching conditions between the different zones are discussed, become very interesting when the algorithm takes advantage of the possibilities of "parallelization" of the computer as in [14].

The Euler equations are derived from the N.S. equations by setting $\vec{H}_V = 0$.

The "thin-layer N.S. equations" considered in [15] require high Reynolds number flows such that viscous effects are restricted near the body. In this case, a body-fitted coordinate system (ξ,η,ζ) is introduced, such that $\zeta = 0$ corresponds to the body surface. Then, the ξ- and η- derivatives (i.e. laterally to the body) are neglected in the viscous stress terms, the heat conduction terms and the viscous dissipation function (i.e. in all dissipative terms). Therefore the thin-layer N.S. equations are deduced from (2.4), where ξ,η and ζ are independent of t, they write as

$$\frac{\partial \bar{W}}{\partial t} + \frac{\partial \bar{E}_I}{\partial \xi} + \frac{\partial \bar{F}_I}{\partial \eta} + \frac{\partial \bar{G}_I}{\partial \zeta} = \frac{\partial \bar{G}_V}{\partial \zeta} . \tag{4.1}$$

A theoretical discussion of these equations, related in particular with the boundary-layer equations is made in [16]. Numerical experiments comparing thin-layer and full N.S. equations were conducted in [17], [18]. For the problems considered in these works, the results were found to be in agreement. On the other hand, the validity of the thin-layer equations is suspected in [19] for unsteady separated flows.

The "parabolized N.S. equations" are derived from (4.1) by assuming the flow to be steady and the existence of a marching (parabolic) direction along the body. Such an approximation requires the inviscid flow outside the viscous layer to be supersonic and the velocity in the marching (streamwise) ξ-direction to be positive. Moreover, the flux term $\bar{E}_I$ in the ξ-direction must be modified in order to obtain a stable marching procedure in the subsonic region. This is done by replacing the pressure p by a constant p_o in the subsonic part of the viscous layer, the constant being the value of the pressure just at the first supersonic point away of the body. If $\hat{E}_I$ denotes the transformed flux term $\bar{E}_I$ in this way, the parabolized N.S. equations write as

$$\frac{\partial \hat{E}_I}{\partial \xi} + \frac{\partial \bar{F}_I}{\partial \eta} + \frac{\partial \bar{G}_I}{\partial \zeta} = \frac{\partial \bar{G}_V}{\partial \zeta} . \qquad (4.2)$$

Derivation and applications of the above parabolized N.S. equations are presented in [20] - [24]. Another way to devise a well posed space-marching procedure was proposed in [25] and applied also in [26]. The limitation of the parabolized N.S. equations consisting in forbidding any upstream influence, in particular separation, has led to the introduction of iterative methods [27], [28].

Finally, it must be pointed out that some other approximations are used : the viscous shock-layer equations for hypersonic blunt body problem [29], or the constant total enthalpy approximation [30], [31] valid within the boundary-layer equations with the Prandtl number equal to one and no wall heat flux [32].

5. SPATIAL DISCRETIZATION

The F.D. method deals with the equations written in the differential form either in cartesian or in curvilinear coordinate systems. The unknowns are the values $W_h = W_{i,j,k}$ of W at the discretization points (i,j,k). Equations for these values are obtained by using finite-difference formulas (based on Taylor series expansions) for approximating the derivatives.

In F.V. and F.E. methods, the computational domain is divided into elementary volume ϑ_h. (For example, quadrangle or triangle in the two-dimensional case).

In F.V. methods, the equations are considered in the integral form (2.1) where ϑ is now replaced by an elementary volume ϑ_h. Then, the approximation of the integrals involved in (2.1) leads to equations for the unknowns W_h that are now the average value of W in the elementary volumes.

A F.E. method makes use of the differential form of the equations in cartesian coordinates. The dependent variable W is now approximated in a volume ϑ_h by means of a polynomial of low degree in general. Then, the application of the Weighted Residuals method (Galerkin-type method) to Eq. (2.3) furnishes the equations for determining the values W_h at the vertices of the elements and possibly at other points (mid-point of the sides, center-point of the element...) according to the degree of the approximation.

We refer to text books (for example [3]) for a more detailed description of the above approximation methods.

6. ADAPTATION OF THE MESH

The question of the mesh is essential in the construction of a numerical method. In most cases, the quality of the results is strongly dependent on the properties of the mesh.

Meshes associated with F.D. or F.V. methods are generally "structured", that is to say the computational domain is divided by surfaces ξ = cte, η = cte, ζ = cte, so that each node is defined as the intersection of these surfaces and is identified by a set of indices (i,j,k). This is not always the case in F.E. methods for which the consideration of "unstructured" meshes permits an accurate representation of the geometry as well as a refined discretization of flow regions where the solution exhibits large variations. The generation of such meshes may be complicated, mainly in the three-dimensional case. We refer to [33] for a presentation of automatic mesh generation techniques and to [34] for some applications to aerodynamical problems.

Obviously, the mesh must be adapted to the geometry. Very often, the mesh is constructed through a body-fitted curvilinear coordinate system. In a very few cases, the geometry is sufficiently simple to be represented by an orthogonal curvilinear coordinate system known analytically in a closed form. Generally, the generation of the mesh is obtained by a less direct method in which the numerics may play an important role. The ways to generate arbitrary curvilinear systems which define the grid are various. We refer to [35] for an extensive review on grid generation techniques. Two categories of curvilinear systems can be distinguished : those where the coordinates are determined by interpolation, and those where these coordinates are determined as the solution of partial differential equations. The well-known technique of conformal mapping belongs to the second category.

Moreover, the mesh must be adapted to the solution in order to permit an accurate representation of the solution in regions of large variation (boundary layer, free shear layer, shock,...) while keeping the number of discretization points as small as possible.

When the location of the regions where the solution varies strongly is known (for example, boundary layer), an "a priori" refinement technique can be used, the simpler in case of structured meshes being the use of stretched coordinates. In [36] a technique of dichotomy was proposed and used with success to refine the mesh in boundary layer or wake regions [10], [13], [37]. The technique is based upon a division of the computation domain into several zones where the mesh size differs by a factor of 2. Such a technique is associated with a systematic procedure of matching the solution between adjacent zones while the solution is advanced in time.

However, when the location of the high gradient regions is not known in advance or when its boundary does not fit well with coordinate lines, a technique of self-adaptation must be employed.

In F.E. methods, this can be done by means of "mesh enrichment" techniques consisting in dividing elements where the accuracy of the solution is considered as insufficient. The criteria for such a division are based on the size of the truncation error or on physical considerations connected for example to the variation of the vorticity, the Mach number etc. [34], [38] - [40]. The corollary of a mesh enrichment procedure would be a "mesh impoverishment" procedure in order to keep the total number of elements within reasonable bounds, but such a technique presents difficulties which do not seem to have been presently solved. Hence, the "mesh movement" technique constitutes an efficient solution to the problem. The method was developed in [41] in association with a F.V. method for solving the N.S. equations in a curvilinear coordinate system, but it can also be employed with F.E. methods using unstructured meshes (see [42] for Euler equations).

This technique is based upon a spring analogy where springs connect adjacent mesh points, and spring constants are a function of the gradient of some quantity (for example, Mach number or internal energy) between the nodes. The number of grid points is kept constant and the points on the boundaries are kept fixed. Then the location of inner grid points is determined (through an iterative procedure) by assuming the points to be in equilibrium. The method is very efficient but, like many adaptation mesh techniques, it requires the determination of various parameters which has to be done for each specific problem.

7. TIME DISCRETIZATION

When the time discretization scheme has to be chosen, the usual question arises concerning its nature : explicit or implicit ? As it is known, each of one of these two types of discretization has its own advantages and drawbacks. The explicit schemes are easily implemented and not expensive in computing time by time cycle because they do not necessitate matrix inversion. On the other hand, the limitation on the size of the time step Δt due to stability requirements can lead to a very large number of cycles in the time integration, in particular when a steady solution is computed. Although, the explicit "Rational Runge-Kutta" scheme [43], [30] permits the use of larger Δt.

The implicit schemes, because of their good stability properties,do not suffer of this stringent limitation on Δt. However, matrices have to be inverted at each step of the time integration. Consequently, such schemes are heavier to implement than explicit ones and they are more time consuming per time cycle.

Moreover, explicit schemes are reputed to be more easily vectorized and hence more efficient on a vector computer. But, the progress in course in the construction of vectorized solution algorithms [44] lead us to think that efficient algorithms can also be devised for implicit methods.

Now, we briefly discuss these two types of schemes.

7.1. Explicit schemes

The simpler one-step explicit scheme writes as

$$W_h^{n+1} = C_h^n \; W_h^n \tag{7.1}$$

where W_h^n refers to the approximate solution at time $n\Delta t$, and C_h^n is a nonlinear operator which may have various expressions according to the method of spatial approximation under consideration. The expression (7.1) is formally valid for F.D., F.V. methods and F.E. methods with mass lumping.

For several reasons connected with stability, accuracy and simplicity of the algebra, multi-step schemes are generally preferred to one-step schemes. In this case intermediate values $W_h^{(r)}$, $r=1,\ldots,p-1$, are computed before to define the final value W_h^{n+1} :

$$\begin{aligned} W_h^{(1)} &= C_h^{(1)} \; W_h^n \\ W_h^{(2)} &= C_h^{(2)} \; W_h^n + \bar{C}_h^{(2)} \; W_h^{(1)} \\ &\cdots\cdots \\ W_h^{n+1} &= C_h^{(p)} \; W_h^n + \bar{C}_h^{(p)} \; W_h^{p-1}. \end{aligned} \tag{7.2}$$

Among these multi-step schemes, we may distinguish two categories :

A) <u>Schemes of type 1</u> : The operators approximating the spatial part of the equations are identical at each step (method of lines). The Braïlovskaya scheme [45] and the Cheng-Allen scheme [46], proposed several years ago belong to this category. The Braïlovskaya scheme is a two-step Runge-Kutta scheme with first order accuracy in time. More recently, multi-step schemes of this type, previously found successful for the Euler equations [47] were adapted to the N.S. equations in [31] , [48] . The "Rational Runge-Kutta" scheme [43] is also a scheme of type 1.

B) <u>Schemes of type 2</u> : The operators approximating the spatial part of the equations are different at each stage of the scheme. The extension of the Richtmyer ("two-step Lax-Wendroff") scheme proposed by Thommen [49] for solving the N.S. equations belongs to this type. It is the same for the well-known MacCormack scheme [50] and for the generalization of the Richtmyer scheme proposed in [3] , [51] and applied to a F.E. method in [52] .

It has been pointed out [3] , [53] that boundary conditions are more accurately introduced into schemes of type 1 than type 2. Moreover, the steady solution given by schemes of type 1 does not depend on Δt, contrary to schemes of type 2. Generally, this dependency on Δt is without importance because Δt

is small due to stability conditions. However, if an implicit smoothing operator is added to the explicit scheme, or if the explicit scheme is used as a part of an implicit scheme as in [11], [54], the size of the allowable Δt is increased and the dependency on Δt could become prejudicial to the accuracy. However, numerical experiments on practical cases [57] show that such a dependency is weak, at least for the Thommen scheme. It can be shown by a theoretical and numerical study of the steady solution given by the Thommen and the MacCormack[50] schemes for the one-dimensional advection-diffusion equation that the error can be strongly minimized by a modification of the scheme near the boundary. More precisely, in case of the Dirichlet problem, the error becomes $0(\Delta x^2, \Delta t \Delta x^2)$ if the first point near a boundary is calculated by the standard 3-point centered scheme. For the Mac Cormack scheme, the combination of the two variants (forward and backward), at the first point ([3], p.59) reduces also the error due to Δt.

7.2. Implicit schemes

The main reason to consider implicit schemes lies in their good stability properties. However, the computing time saved by the use of large Δt must not be lost by a too large computational effort needed at each step of the time integration. This lead us to avoid the use of any iterative procedure which could be associated with the nonlinearity of the N.S. equations. Therefore, some "linearization" procedure has to be employed in order to obtain a scheme depending linearly on W_h^{n+1}. Moreover, because the system defining W_h^{n+1} is solved at each time cycle, the associated matrix must have a structure leading to an efficient solution algorithm.

Let us consider the equation (2.5) discretized with respect to time according to the simple first-order scheme

$$W^{n+1} - W^n - \Delta t \; A(W^{n+1}) = 0. \tag{7.3}$$

More accurate schemes may be constructed in the usual ways : Crank-Nicolson, three-level backward, etc. (see [3], [55]). According to [55] - [57], the term $A(W^{n+1})$ can be "linearized" by means of a Taylor series expansion, so that (7.3) is replaced by

$$(I+\Delta t \; \Lambda_h^n)(W_h^{n+1} - W_h^n) = \Delta t \; A_h(W_h^n) \tag{7.4}$$

where Λ_h and A_h are discretized operators. At steady state, $W_h^{n+1} = W_h^n = W_h$, this equation (7.4) reduces to

$$A_h(W_h) = 0 \tag{7.5}$$

that is a discretized version at the steady N.S. equations. This remark leads to the construction of schemes for calculating steady solutions more efficient than (7.4), by modifying the operator Λ_h^n in order to simplify the solution of the associated algebraic system. For example, some terms (like those arising from mixed derivatives) can be neglected, or some of the matrices can be replaced by diagonal matrices [58], [59].

Also, it may be interesting to replace $A_h(W_h^n)$ in the R.H.S. of Eq. (7.4) by another approximation leading to better properties (stability, simplicity of the implementation,...). This is done in [11] where the solution at each time cycle is obtained in two stages :

A) <u>Explicit stage</u> : A provisional solution $\hat{W}_h^{n+1}$ is calculated from a predictor-corrector scheme of Thommen-type.

B) <u>Implicit stage</u> : This provisional solution is stabilized by application of the implicit operator defined in the L.H.S. of (7.4), that gives the final value W_h^{n+1} :

$$(I+\Delta t\ \Lambda_h^n)(W_h^{n+1} - W_h^n) = \hat{W}_h^{n+1} - W_h^n \tag{7.6}$$

so that the steady solution is the solution of the steady Thommen-type scheme.

In F.D. or F.V. methods, the solution of (7.4) or (7.6) can be obtained by means of a generalized ADI procedure (also called "Approximate factorization"). This procedure needs the operator Λ_h to be split into one-dimensional operators

$$\Lambda_h = \Lambda_x + \Lambda_y + \Lambda_z \tag{7.7}$$

then the ADI algorithm is

$$\begin{aligned}(I+\Delta t\ \Lambda_x^n)\ \Phi_h^{(1)} &= \Delta t\ A_h(W_h^n)\\ (I+\Delta t\ \Lambda_y^n)\ \Phi_h^{(2)} &= \Phi_h^{(1)}\\ (I+\Delta t\ \Lambda_z^n)\ \Phi_h^{(3)} &= \Phi_h^{(2)}\\ W_h^{n+1} &= W_h^n + \Phi_h^{(3)}.\end{aligned} \tag{7.8}$$

In case of 3-point approximation, only block-tridiagonal matrices have to be inverted at each step of the algorithm. Even, it is possible, thanks to some simplification of Λ_x, Λ_y and Λ_z and application of a diagonalization process to reduce the algorithm to the solution of scalar tridiagonal matrices [59], [60].

It must be pointed out that the equation resulting from the algorithm (7.8) differs from the equation (7.4) by a term $O(\Delta t^2)$. This error term can forbid the use of a very large Δt. This remark leads to the construction of "unfactored" methods in which the equation (7.4) is solved by a relaxation method. For instance, in [61], a single double-sweep forward-backward of a line Gauss-Seidel iteration at each time cycle was found sufficient.

8. ARTIFICIAL VISCOSITY

In high Reynolds number flows, it exists regions where the flow is practically inviscid. In such a case the scheme must be able to correctly solve the Euler equations. In particular, it must be able to capture shocks without spurious oscillations. This can be done by an appropriate differencing of the convective terms, for example by using the flux vector splitting approach [62] .

Another way is to add artificial viscosity terms to the scheme. The artificial viscosity must be sufficient to stabilize the solution and to damp the possible oscillations but its effect in the viscous regions of the flow must be very weak compared to the real viscous effects. Various types of artificial viscosity are used is viscous flow computations. We shall present some of the most employed. Let us denote by

$$D(W) = D_x(W) + D_y(W) + D_z(W) \tag{8.1}$$

the damping term whose the discretized version is added to the scheme according to

$$W_{hd}^{n+1} = W_h^{n+1} + D_h(W_h^{n+1}) , \tag{8.2}$$

where W_{hd}^{n+1} is the damped solution associated with W_h^{n+1} given by the scheme and let us consider the expression for $D_x(W)$ only :

A) Second-order nonlinear viscosity

This viscosity is particularly adapted to the computation of shock wave. It has been devised by Lax and Wendroff [63] , but it is generally used in a simplified version :

$$D_x = \kappa \, \Delta t \, \Delta x^q \frac{\partial}{\partial x}\left(Q \frac{\partial W}{\partial x}\right) \tag{8.3}$$

where κ is a positive constant chosen as small as possible. Possible choices for Q and q are :

$$Q = \text{diagonal matrix with entries} = |\partial W_k/\partial x| , \tag{8.4}$$
$$W_k = \text{components of } W,$$
$$q = 2$$

$$Q = \frac{|u| + c}{p} \left|\frac{\partial^2 p}{\partial x^2}\right| , \quad q = 3 \qquad \text{(Ref. [64])} \tag{8.5}$$

where c is the sound speed and p is the pressure.

B) Fourth-order linear viscosity

In [65] , Gnoffo has employed the fourth-order damping term

$$D_x = - \kappa \, \Delta x^4 \frac{\partial^4 W}{\partial x^4} \tag{8.6}$$

which is now currently used (see [3]).

C) <u>Blended 2nd-order/4th-order viscosity</u>

This viscosity term modelled for the Euler equations [47] is applied to the N.S. equations in [31], [48]. It can be expressed as

$$D_x = D_x^{(2)} + D_x^{(4)} \qquad (8.7)$$

where

$$D_x^{(2)} = \kappa^{(2)} \Delta x^4 \frac{\partial}{\partial x} \left(Q \frac{\partial W}{\partial x}\right) \qquad (8.8)$$

with Q given by (8.5) and

$$D_x^{(4)} = \begin{cases} -\Delta x^4 \frac{\partial}{\partial x} \left[(\kappa^{(4)} - \Delta x^2 Q) \frac{\partial^3 W}{\partial x^3} \right] & \text{if } \kappa^{(4)} > \Delta x^2 Q \\ 0 & \text{if } \kappa^{(4)} \le \Delta x^2 Q \end{cases} \qquad (8.9)$$

where $\kappa^{(2)}$ and $\kappa^{(4)}$ are positive constant. This viscosity term has the property to reduce to a nonlinear second-order viscosity term in shock regions. In regions of smooth flow the dissipative term is $0(\Delta x^4/\Delta t)$ while it becomes $0(\Delta x^2/\Delta t)$ in the neighborood of a shock.

9. ACCELERATION OF CONVERGENCE

When steady flows are computed through the unsteady Navier-Stokes equations by considering the asymptotic solution as $t \to \infty$, it is important to reach this steady solution, defined by sufficiently small values of the time-derivatives, as fast as possible.

The consideration of implicit schemes, which permit large time steps, gives a solution to this question. But other various ways are possible.

First, the use of an unsteady system of equations different from the exact unsteady N.S. equations may improve the convergence towards the steady state, for example by allowing a larger time step than this associated with the exact equations [66].

Another way is to use at each time cycle a spatially varying time step rather than a constant one. This technique, currently, called "local time step technique", consists in using at each discretization point or elementary cell, a variable Δt_h defined from the maximum local value Δt_{cr} given by the stability criterion :

$$\Delta t_h = \sigma \, \Delta t_{cr} \qquad (9.1)$$

where σ is a safety factor which may depend on space variables in order to avoid too large variations of Δt_h induced by large variations of the mesh.

When applied to an implicit scheme, the critical time step Δt_{cr} may be derived from the associated explicit scheme by setting $\Lambda_h^n = 0$ in Equation (7.4).

This technique, as well as the previous one, belongs to the general pseudo-unsteady approach discussed in [67].

Finally, we note the use of multiple-grid techniques which have been recently found successful for the solution of the N.S. equations [68] - [71].

REFERENCES

[1] PEYRET R. and VIVIAND H. "Computation of viscous compressible flows based on the Navier-Stokes equations", AGARDograph No. 212, 1975.

[2] HOLLANDERS H. and VIVIAND H. "The numerical treatment of compressible high Reynolds number flows", in "Computational Fluid Dynamics", Von Karman Institute Book, Vol.2, p. 1-65, Hemisphere Publ., 1980.

[3] PEYRET R. and TAYLOR T.D. "Computational Methods for Fluid Flow", Springer-Verlag, 1983.

[4] LOMAX H. and MEHTA U.B. "Some physical and numerical aspects of computing the effect of viscosity on fluid flow", in "Computational methods in viscous flows", W.G.Habashi, Ed., p.1-50, Pineridge Press, Swansea, 1984.

[5] SHANG J.S. "An assessment of numerical solutions of the compressible Navier-Stokes equations", J. Aircraft, 22, 353-370, 1985.

[6] VIVIAND H. "Formes conservatives des équations de la Dynamique des Gaz", La Recherche Aérospatiale, No.1974-1, 65-66, 1974.

[7] FLETCHER C.A.J. "Computational Galerkin methods", Springer-Verlag, New-York, 1984.

[8] GUSTAFSSON B. and SUNDSTROM A. "Incompletely parabolic problems in Fluid Dynamics", SIAM J. Appl. Math., 35 , 343-357, 1978.

[9] OLIGER J. and SUNDSTROM A. "Theoretical and practical aspects of some initial boundary value problems in Fluid Dynamics", SIAM J. Appl. Math., 35, 419-446, 1978.

[10] CAMBIER L. "Computation of viscous transonic flows using an unsteady type method and a zonal grid refinement technique" in present volume, pp. 105-122.

[11] HOLLANDERS H., LERAT A. and PEYRET R. "Three-Dimensional calculation of transonic viscous flows by an implicit method", AIAA J., 23, 1670-1678, 1985.

[12] HOLLANDERS H. and RAVALASON W. "Résolution des équations de Navier-Stokes en fluide compressible par une méthode implicite", 22ème colloque Aérodynamique Appliquée, Lille, 13-15 Novembre 1985.

[13] CAMBIER L., GHAZZI W., VEUILLOT J.P. and VIVIAND H. "A multi-domain approach for the computation of viscous transonic flows", in "Computational Methods for viscous flows", W.G. Habashi Ed., p.513-539, Pineridge Press, Swansea, 1984.

[14] HOLST T.L., THOMAS S.D., KAYNAK U., GUNDY K.L., FLORES J. and CHADERJIAN N.M. "Computational aspects of zonal algorithms for solving the compressible Navier-Stokes equations in three-dimensions", Intern. Symp. Computational Fluid Dynamics-Tokyo, Sept. 9-12, 1985, p. 305-316.

[15] PULLIAM T.H. and STEGER J.L. "Implicit finite-difference simulations of three-dimensional compressible flow", AIAA J., 18, 159-167, 1980.

[16] BLOTTNER F.G. "Significance of the thin-layer Navier-Stokes equations", Proceed. Third Symp. on Numerical and Physical Aspects of Aerodynamic Flow, Long Beach, Jan. 21-24, 1985. To be published by Springer-Verlag.

[17] DEGANI D. and STEGER J.L. "Comparison between Navier-Stokes and thin-layer computation for separated supersonic flow", AIAA J., 21, 1604-1607, 1983.

[18] MULLER B. "Navier-Stokes solution for hypersonic flow over an indented nosetip", AIAA paper 85-1504, 1985.

[19] CHYU W.J. and KUWAHARA K. "Computations of transonic flow over an oscillating airfoil with shock-induced separation, AIAA paper 82-350, 1982.

[20] RUBIN S.G. and LIN T.C. "A numerical method for three-dimensional viscous flow : application to the hypersonic leading edge", J. Comput. Phys. 9, 339-364, 1972.

[21] SCHIFF L.B. and STEGER J.L. "Numerical simulation of steady supersonic viscous flow", AIAA J., 18, 1421-1430, 1980.

[22] LI C.P. "A computational procedure for supersonic flows governed by the parabolic Navier-Stokes equations", J. Comput. Phys., 35, 356-380, 1980.

[23] RIZK Y.M., CHAUSSEE D.S. and Mc RAE D.S. "Computation of hypersonic viscous flow around three-dimensional bodies at high angles of attack", AIAA paper 81-1261, 1981.

[24] RAI M.M., CHAUSSEE D.S. and RIZK Y.M. "Calculation of viscous supersonic flow over finned bodies", AIAA paper 83-1667, 1983.

[25] VIGNERON Y.C., RAKICH J.V. and TANNEHILL J.C. "Calculation of supersonic flow over delta wings with sharp leading edges", AIAA paper 78-1137, 1978.

[26] LAWRENCE S.L., TANNEHILL J.C. and CHAUSSEE D.S. "Application of the implicit Mac Cormack scheme to the Parabolized Navier-Stokes equations", AIAA J., 22, 1755-1763, 1984.

[27] RAKICH J.V. "Interactive PNS method for attached flows with upstream influence", AIAA paper 83-1955, 1983.

[28] LOMBARD C.K., VENKAPATHY E. and BARDINA J. "Universal single level implicit algorithm for Gasdynamics", AIAA paper 84-1533, 1984.

[29] GUPTA R.N., GNOFFO P.A. and MACCORMACK R.W. "Viscous Shock-layer flowfield analysis by an explicit-implicit method", AIAA J., 23, 723-732, 1985.

[30] BRISTEAU M.O., MANTEL B., GLOWINSKI R., PERIAUX J., POULETTY C. and SING G.H. "Implicit and semi-implicit methods for the compressible Navier-Stokes equations", Proc. Sixth GAMM Conf. Numerical Methods in Fluid Mechanics, Göttingen, Sept. 1985, Vol.13, Vieweg.

[31] SWANSON R.C. and TURKEL E. "A multistage time-stepping procedure for the Navier-Stokes equations", AIAA paper 85-0035, 1985.

[32] LAGERSTROM P.A. "Laminar flow theory" in "Theory of Laminar Flow", F.K. Moore, Ed., p. 20-285, Princeton Univ. Press, 1964.

[33] GEORGE P.L. "Modulef : Génération automatique de maillages", Collection didactique, INRIA, Rocquencourt, 1986.

[34] POULETTY C. "Génération et optimisation de maillages en éléments finis. Application à la résolution de quelques équations en Mécanique des Fluides", Thèse Docteur-Ingénieur, Génie industriel - Analyse Numérique, Ecole Centrale des Arts et Manufactures, Paris, 1985.

[35] THOMPSON J.F. "Grid generation technique in Computational Fluid Dynamics", AIAA J., 22, 1505-1523, 1984.

[36] VIVIAND H. and GHAZZI W. "Numerical solutions of the compressible Navier-Stokes equations at high Reynolds number with application to the blunt body problem", Lecture Notes in Physics", Vol.59, p. 434-439, Springer-Verlag, 1976.

[37] LE BALLEUR J.C., PEYRET R. and VIVIAND H. "Numerical studies in high Reynolds number Aerodynamics", Computers and Fluids, 8, 1-30, 1980.

[38] PALMERIO B., BILLEY V., DERVIEUX A. and PERIAUX J. "Self-adaptive mesh refinements and finite element methods for solving the Euler equations", ICFD Conference on "Numerical Methods for Fluid Dynamics", Reading, U.K., April 1-4, 1985.

[39] LÖHNER R., MORGAN K. and ZIENKIEWICZ O.C. "An adaptive finite element method for high speed compressible flow", 9th International Conference on Numerical Methods in Fluid Dynamics. Lecture Notes in Physics, Vol. 218, p. 388-392, Springer-Verlag, 1985.

[40] LÖHNER R., MORGAN K., PERAIRE J., ZIENKIEWICZ O.C. and KONG L. "Finite element methods for compressible flow", ICFD Conf. on "Numerical Methods for Fluid Dynamics", Reading, U.K., April 1-4, 1985.

[41] GNOFFO P.A. "A finite-volume, adaptive grid algorithm applied to planetary entry flowfields", AIAA J., 21, 1249-1254, 1983.

[42] PALMERIO B. and DERVIEUX A. "Application of a FEM moving node adaptive method to accurate shock capturing", 1st Intern. Conf. on Numerical Grid Generation in Computational Fluid Dynamics", Landshut, Germany, July 14-17, 1986.

[43] SATOFUKA N. "Unconditionally stable explicit method for the numerical solutions of the compressible Navier-Stokes equations", Proc. Fifth GAMM Conf. on Numerical Methods in Fluid Mech., Rome, October 5-7, 1983, M.Pandolfi and R. Piva, Eds., p.291-298, Vieweg, Braunschweig, 1984.

[44] ORTEGA J.M. and VOIGT R.G. "Solution of partial differential equations on vector and parallel computers", ICASE Report No. 85-1, January 1985.

[45] BRAILOVSKAYA I.Yu "A difference scheme for numerical solution of the two-dimensional nonstationary Navier-Stokes equations for a compressible gas". Soviet Physics-Doklady, 10, 107-110, 1965.

[46] ALLEN J.S. and CHENG S.I. "Numerical solutions of the compressible Navier-Stokes equations for the laminar near wake", Phys. Fluids, 19, 37-52, 1970.

[47] JAMESON A., SCHMIDT W. and TURKEL E. "Numerical Solutions of the Euler equations by finite volume methods using Runge-Kutta time-stepping schemes", AIAA paper 81-1259, 1981.

[48] HAASE W., WAGNER B. and JAMESON A. "Development of a Navier-Stokes method based on a finite-volume technique for the unsteady Euler equations", Proc. Fifth GAMM Conf. in Numerical methods in Fluid Mech., Rome, October 5-7, 1983, M.Pandolfi and R.Piva Eds., p. 99-107, Vieweg, Braunschweig, 1984.

[49] THOMMEN H.U. "Numerical integration of the Navier-Stokes equations", ZAMP, 17, 369-384, 1966.

[50] MACCORMACK R.W. "The effect of viscosity in hypervelocity impact cratering", AIAA paper 69-354, 1969.

[51] LERAT A. et PEYRET P. "Propriétés dispersives et dissipatives d'une classe de schémas aux différences pour les systèmes hyperboliques non-linéaires", La Recherche Aerospatiale, n° 1975-2, p. 61-79, 1975.

[52] ANGRAND F. "Viscous perturbation for the compressible Euler equations. Application to the numerical simulation of compressible viscous flows", in present volume, pp. 69-84.

[53] CHENG S.I. "A critical review of the numerical solution of Navier-Stokes equations", in "Progress in Numerical Fluid Dynamics", Lecture Notes in Physics, Vol.41, p. 78-225, Springer-Verlag, 1975.

[54] MACCORMACK R.W. "A numerical method for solving the equations of compressible viscous flows", AIAA J., 20, 1275-1281, 1982.

[55] BEAM R.M. and WARMING R.F. "An implicit factored scheme for the compressible Navier-Stokes equations", AIAA J., 16, 393-402, 1978.

[56] LINDEMUTH I. and KILLEEN J. "Alternating direction implicit techniques for two-dimensional magnetohydrodynamic calculations", J. Compt. Phys., 13, 181-208, 1973.

[57] BRILEY W.R. and Mc DONALD H. "Solution of the three-dimensional compressible Navier-Stokes equations by an implicit technique", Lecture Notes in Physics, Vol.35, p. 105-110, Springer, 1975.

[58] LERAT A. and SIDES J. "A new finite-volume method for the Euler equations with applications to transonic flows", in "Numerical Methods in Aeronautical Fluid Dynamics", P.L. Roe Ed., p. 245-288, Academic Press, 1982.

[59] LI C.P. "A finite-difference method for solving unsteady flow problems", AIAA J., 23, 659-668, 1985.

[60] CHAUSSEE D.S. and PULLIAM T.H. "Two-dimensional inlet simulation using a diagonal implicit algorithm", AIAA J., 19, 153-159, 1981.

[61] MACCORMACK R.W. "Current status of numerical solutions of the Navier-Stokes equations", AIAA paper 85-0032, 1985.

[62] STEGER J.L. and WARMING R.F. "Flux vector splitting of the inviscid gas dynamic equations with application to finite-difference methods", J. Comput. Phys., 40, 263-293, 1981.

[63] LAX P.D. and WENDROFF B. "Systems of conservation laws", Comm. Pure and Appl. Math., 13, 217-237, 1960.

[64] MACCORMACK R.W. and BALDWIN B.S. "A numerical method for solving the Navier-Stokes equations with application to shock-boundary layer interactions", AIAA paper 75-1, 1975.

[65] GNOFFO P.A. "Solutions of the Navier-Stokes equations for supersonic flow over blunt bodies in a generalized coordinate system", M.S. Thesis, Polytechnic Institute of Brooklyn, 1974.

[66] PEYRET R. and VIVIAND H. "Calcul de l'écoulement d'un fluide visqueux autour d'un obstacle de forme parabolique", Lecture Notes in Physics, Vol. 19, p. 222-229, Springer-Verlag, 1973.

[67] PEYRET R. and VIVIAND H. "Pseudo-unsteady methods for inviscid of viscous flow computation", in "Recent Advances in the Aerospace Sciences", C. Casci, Ed., p.41-71, Plenum Press, 1985.

[68] DAVID R.L., NI R.H. and BOWLEY W.W. "Prediction of compressible, laminar viscous flows using a time-marching control volume and multiple-grid technique", AIAA J., 22 1573-1581, 1984.

[69] JOHNSON G.M. "Convergence acceleration of viscous flow computations", NASA TM 83039, October 1982.

[70] CHIMA R.W. and JOHNSON G.M. "Efficient solution of the Euler and Navier-Stokes equations with a vectorized multiple-grid algorithm", AIAA J., 23, 23-32, 1985.

[71] JAMESON A., MARTINELLI L. and GRASSO F. "A multigrid method for the Navier-Stokes equations", AIAA paper 86-0208, January 1986.

EXPERIMENTAL FLOWFIELDS AROUND NACA 0012 AIRFOILS LOCATED IN SUBSONIC AND SUPERSONIC RAREFIED AIR STREAMS

J. Allègre, M. Raffin, J.C. Lengrand*

S.E.S.S.I.A., 75116-Paris, France

*Laboratoire d'Aérothermique du C.N.R.S., 92190-Meudon, France

INTRODUCTION

During the past decade, significant progress has been achieved in the development of numerical methods, increasing the possibility of aerodynamic calculations. These methods are a valuable tool for everyone who tries to understand specific problems or wants to develop new projects.

However, even for sophisticated methods, it is necessary to verify the validity of chosen hypotheses and, consequently, to establish comparisons, for some typical configurations, between theoretical flowfields and experimental flowfields as obtained in testing facilities. That is the purpose of the present investigation with presentation of a number of experimental flowfields which could be considered as references to offer further comparisons with available theoretical models.

TEST CONDITIONS

At low Reynolds numbers, flow structures are quite satisfactory to enable comparisons between theoretical and experimental flowfields. As a matter of fact, rarefaction conditions conduct to important shock thicknesses and to large boundary layers, making easier the flow probing. In addition, rarefied flow surveys may be performed by means of an electron beam. One should recall that the electron beam gives local density measurements without disturbing the investigated flow. The measurement technique consists of emitting a high energy electron beam right across the flow in order to produce ionization and excitation of collided molecules or atoms of the tested gas ; by recovering states of lower energy, these molecules or atoms emit photons. Following the excitation process, locally, the intensity of emitted light is a function of the molecular concentration of the gas. The recorded light intensity allows to give access to the value of local density. The electron beam may also be used as a means of flow visualization.

In France, at the present time, experimental aerodynamic investigations of rarefied gases are mainly performed in the SR3 facility of "laboratoire d'Aérothermique" at Meudon. Investigations described in the present paper allow to obtain global description of aerodynamic flowfields.

Test conditions correspond to subsonic and supersonic rarefied air flows characterized by Mach numbers of 0.8 and 2 and respective Reynolds numbers of 73 and 106 as referred to the airfoil chord.

FLOWFIELD PRESENTATION IN THE SUBSONIC REGIME

The NACA 0012 airfoil, 4 cm long, is located in the air stream characterized by a uniform Mach number of 0.8 over a diameter of the order of 10 cm and a Reynolds number of 73 calculated on the airfoil chord. The wall temperature of the airfoil, equal to the stagnation temperature of the flow, is about 290 K.

Transverse flow surveys have been performed along fifty cross sections distributed from the abscissa located at $3/10^{th}$ of the chord, ahead of the leading edge, up to the abscissa located at $1/10^{th}$ of the chord, downstream of the trailing edge. Transverse density distributions allow to draw lines of equal density around the airfoil. Density distributions are presented in figures 1 and 2 respectively for incidences of 0° and 20°. For the 10° incidence, the overall density flowfield is presented in figure 3 for a flow region extending from $3/10^{th}$ of the chord, ahead of the leading edge, up to $8/10^{th}$ of the chord, downstream of the trailing edge.

FLOWFIELD PRESENTATION IN THE SUPERSONIC REGIME

The air flow, issued from an axisymmetric contoured nozzle, is characterized by a uniform Mach number of 2, over a diameter of about 16 cm, and a Reynolds number of 106 calculated on the airfoil chord. As indicated for previous configurations in subsonic regime, transverse density surveys allowed to obtain lines of equal density all around the airfoil. Density distributions are shown in figures 4 and 5 for respective airfoil incidences of 0° and 20° ; density distribution is also presented in figure 6 for the intermediate 10° incidence.

An attempt of comparison has been made between experimental flowfields, obtained through present wind-tunnel tests, and calculated flowfields. Among available methods on the resolution of Navier-Stokes equations in compressible flows (1-3), ONERA program (2) was slightly modified in order to be suitable with CNRS computer of CIRCE. For the present paper, the comparison between experimental and theoretical flowfields is limited to the zero incidence airfoil configuration.

The theoretical density flowfield presented in figure 7 corresponds to wind-tunnel test conditions, i.e. flow Mach number of 2 and Reynolds number of 106 based on the airfoil chord. Hypotheses of no slip conditions and of adiabatic wall have been assumed. In order to provide easier comparisons between experimental and theoretical density flowfields presented respectively in figures 4 and 7, lines representative of a limited number of density values have been drawn in figure 8. Continuous lines and discontinuous lines correspond respectively to calculated density and to measured density. An acceptable agreement appears between experimental and calculated density contours in the region surrounding the wall of the NACA 0012 airfoil. Nevertheless, ahead of the leading edge or across the shock, observed differences show that the experimental shock, as deduced from density surveys, is pushed backwards, in the stream direction, as compared with its theoretical location. In order to materialize the theoretical or experimental shock location, lines of constant density, corresponding to 1.2 time the infinite density value, are drawn in figure 9 ; the upstream experimental limit of the shock merges together with the shock location as deduced from flow visualization obtained by means of the electron beam. On the same figure are indicated lines of maximum values for the calculated and experimental density.

CONCLUSIONS

Experimental density flowfields around NACA 0012 airfoils have been presented for subsonic and supersonic conditions characterized by Mach numbers of 0.8 and 2 and respective Reynolds numbers of 73 and 106 based on the airfoil chord.

In the supersonic regime, comparisons between experimental density flowfields and theoretical density distributions, deduced from Navier-Stokes calculation, point out significant discrepancies between test results and theoretical predictions. Such a difference may also be noticed between experimental velocities around the airfoil and theoretical values deduced from Navier-Stokes equations in the hypothesis of no slip conditions (4). In fact, the noticed difference could be explained, at least partly, by the fact that a slip velocity does exist along the airfoil surface. The present flow conditions correspond effectively to the slip region as defined on the Mach-Reynolds representation of figure 10. At the limit of continuum flow conditions and for transition flow regimes, theoretical approaches to quantify slip conditions don't allow to calculate with accuracy the slip velocity ; this last one depends on local rarefied conditions but also on reflexion and accommodation coefficients. An investigation has been initiated in order to quantify slip effects and corresponding results should be available for the 15th Rarefied Gas Dynamics Symposium to be held on June 86. Thus, theoretical determinations of flowfields should also take into account expected slip conditions.

Acknowledgements

The authors gratefully acknowledge the financial support of "Direction des Recherches et Etudes Techniques" and H. Viviand from ONERA for his valuable help.

REFERENCES

(1) PEYRET, R., VIVIAND, H. : "Computation of viscous compressible flows based on Navier-Stokes equations", AGARD-AG 212, 1975.

(2) VIVIAND, H., GHAZZI, W. : "Solution numérique des équations de Navier-Stokes en compressible à grands nombres de Reynolds, avec application au problème du corps émoussé", ONERA, T.P. n° 1977-17.

(3) BRISTEAU, M.O., et al. : "Finite element methods for the compressible Navier-Stokes equations", AIAA Computational Fluid Dynamics Conference, Danvers, 13-15 july 1983.

(4) ALLEGRE, J., LENGRAND, J.C., RAFFIN, M. : "Champs d'écoulements raréfiés compressibles autour d'un profil NACA 0012", SESSIA Report 894/82.380, nov. 1982.

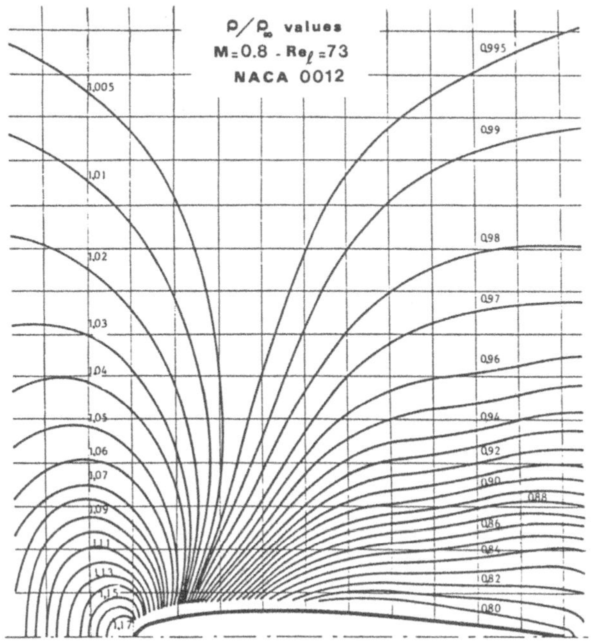

Fig. 1

Experimental density flowfield (i=0°)

Fig. 2

Experimental density flowfield (i=20°)

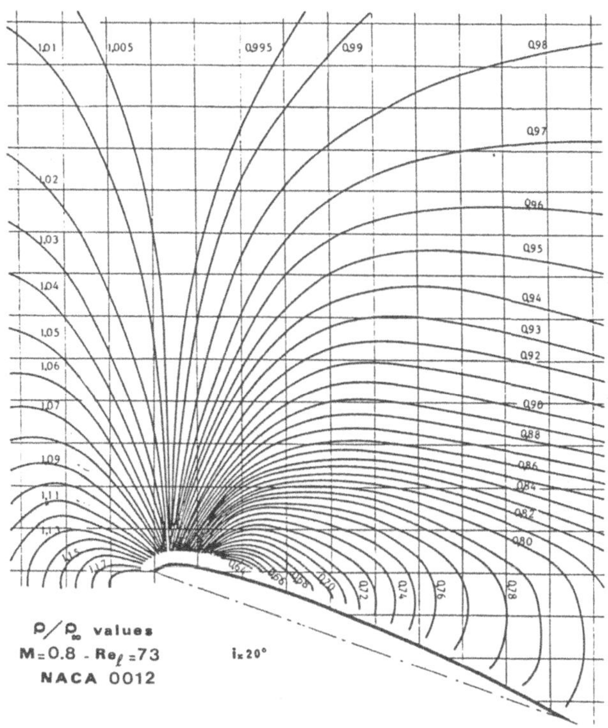

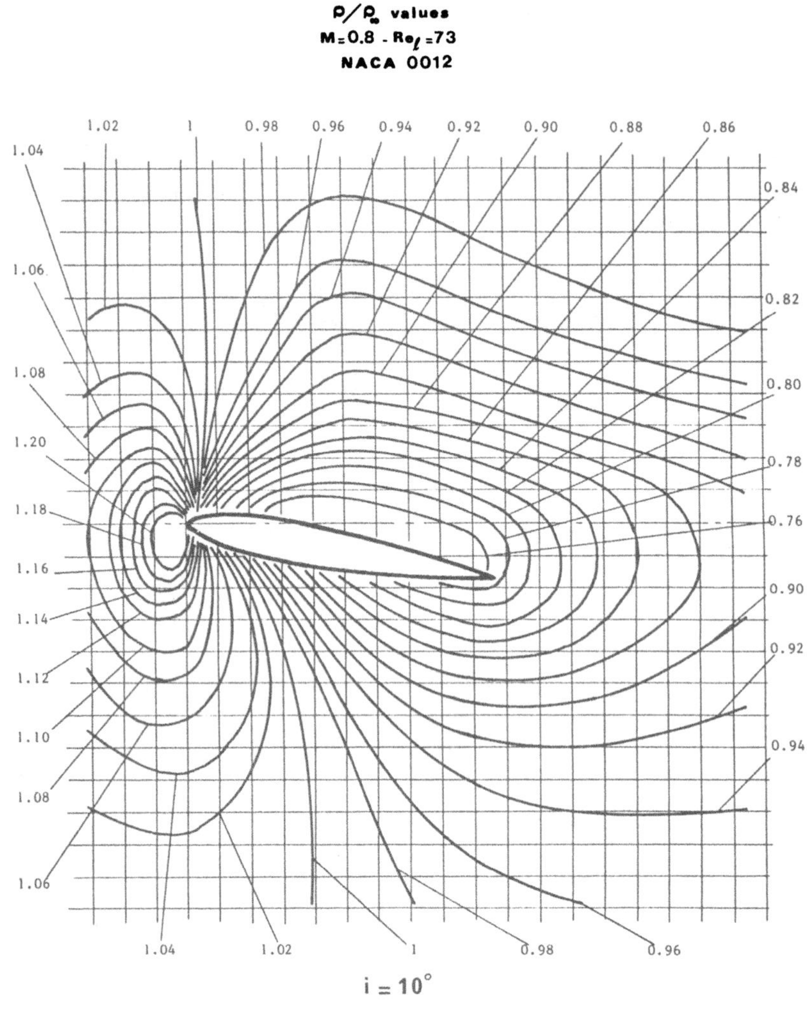

Fig. 3
Experimental density flowfield (i=10°)

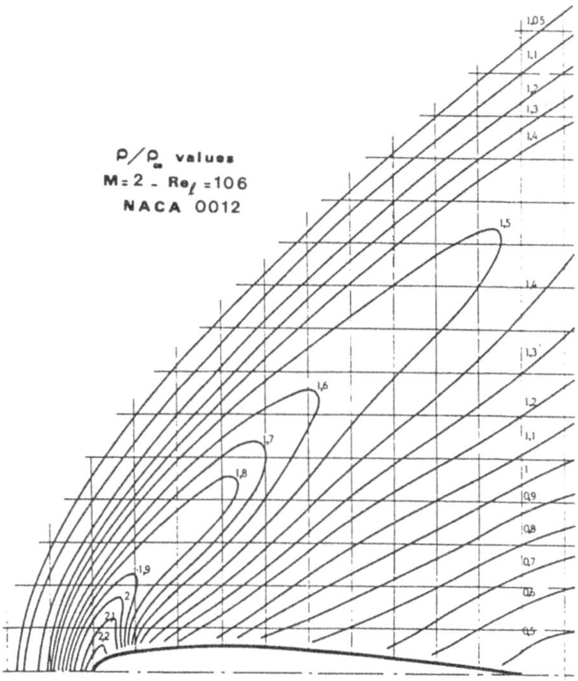

Fig. 4
Experimental density flowfield (i=0°)

Fig. 5
Experimental density flowfield (i=20°)

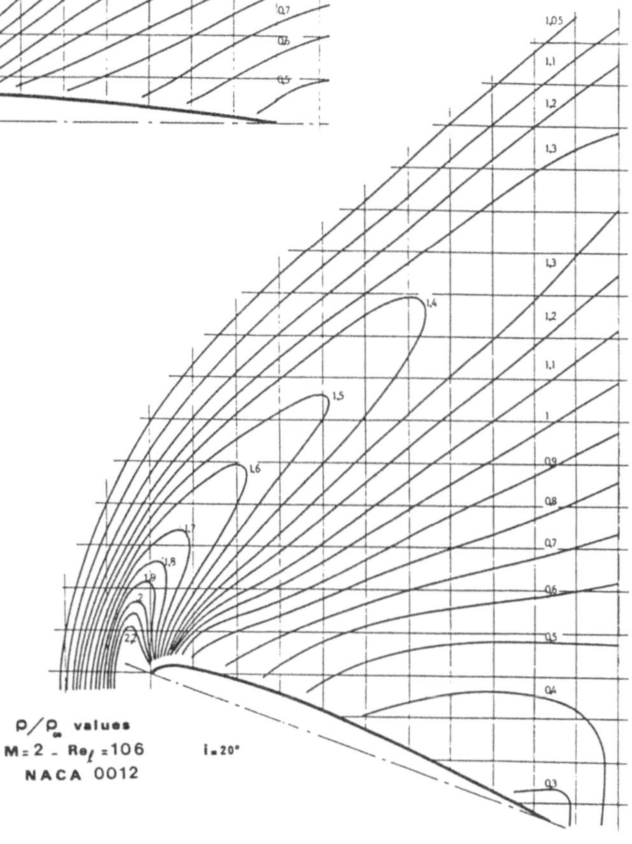

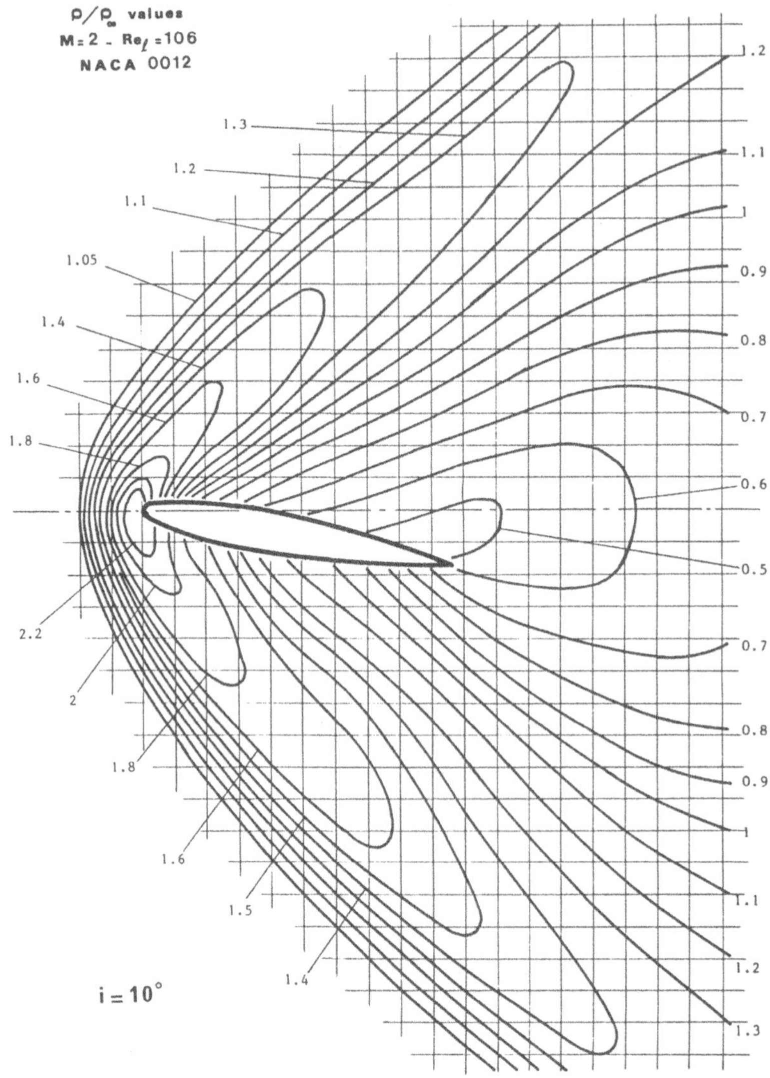

Fig. 6
Experimental density flowfield (i=10°)

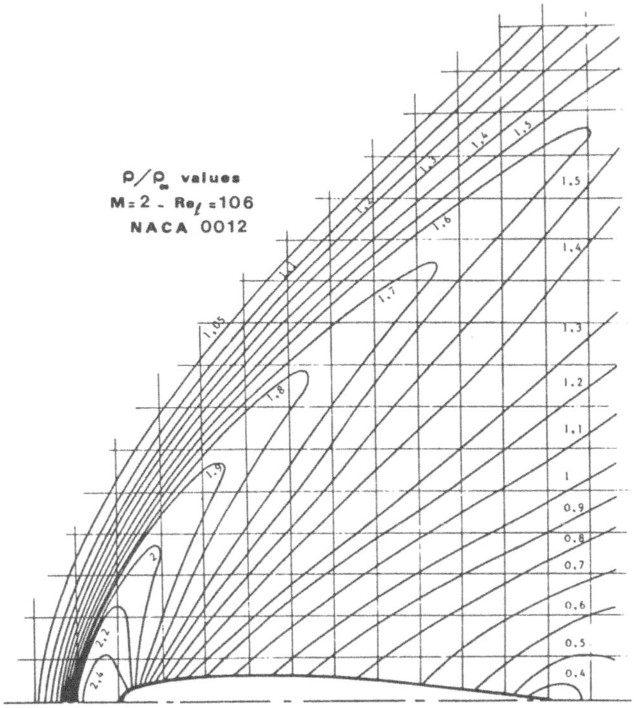

Fig. 7

Theoretical density flowfield (i=0°) (Navier-Stokes, no slip conditions)

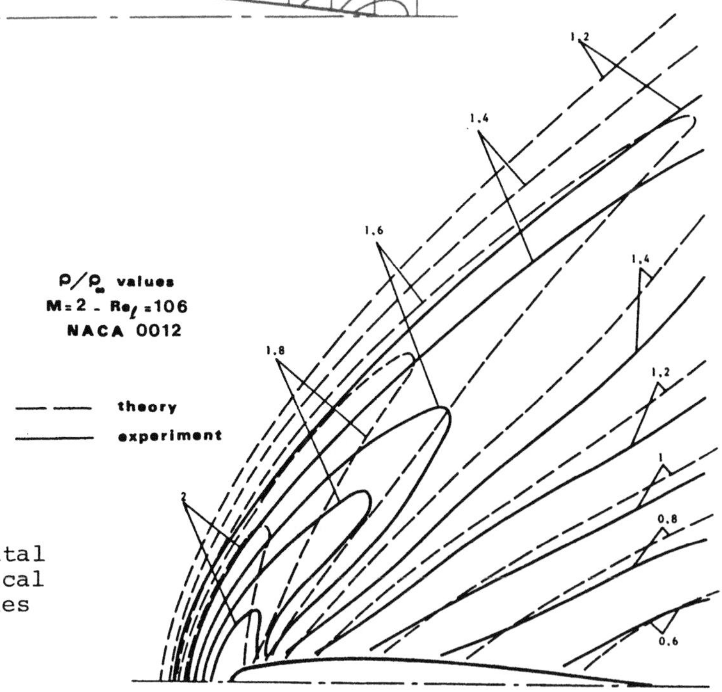

Fig. 8

Experimental and theoretical density values (i=0°)

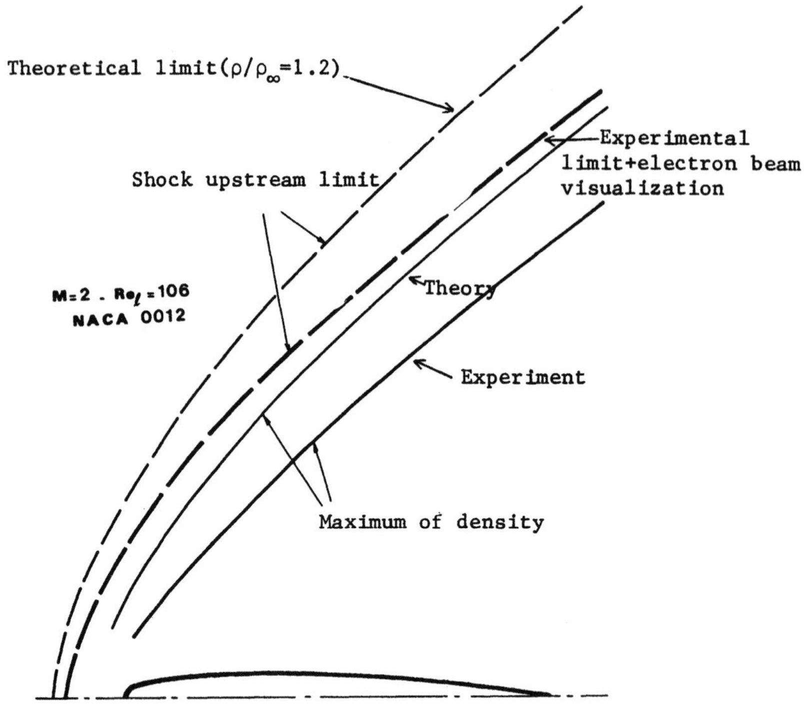

Fig. 9 - Shock location

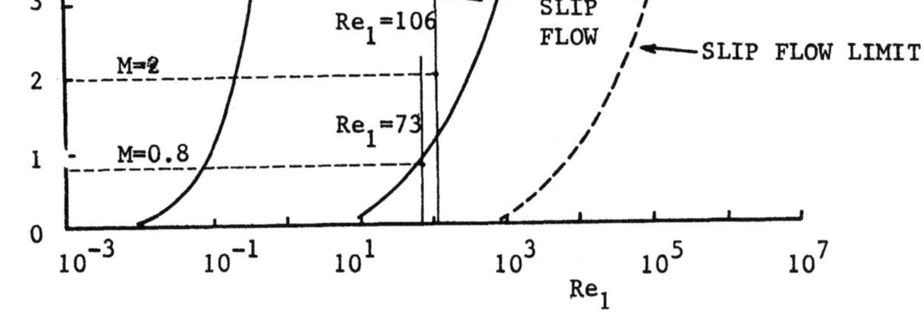

Fig. 10 - Flow regimes

VISCOUS PERTURBATION FOR THE COMPRESSIBLE EULER EQUATIONS. APPLICATION TO THE NUMERICAL SIMULATION OF COMPRESSIBLE VISCOUS FLOWS.

F. ANGRAND
INRIA - ROCQUENCOURT BP 105 78153 LE CHESNAY CEDEX - FRANCE

A new method for solving Navier-Stokes equations is introduced issued from an Euler solver ([1], [2], [4]). The spatial approximation is a Galerkin Finite Element Method which handles non-structured meshes. A pseudo unsteady approach is used in which the time stepping is a two step Lax-Wendroff explicit scheme.

I. EQUATIONS

The compressible Navier Stokes system in 2-D conservative form can be written as follows :

$$\begin{cases} \dfrac{\partial U}{\partial t} + \dfrac{\partial F}{\partial x} + \dfrac{\partial G}{\partial y} = \dfrac{1}{Re}\left(\dfrac{\partial R}{\partial x} + \dfrac{\partial S}{\partial y} \right) \\ + \text{ boundary and initial conditions} \end{cases} \tag{1}$$

with

$$U = \begin{cases} \rho \\ \rho u \\ \rho v \\ e \end{cases} \qquad F(U) = \begin{cases} \rho u \\ \rho u^2 + p \\ \rho uv \\ u(e+p) \end{cases} \quad , \; G(U) = \begin{cases} \rho v \\ \rho uv \\ \rho v^2 + p \\ v(e+p) \end{cases}$$

$$R = \begin{cases} 0 \\ \dfrac{4}{3}\dfrac{\partial u}{\partial x} - \dfrac{2}{3}\dfrac{\partial v}{\partial y} \\ \dfrac{\partial u}{\partial y} + \dfrac{\partial v}{\partial x} \\ u\left(\dfrac{4}{3}\dfrac{\partial u}{\partial x} - \dfrac{2}{3}\dfrac{\partial v}{\partial y}\right) + v\left(\dfrac{\partial u}{\partial y} + \dfrac{\partial v}{\partial x}\right) + \dfrac{\gamma}{Pr}\dfrac{\partial \varepsilon}{\partial x} \end{cases} \tag{2}$$

$$S = \begin{cases} 0 \\ \dfrac{\partial u}{\partial y} + \dfrac{\partial v}{\partial x} \\ \dfrac{4}{3}\dfrac{\partial v}{\partial y} - \dfrac{2}{3}\dfrac{\partial u}{\partial x} \\ u\left(\dfrac{\partial u}{\partial y} + \dfrac{\partial v}{\partial x}\right) + v\left(\dfrac{4}{3}\dfrac{\partial v}{\partial y} - \dfrac{2}{3}\dfrac{\partial u}{\partial x}\right) + \dfrac{\gamma}{Pr}\dfrac{\partial \varepsilon}{\partial y} \end{cases} .$$

Moreover, we have

$$\varepsilon = \frac{1}{\gamma-1}\,\frac{p}{\rho}\,, \quad \gamma = 1.4$$

$$e = \rho\varepsilon + \rho\,\frac{u^2+v^2}{2}\,. \qquad (3)$$

ρ, u, v, p and e denote respectively the density, components of the velocity, pressure and total specific energy of the fluid. Re is the Reynolds number and Pr the Prandtl number.

II. DISCRETIZATION

When the Reynolds number Re increases and go to infinity, we obtain the Euler system. It's the reason why we use an extension of an Euler solver. We present thereafter an unsteady two-step Ritchmyer-like scheme which extends the 1-D S_β^α scheme proposed by Lerat-Peyret [3], to solve Euler equations. Then the strategy consists of adding explicitly the viscous terms of the Navier-Stokes equations and taking into account appropriate wall boundary conditions.
Let us consider first a triangulation $\mathcal{T}_h$ of the discretized computational domain Ω occupied by the fluid,

$$\mathcal{T}_h = \{\text{set of triangles } T\},$$

and the standard space V_h,

$$V_h = \{\phi \in C^0(\bar{\Omega}) \mid \phi|_T \in P_1 \ \forall T \in \mathcal{T}_h\}$$

where P_1 denotes the set of polynomials of degree ≤ 1.
We apply the following algorithm.
Knowing $U^n \in V_h^4$, value of U at time $n\Delta t$, we compute U^{n+1} by :

Step 1 : **Predictor.**

$$\forall\, T \in \mathcal{T}_h$$

$$U^P(T) = \frac{1}{\text{aire}(T)}\left\{ \iint_T U^n dx\,dy - \alpha\,\Delta t \int_{\partial T} [F(U^n)\nu_x + G(U^n)\nu_y]\,d\sigma \right\}.$$

The resulting predicted value U^P is a constant function on each triangle.

Step 2 : **Corrector**

$$U^{n+1} \in V_h^4 \text{ and } \forall\phi \in V_h^4 \quad \forall k = 1,\dots,4$$

$$\iint_\Omega \Sigma_0 \frac{U_k^{n+1}-U_k^n}{\Delta t} \Sigma_0\, \phi_k\, dx\,dy = \qquad \langle 1\rangle$$

$$\beta_1 \iint_\Omega \left(F_k(U^n)\frac{\partial\phi_k}{\partial x} + G_k(U^n)\frac{\partial\phi_k}{\partial y}\right) dx\,dy \qquad \langle 2\rangle$$

$$+ \beta_2 \iint_\Omega (F_k(U^p) \frac{\partial \phi_k}{\partial x} + G_k(U^p) \frac{\partial \phi_k}{\partial y}) \, dx \, dy \qquad \langle 3 \rangle$$

$$- \int_{\partial\Omega} \phi_k (F_k(\bar{U})\nu_x + G_k(\bar{U})\nu_y) \, d\sigma \qquad \langle 4 \rangle$$

$$+ \chi \iint_\Omega \langle f(U^n) \, . \, \nabla U_k^n, \nabla \phi_k \rangle \, dx \, dy \qquad \langle 5 \rangle$$

$$+ \frac{1}{Re} \int_{\partial\Omega} \phi_k (R_k \nu_x + S_k \nu_y) \, d\sigma \qquad \langle 6 \rangle$$

$$- \frac{1}{Re} \iint_\Omega (R_k \frac{\partial \phi_k}{\partial x} + S_k \frac{\partial \phi_k}{\partial y}) \, dx \, dy \qquad \langle 7 \rangle$$

$\nu = (\nu_x, \nu_y)$ is the outward unit vector normal to the boundary of the domain. Σ_o is the mass lumped operator which gives a diagonal matrix to inverse.

$$\alpha = 1 + \frac{\sqrt{5}}{2} , \ \beta_1 = \frac{2\alpha - 1}{2\alpha}, \ \beta_2 = \frac{1}{2\alpha} .$$

These parameters are optimal in the one dimensional case see [3].
The above two-step explicit scheme is stable under a classical Courant condition [1].
Term <4> contains the physical boundary conditions (except on the body). They are taken into account through the vector $\bar{U}$ which is computed from quantities depending on the interior value U and quantities determined by physical boundary conditions. When dealing with a slip condition, we only need the pressure of $\bar{U}$ to compute the flux, we choose to set this pressure equal to the previous value of the pressure on the boundary. With this procedure, the condition is only weakly satisfied.
For the wall boundary, we extrapolate pressure from internal values using momentum equations. With the temperature, given on the body, the density and the energy of the fluid can be thereby determined. With this wall boundary condition, <6> and <7> are the perturbation of our Euler code.
Finally <5> is an artifical viscosity term with parameter χ. For simplicity, we choose a function f(U), constructed along x, y directions

$$f(U) = \begin{bmatrix} f^x & 0 \\ 0 & f^y \end{bmatrix}$$

and f^x(resp f^y) is proportional to the square of a local Δx (resp Δy) space increment evaluation and involves first derivatives of quantities such as pressure, density, velocity.

III. NUMERICAL RESULTS

II.1 Mandatory cases

These calculations are first experiments of this Navier-Stokes code. So we have only solved mandatory cases. The calculations have been carried out on the Cray 1S computer of CCVR. For all test cases, we have the same triangulation T_1 : 2354 nodes, 4568 triangles (Figures 1, 2). The computational domain is inscribed in the rectangle :

$$- 5 \le x \le 8$$
$$- 8.5 \le y \le 8.5 .$$

Smoothing parameter χ is equal to zero except for A3 where χ is equal to .2. For every calculation we stop at iteration n = 4000 (about 1500 s on CRAY 1.S). The required storage is about 54 000 words. The CFL number is equal to .7. In Figures 3-4-5-6, we present pressure coefficients for mandatory cases. Figure 7-8-9-10 correspond to skin friction coefficient and Figures 11-12-13-14 are iso-density lines for these cases. In Table 1, we give the value of the residu, and lift, drag and momentum coefficients.

Table 1

	A1	A2	A3	A5
$\lvert\frac{\partial\rho}{\partial t}^n\rvert_{L^2} / \lvert\frac{\partial\rho}{\partial t}^o\rvert_{L^2}$	$.3\ 10^{-5}$	$.4\ 10^{-4}$	.97 10-8	$.25\ 10^{-7}$
C_L total	0.5916	0.517	0.367	$0.16\ 10^{-4}$
C_D total	0.581	0.243	0.412	0.186
C_M	0,233	0.166	0.1817	$0.6\ 10^{-6}$
C_L pressure	0.621	0.52	0.394	$0.14\ 10^{-4}$
C_D pressure	0.218	0.17	0.179	$0.84\ 10^{-1}$

For case A6 it seems that the solution is unsteady.

III.2 Some investigation work on case A5

The geometric flexibility of triangular elements is well suited for refinement processes. This has been studied by B. Palmerio [5] and C. Pouletty [6], [7]. The basic idea is, starting from a solution on an initial mesh, to improve locally the accuracy of this solution by refining the mesh according to criteria which are functions of the known solutions. Then the mesh is depending not only of the geometry or of an a priori knowledge of the critical areas of the computational domain, but is also influenced a posteriori by the solution itself.

We have to find criteria which are able to localize not only discontinuities but also boundary layers and wakes.

For case A5, the adapted meshes are constructed via the criterion C :

$$C = \frac{(V.\nabla M + V\wedge\nabla M)}{|V.\nabla M + V\wedge\nabla M|} \times L$$

V : Vector of components u and v
M : Local Mach number
L : Distance between the local point and the leading edge
$||$: L^∞ norm.

New points are added in the regions where the criterion C is greater than a given threshold value. We perform sequentially computations on ever better adapted meshes. Such sequence of locally nested grids is presented on Figure 15. The corresponding Mach contours of the computations with an increasing stability of the lines and also a better capture of the boundary layer can be observed on Figure 16.

No important improvement can be seen between the results on the second and the thrird grid. But for the skin friction coefficient, the value continues to increase at the leading edge (Figure 17).

We can remark that the second mesh has less nodes than the triangulation T_1, but the density of nodes is more appropriate to the characteristics of the flow and therefore we obtain more accurate results.

The CFL number used is .7, there is no artifical viscosity added. After some vectorization work [8], we obtain some improvement of the CPU time given in Table 2.

Table 2

number of nodes of the triangulation	n = number of iterations	CPU time CRAY 1 seconds	$\left\|\frac{\partial\rho}{\partial t}^n\right\|_{L^2} / \left\|\frac{\partial\rho}{\partial t}^o\right\|_{L^2}$	C_D total
1006	1500	89,7	10^{-3}	0.303
2282	4500	619.9	10^{-5}	0.2107
3330	3000	617.8	10^{-4}	0.2184

CONCLUSION

We have presented in this paper a **finite element** method for solving compressible viscous flow problems modelled by the unsteady Navier-Stokes equations, written in **conservative** form. The main numerical ingredients are a centered **Lax Wendroff** type scheme and **self adaptive mesh refinement** techniques for accuracy improvement.

This method seems to be well adapted for the moderate Reynolds numbers considered here, however in the range of spatial applications (high Mach number and high angle of attack) there is a lot of work to do in order to obtain similar results ; specially generalized Euler solvers based on space centered schemes with 4th order artificial viscosity or upwinded characteristic variables are highly suggested.

REFERENCES

[1] ANGRAND F., DERVIEUX A., "Some explicit triangular finite element schemes for the Euler equations" in International Journal for Numerical Methods in Fluids, Vol. 4, 749-764 (1984).

[2] ANGRAND F., DERVIEUX A., BOULARD V., PERIAUX J., VIJAYASUNDARAM G., "Transonic Euler simulations by means of finite element explicit shemes" AIAA paper, 83-1924 (1983).

[3] LERAT A., PEYRET R., "Sur le choix de schémas aux différences du second ordre fournissant des profils de chocs sans oscillation" Comptes rendus Acad. Sc. Paris, Série A, 277, 363-366 (1973).

[4] BILLEY V., "Résolution des équations d'Euler par des méthodes d'éléments finis, application aux écoulements 3-D de l'aérodynamique" Thèse Paris VI (1984).

[5] POULETTY C., "Générations et optimisations de maillages en éléments finis par déformations et raffinements automatiques. Application à la simulation numérique de quelques phénomènes de l'Aérodynamique" Thèse de Docteur Ingenieur, Ecole Centrale, Dec. 85.

[6] PALMERIO B., "Self-Adaptative F.E.M. algorithms for the Euler equations" Rapport INRIA, N° 338 (1984).

[7] PALMERIO B., BILLEY V., DERVIEUX A., PERIAUX J., "Self adaptive mesh refinements and finite element methods for solving the Euler equations" Proceedings of the ICFD Conference on 'Numerical methods for fluid dynamics', Reading 1985.

[8] ANGRAND F., ERHEL J., "Vectorized finite element codes for compressible flows". Proceedings of the 6th International Symposium Finite Element Methods in flow problems, Antibes juin 86.

ACKNOWLEDGEMENTS

The access to CRAY 1 was provided by the CCVR Committee.

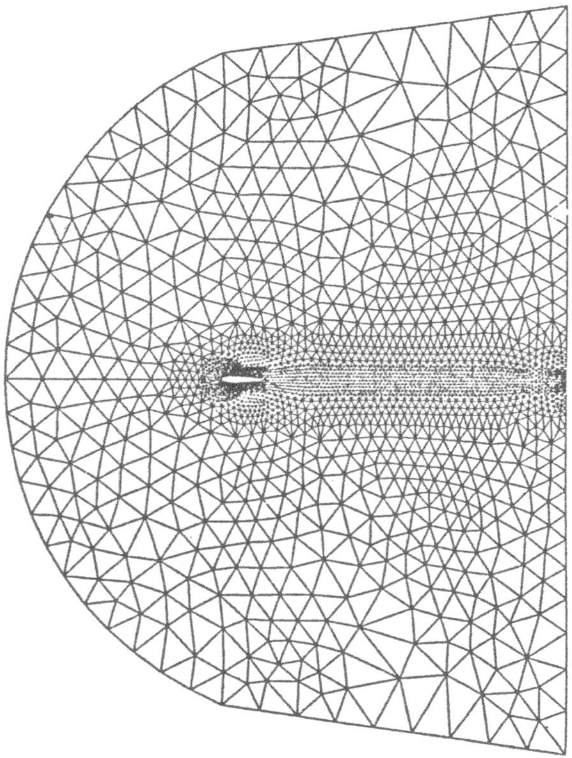

Figure 1 Global view of the mesh

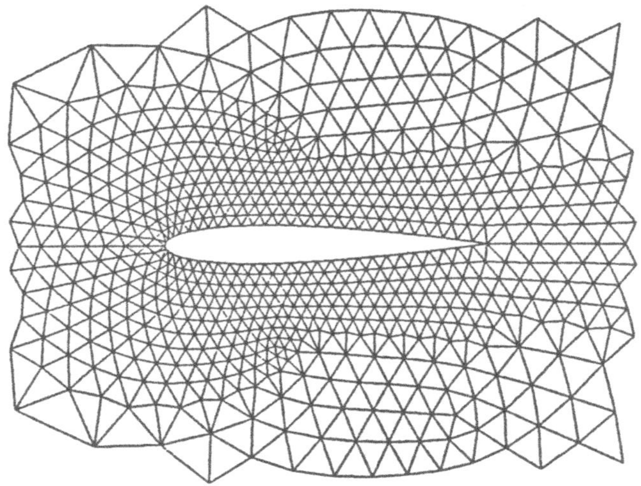

NACA0012 NODES = 2354 ELEMENTS = 4568

Figure 2 An enlargment of the mesh

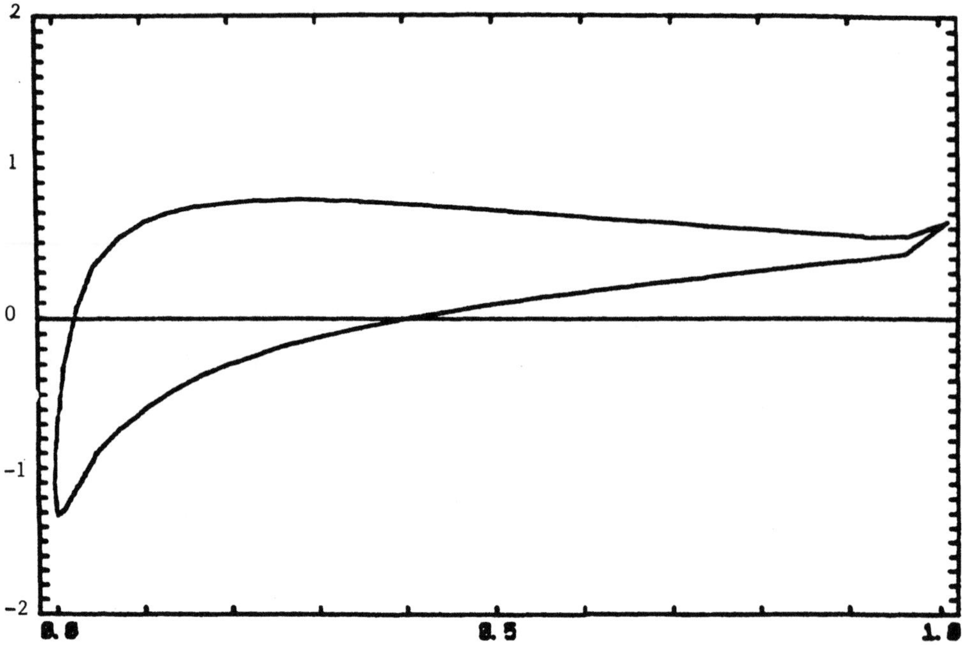

Figure 3 : A1 : M = .8 ALP = 10 RE = 73. Pressure coefficient

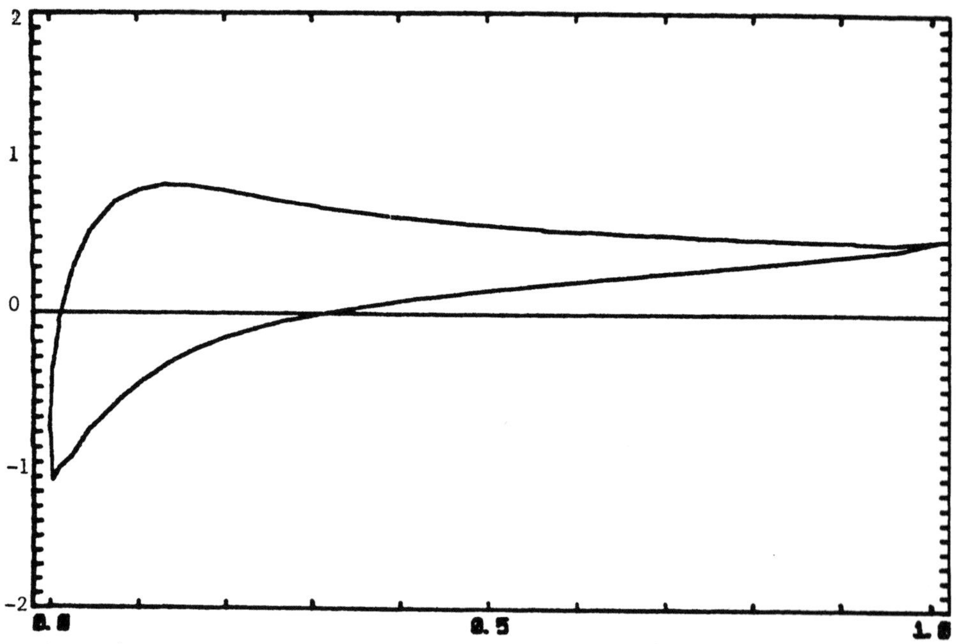

Figure 4 : A2 : M = .8 ALP = 10 RE = 500. Pressure coefficient

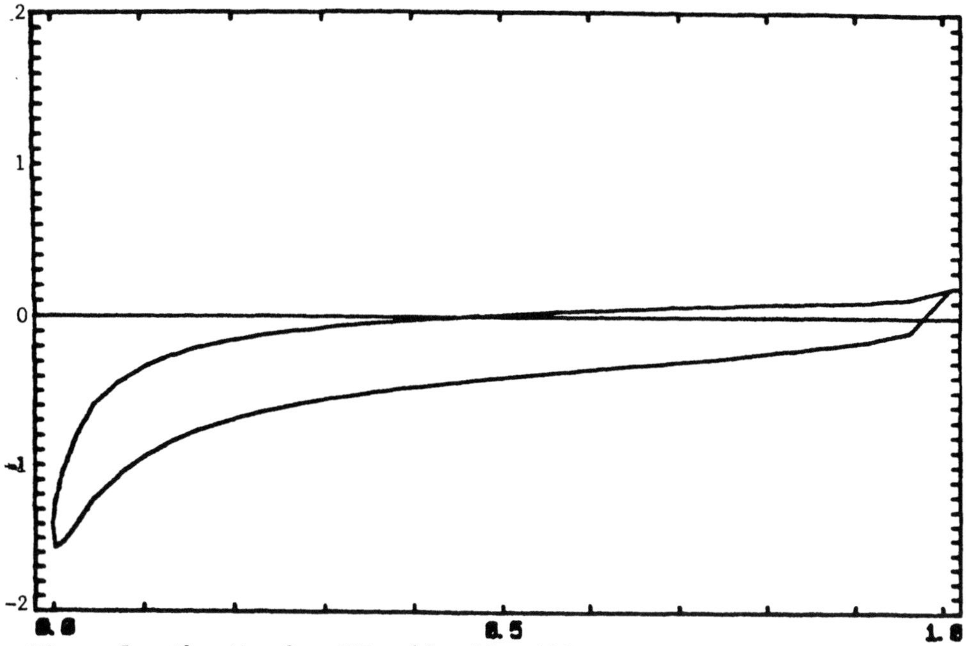

<u>Figure 5</u> : A3 : M = 2. ALP = 10. RE = 106. Pressure coefficient

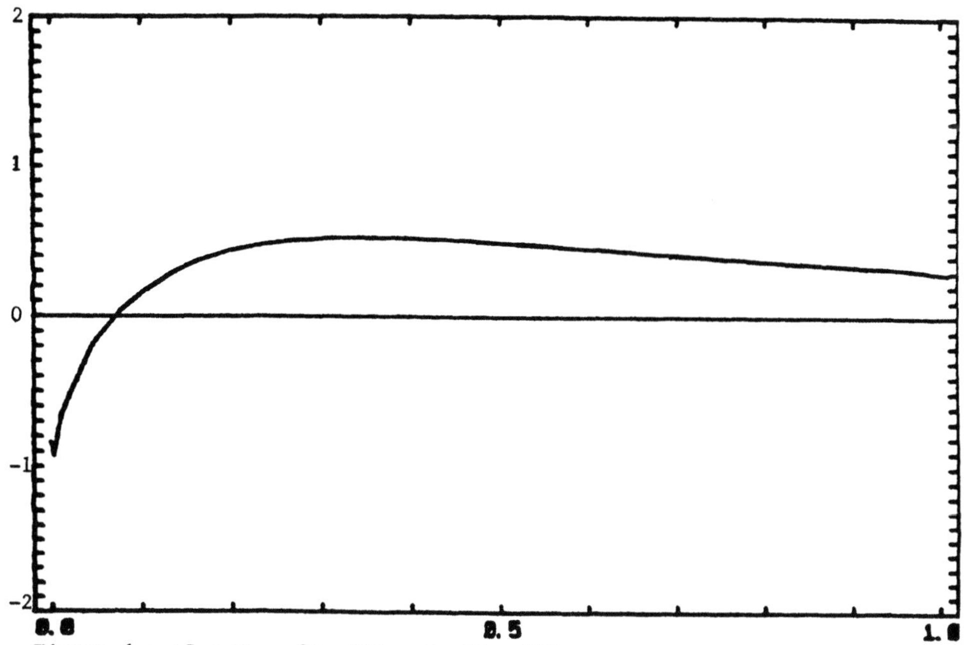

<u>Figure 6</u> : A5 : M = .85 ALP = 0 RE = 500. Pressure coefficient

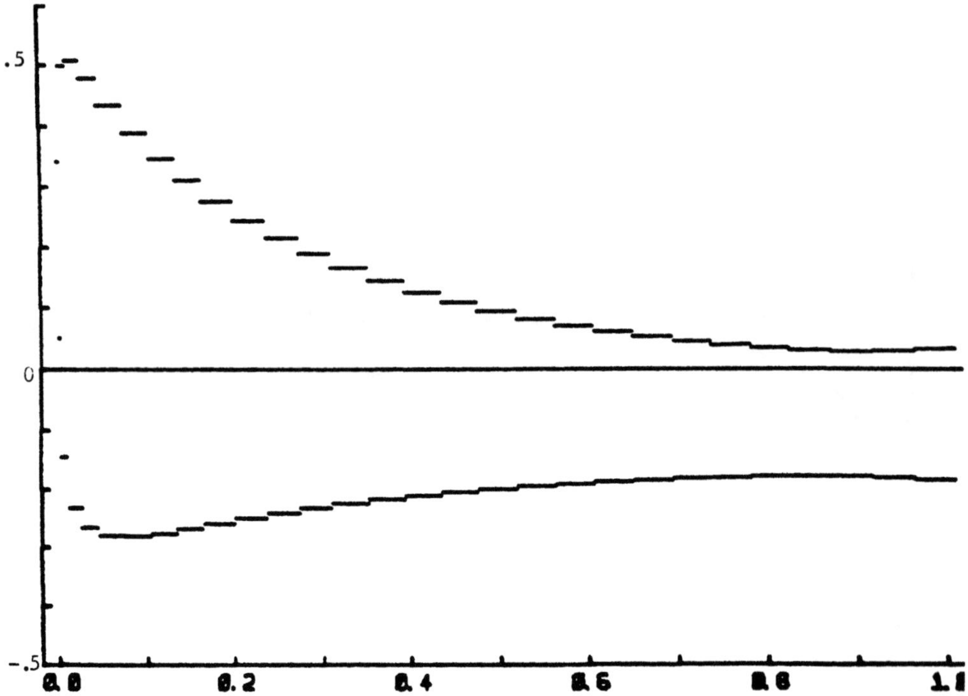

<u>Figure 7</u> : A1 : M = .8 ALP = 10. RE = 73. SKIN FRICTION COEFFICIENT

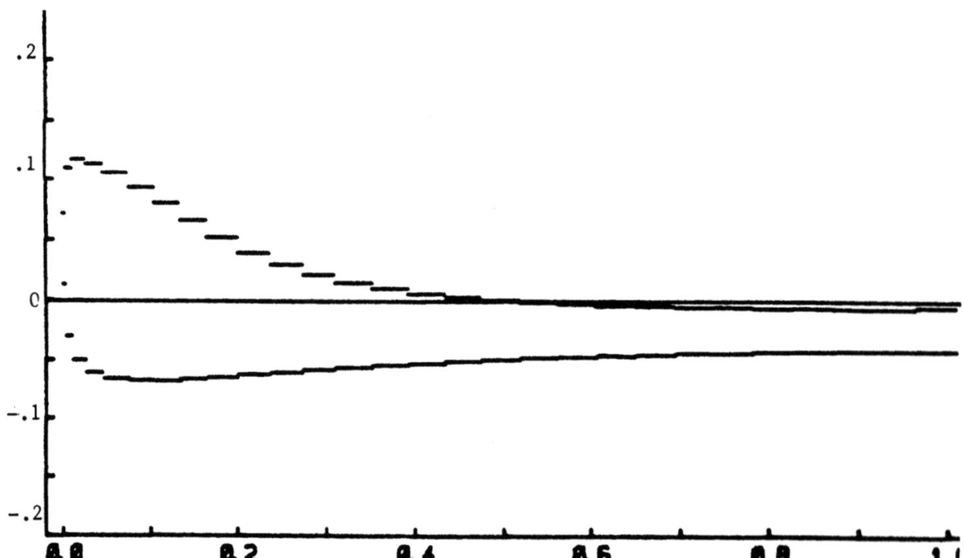

<u>Figure 8</u> : A2 : M = .8 ALP = 10. RE = 500. SKIN FRICTION COEFFICIENT

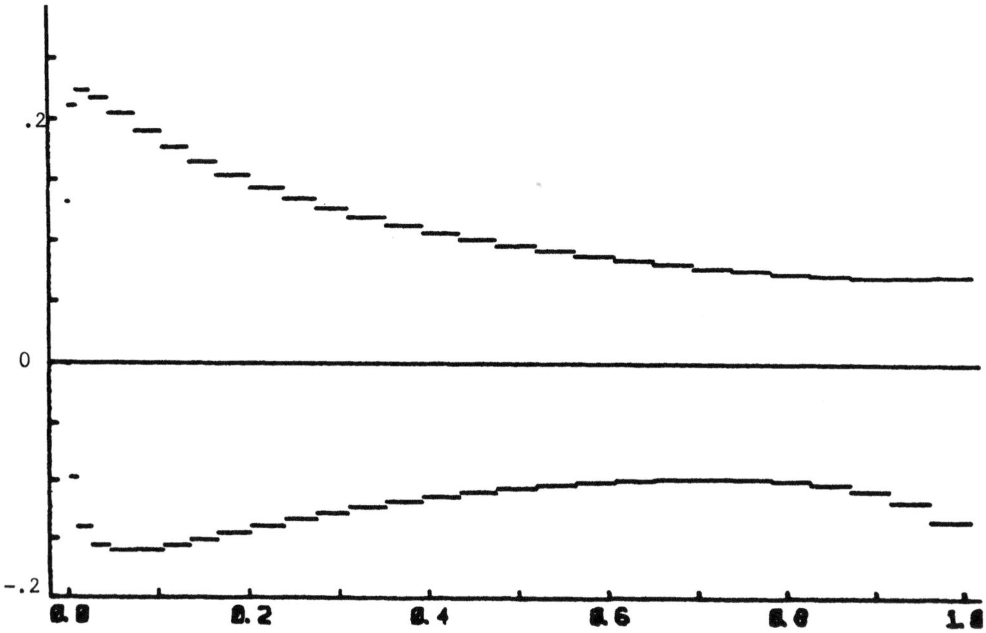

Figure 9 : A3 : M = 2. ALP = 10. RE = 106. SKIN FRICTION COEFFICIENT

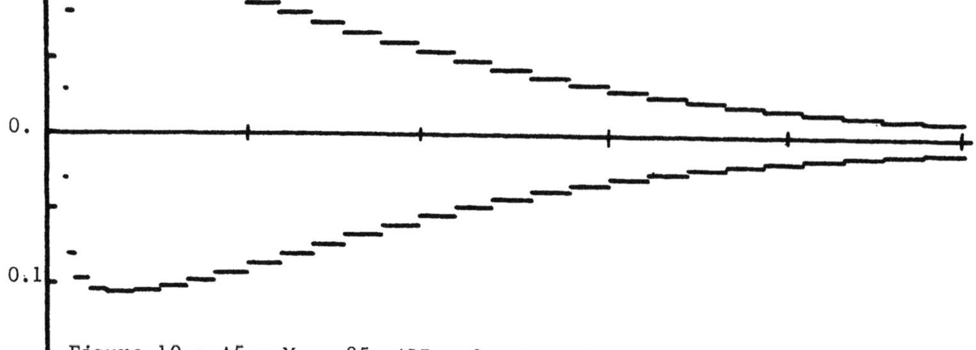

Figure 10 : A5 : M = .85 ALP = 0. RE = 500. SKIN FRICTION COEFFICIENT

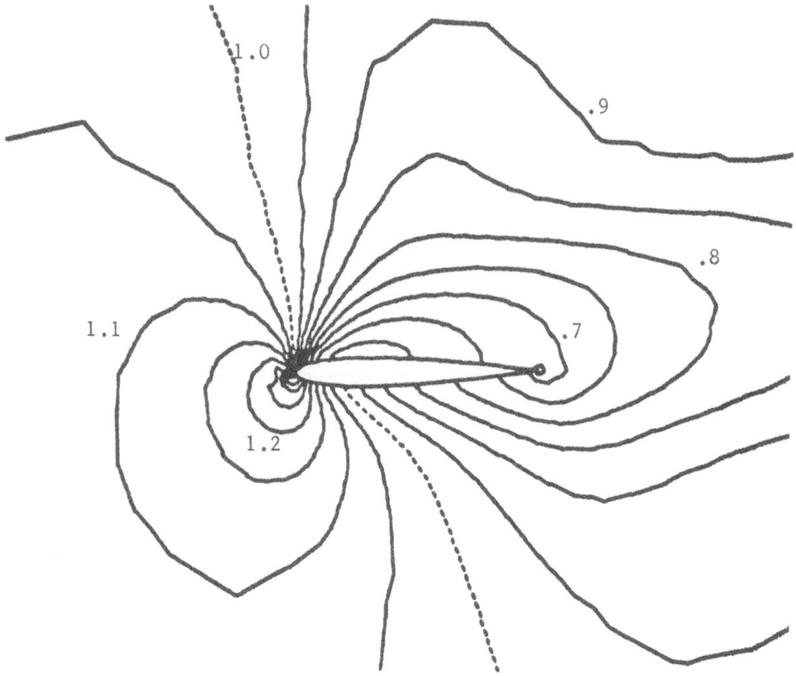

Figure 11 : A1 : M = .8 ALP = 10. RE = 73. Iso- ρ/ρ_∞ lines ($\delta\rho/\rho_\infty$ = .05).

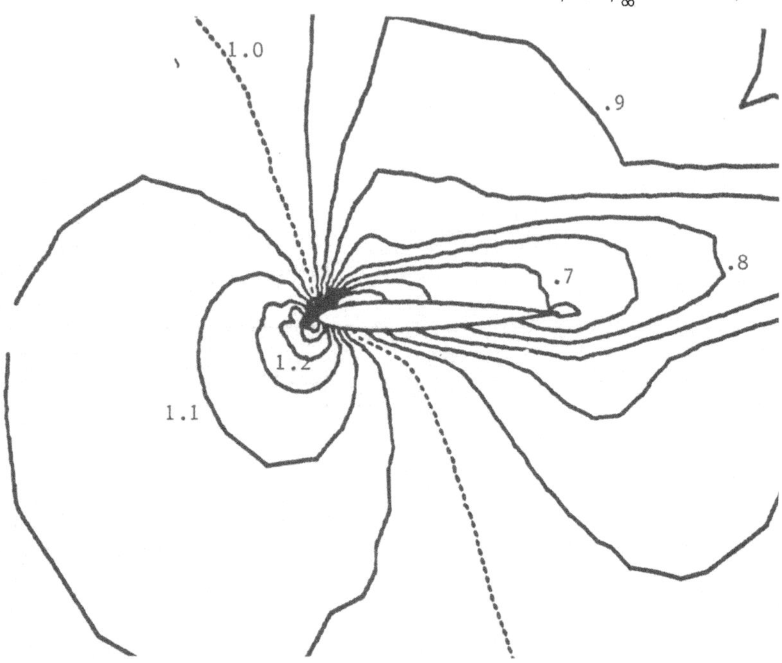

Figure 12 : A2 : M = .8 ALP = 10. RE = 500 iso-ρ/ρ_∞ lines ($\delta\rho/\rho_\infty$ = .05).

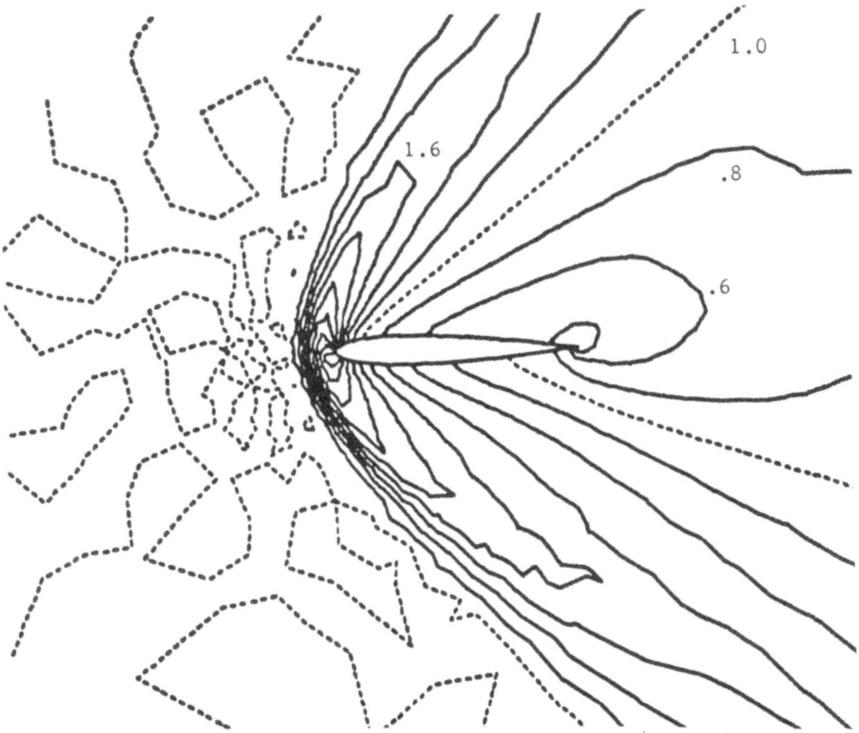

Figure 13 :A3 : M = 2. ALP = 10. RE = 106. Iso-ρ / ρ_∞ lines ($\delta\rho$ / ρ_∞ = .2).

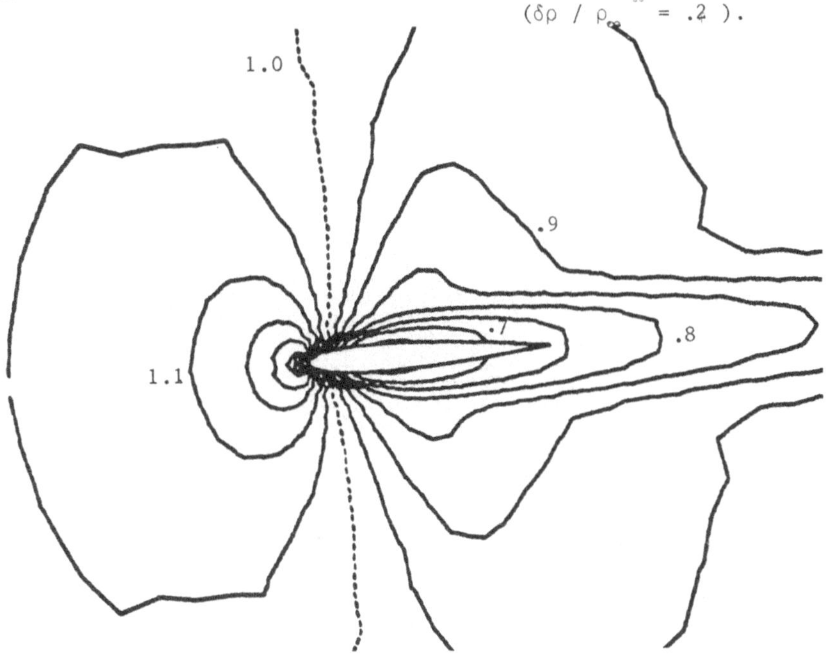

Figure 14 : A5 : M = .85 ALP = 0. RE = 500. Iso-ρ / ρ_∞ lines ($\delta\rho$ / ρ_∞ = .05).

Nodes :

1006

Elements :

1916

Nodes :

2282

Elements :

4428

Nodes :

3330

Elements :

6456

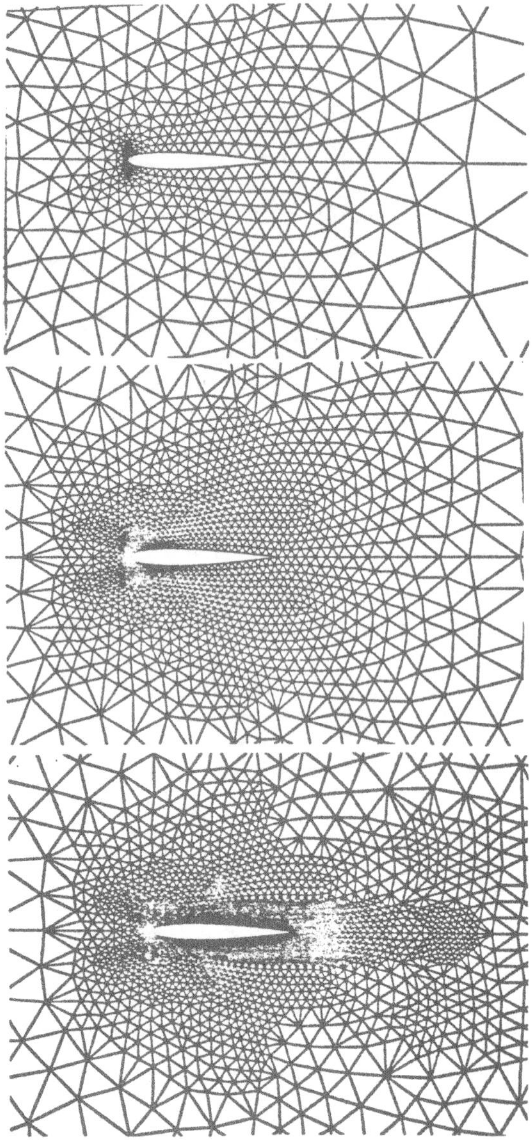

Figure 15

Initial and adapted meshes around the NACA0012 airfoil.

$M_{\infty} = 0.85$, Re = 500.

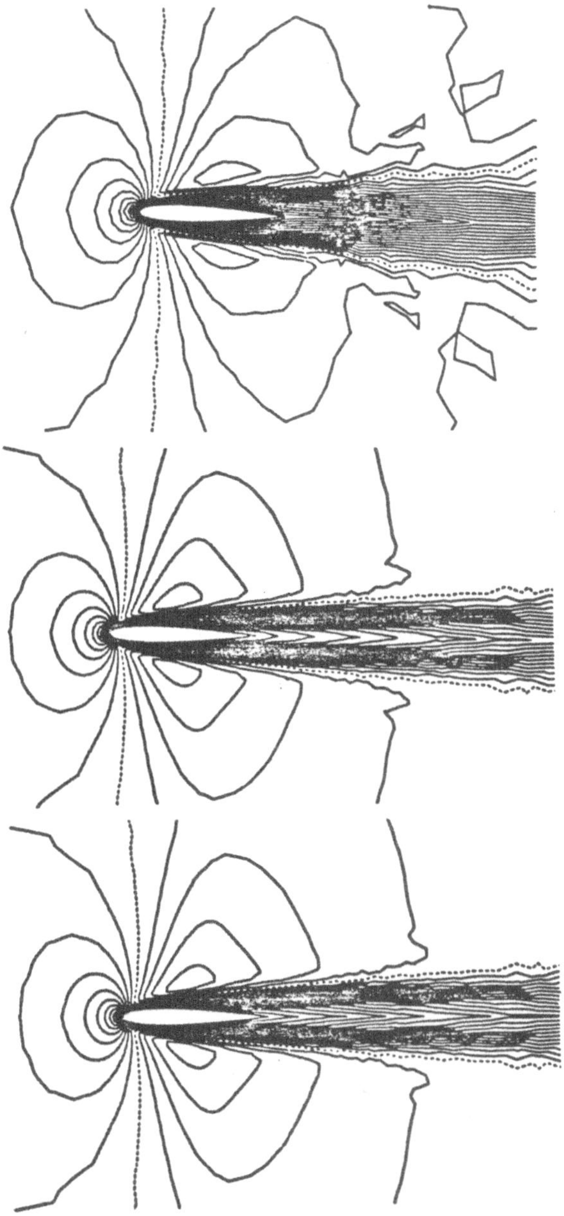

Nodes :
1006
Elements :
1916

Nodes :
2282
Elements :
4428

Nodes :
3330
Elements :
6456

Figure 16

Mach contours on the initial and adapted meshes.
$M_\infty = 0.85$, Re = 500.

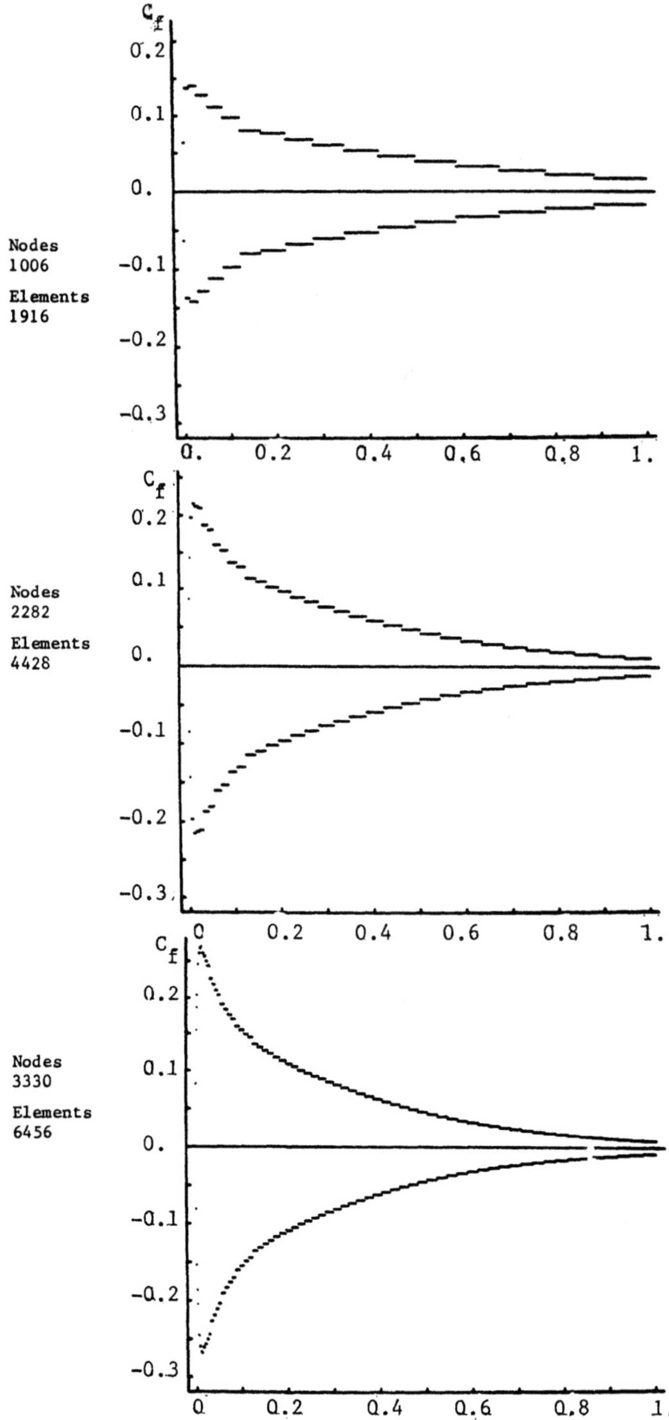

Figure 17 : Skin friction coefficient on the initial and adapted meshes M_∞ = .85 Re = 500.

SOLUTION OF THE COMPRESSIBLE NAVIER-STOKES EQUATIONS BY LEAST-SQUARES AND FINITE ELEMENT METHODS

by

M.O.BRISTEAU
INRIA, B.P.105, 78153 LE CHESNAY CEDEX, France
R.GLOWINSKI
INRIA and UNIVERSITY OF HOUSTON, Texas, U.S.A.
B.MANTEL, J.PERIAUX, C.POULETTY
AMD/BA, B.P.300, 92214 ST CLOUD, France

1. INTRODUCTION

We discuss in this paper two methods for the numerical simulation of compressible viscous flows, associated with two types of mathematical modelling assuming that the total enthalpy is constant (M1) or not (M2).

These methods are based on the following ingredients :

i) P_1-finite elements for space discretization. This approximation is well suited to deal with complex geometries and recent studies on self adaptive mesh refinement ([1] and Ref. therein), enhance its interest.

ii) semi-implicit (M1) or implicit (M2) schemes in order to avoid too restrictive time step limitations.

iii) least-squares formulation to identify the non-linearities and to decouple the primitive variables occuring in the linear systems.

iv) centered approximation without artificial viscosity to provide, for low Reynolds numbers, references for further comparisons with results using other approximations.

In this approach, we generalize tools which have been developped for the solutions of the incompressible Navier-Stokes equations [2] .

THE GOVERNING EQUATIONS

Let $\Omega \subset \mathbb{R}^N$ be the flow domain and Γ be its boundary. We suppose the flow is governed by the following set of Navier-Stokes equations, (M2) written in non-conservative form :

$$\frac{\partial \rho}{\partial t} + \nabla . \rho \mathbf{u} = 0 \qquad (1)$$

$$\rho \frac{\partial \mathbf{u}}{\partial t} + \rho(\mathbf{u}.\nabla)\mathbf{u} + (\gamma-1)\nabla \rho T = \frac{1}{Re}[\Delta \mathbf{u} + \frac{1}{3}\nabla(\nabla.\mathbf{u})] \qquad (2)$$

$$\rho \frac{\partial T}{\partial t} + \rho \underset{\sim}{u} . \underset{\sim}{\nabla} T + (\gamma-1)\rho T \underset{\sim}{\nabla} . \underset{\sim}{u} = \frac{1}{Re} [\frac{\gamma}{Pr} \Delta T + F(\underset{\sim}{\nabla} \underset{\sim}{u})] \tag{3}$$

with the pressure given by the perfect gas law :

$$p = (\gamma-1) \rho T , \tag{4}$$

where

ρ,u,T denote respectively the non dimensionalized density, velocity and temperature variables,

$$F(\underset{\sim}{\nabla} \underset{\sim}{u}) = \frac{4}{3} [(\frac{\partial u_1}{\partial x_1})^2 + (\frac{\partial u_2}{\partial x_2})^2 - \frac{\partial u_1}{\partial x_1} \frac{\partial u_2}{\partial x_2}] + (\frac{\partial u_1}{\partial x_2} + \frac{\partial u_2}{\partial x_1})^2 \tag{5}$$

with $\underset{\sim}{u} = (u_1, u_2)$ for N=2.

An usual simplification is provided by the following constant total enthalpy assumption (M1) :

$$T + \frac{|\underset{\sim}{u}|^2}{2} = H_o \tag{6}$$

where H_o is defined from the boundary conditions.

If we suppose that (6) holds instead of (3) then (2) becomes :

$$\rho \frac{\partial \underset{\sim}{u}}{\partial t} + \rho (\underset{\sim}{u} . \underset{\sim}{\nabla}) \underset{\sim}{u} - \frac{\gamma-1}{2} \underset{\sim}{\nabla} (\rho |\underset{\sim}{u}|^2) + \frac{\gamma-1}{\gamma} H_o \underset{\sim}{\nabla} \rho = \frac{1}{Re} [\Delta \underset{\sim}{u} + \frac{1}{3} \underset{\sim}{\nabla} (\underset{\sim}{\nabla} . \underset{\sim}{u})] . \tag{7}$$

BOUNDARY AND INITIAL CONDITIONS

We consider an external flow around a NACA0012 airfoil ; the domain of computation is described in Figure 1.

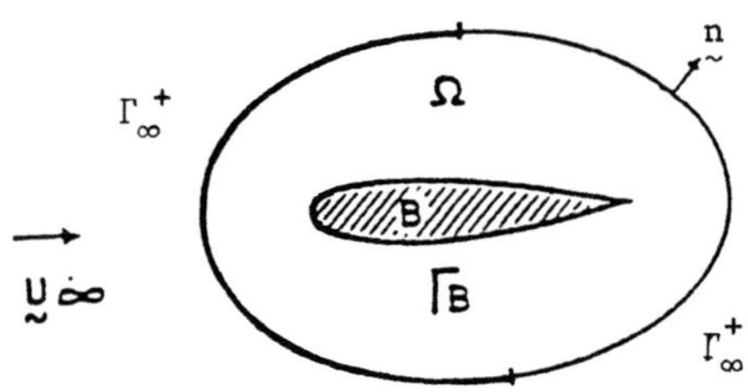

Figure 1

Let Γ_∞ be the external remote boundary of the domain, we introduce :

$$\Gamma_\infty^- = \{x | x \in \Gamma, \ \underset{\sim}{u}_\infty . \underset{\sim}{n} < 0\} \tag{8}$$

$$\Gamma_\infty^+ = \Gamma \setminus \Gamma_\infty^- \tag{9}$$

where u_∞ denotes the free stream velocity and $\underset{\sim}{n}$ the unit vector of the outward normal at Γ .

On the inflow boundary Γ_∞^-, we prescribe :

$$\underset{\sim}{u} = \underset{\sim}{u}_\infty = \begin{pmatrix} \cos\alpha \\ \sin\alpha \end{pmatrix}, \ \alpha \text{ angle of attack} \tag{10}$$

$$\rho = 1 \ , \tag{11}$$

$$T = T_\infty = \frac{1}{\gamma(\gamma-1)M_\infty^2} \ , \tag{12}$$

while on the outflow boundary Γ_∞^+, we prescribe Neumann boundary conditions.

On the wall Γ_B, we use the following conditions :

$$\underset{\sim}{u} = \underset{\sim}{0} \ , \tag{13}$$

$$T = T_B = T_\infty \ (1 + \frac{\gamma-1}{2} M_\infty^2) \ . \tag{14}$$

Finally, as problems (1),(2),(3) or (1),(6) are time dependent, initial conditions have to be added :

$$\rho(x,0) = \rho_o(x) \ , \tag{15}$$

$$\underset{\sim}{u}(x,0) = \underset{\sim}{u}_o(x) \ , \tag{16}$$

$$T(x,0) = T_o(x) \ . \tag{17}$$

TRANSFORMATION OF THE EQUATIONS

In order to extend the methodology previously used for solving the incompressible Navier-Stokes equations [2] , [3] (or the so-called shallow water equations [4]), we perform the following change of variable :

$$\sigma = \text{Log}\ \rho \tag{18}$$

With this new variable, the system (1),(2),(3) becomes :

$$\frac{\partial\sigma}{\partial t} + \underset{\sim}{\nabla}.\underset{\sim}{u} + \underset{\sim}{u}.\underset{\sim}{\nabla}\sigma = 0 \ , \tag{19}$$

$$\frac{\partial \underset{\sim}{u}}{\partial t} + (\underset{\sim}{u}.\underset{\sim}{\nabla})\underset{\sim}{u} + (\gamma-1)(T\underset{\sim}{\nabla}\sigma + \underset{\sim}{\nabla}T) = \frac{e^{-\sigma}}{Re}(\Delta\underset{\sim}{u} + \frac{1}{3}\underset{\sim}{\nabla}(\underset{\sim}{\nabla}.\underset{\sim}{u})) \ , \tag{20}$$

$$\frac{\partial T}{\partial t} + \underset{\sim}{u}.\underset{\sim}{\nabla}T + (\gamma-1)T\ \underset{\sim}{\nabla}.\underset{\sim}{u} = \frac{e^{-\sigma}}{Re}(\frac{\gamma}{Pr}\Delta T + F(\underset{\sim}{\nabla}\underset{\sim}{u})) \ , \tag{21}$$

while the equation (7) is replaced by :

$$\frac{\partial \underset{\sim}{u}}{\partial t} + (\underset{\sim}{u}.\underset{\sim}{\nabla})\underset{\sim}{u} - \frac{\gamma-1}{2\gamma}(\nabla|\underset{\sim}{u}|^2 + |\underset{\sim}{u}|^2\underset{\sim}{\nabla}\sigma) + \frac{\gamma-1}{\gamma}H_o\underset{\sim}{\nabla}\sigma = \frac{e^{-\sigma}}{Re}[\Delta\underset{\sim}{u} + \frac{1}{3}\underset{\sim}{\nabla}(\underset{\sim}{\nabla}.\underset{\sim}{u})] \tag{22}$$

TIME DISCRETIZATION

For system (19),(22) with constant total enthalpy assumption, we introduce a semi-implicit scheme using an operator splitting [5] while for the full set of equations (19)-(21), we apply a fully implicit scheme [7].

The two methods for the solution of the associated time discretized systems rely on a generalized Stokes solver and a least squares solution of a non-linear system.

The generalized Stokes problem which is also found in the shallow-water equations [4], is solved by a direct Stokes solver [5] introduced by Glowinski-Pironneau [6] and which reduces the problem to a cascade of Poisson problems.

The non-linear system is solved via a least squares formulation [5] and preconditioned conjugate gradient algorithms [8].

SPACE DISCRETIZATION

The space discretization for solving (M1) or (M2) is achieved by a Lagrange finite element approximation.
If we denote :

$\mathcal{T}_h$: a standard finite element triangulation of Ω,

P_k : space of polynomials of degree $\leq k$,

the variables σ,u,v,T are approximated by functions belonging to appropriate subspaces of

$$H_h^1 = \{\phi_h \mid \phi_h \in C^0(\bar{\Omega}),\quad \phi_h|_T \in P_1,\ \forall T \in \mathcal{T}_h\}$$

defined from the boundary conditions.

NUMERICAL RESULTS

Computations related to five test cases of problem A, using the two above models M1 and M2 and the associated semi-implicit or implicit algorithms have been performed on IBM3090.

Triangulations

Several triangulations are used for the different test cases. Cases A1 and A5 are computed from grids obtained by the MODULEF Generator [9] while cases A2, A3, A6 use locally refined triangulations, generated by self-adaptive algorithms [1],[10]. Physical criteria based on vorticity production and transport of the Mach number, identifying the main features of the flow (boundary layers and wakes, shocks and stagnation points), provide local enrichment of the mesh by extranodes and thereby sensitive improvement of the solution. Enlargements of the adapted triangulations for cases A2, A3, A5, A6 are presented on Figures 2-5.

Domains of computation

The computational domain occupied by the fluid is surrounded by rectangular limits sized as :

$-6 \leq x \leq 7$
$-10 \leq y \leq 10$ for cases A1, A3

$-5 \leq x \leq 8$
$-8,5 \leq y \leq 8,5$ for cases A2, A5, A6

Steady solutions :

Steady state solutions independent of the timestep Δt have been obtained for cases A1, A2, A3, A5, A6. Different time step sizes are selected depending of the stiffness of the problem . Convergence histories show that two or three orders of magnitude are necessary to reach a steady solution, the solutions presented have been obtained with :

Δt = .2 and 40 time steps for A2, A3,
Δt = .2 and 60 time steps for A1, A5,
Δt = .05 and 180 time steps for A6.

Comments

Figures 8, 11, 14, 17, 20 show the density lines obtained by model M2 for the five test cases considered ; and for cases A1 and A3, these numerical results can be compared with the experimental results presented by J. Allegre, J.C. Lengrand, M. Raffin.

The pressure distributions (resp. skin friction coefficients) on the airfoil obtained by the two models are compared on Figures 6, 9, 12, 15, 18 (resp. Fig. 7, 10, 13, 16, 19). The good agreement of the two results shows that the assumption of constant total enthalpy does not affect these coefficients. However, as expected, the comparison of heat flux coefficients for cases A3 and A6 (Fig. 21-22) shows an important discrepancy.

The values of the skin friction coefficient could be improved by adding nodes near the body as shown by F. Angrand. Values of lift, drag and momentum coefficients are presented on Table 1 for model M1 and Table 2 for model M2. The C.P.U. time on IBM 3090 is given in Table 2 for model M2. The constant enthalpy assumption does not provide sensitive C.P.U. time saving but core memory saving.

Table 1 : Integral values computed with model M1.

cases	A1	A2	A3	A5	A6
Nodes	3114	3060	2646	2354	4154
C_L	.544	.449	.324	10^{-5}	10^{-5}
C_D	.604	.252	.443	.179	.095
C_M	.212	.129	.165	10^{-5}	10^{-4}

Table 2 : Integral values computed with model M2 and C.P.U. time.

cases	A1	A2	A3	A5	A6
Nodes	3114	3060	2646	2354	4154
C_L	.551	.461	.331	10^{-5}	10^{-5}
C_D	.608	.257	.443	.179	.101
C_M	.216	.133	.171	10^{-5}	10^{-4}
CPU_{min}	118	159	192	54	280

For case A2, the separation point is located at $x_s = 0.45$.

FURTHER COMMENTS AND CONCLUSION

A finite element methodology using semi or full implicit schemes for time integration and least squares/conjugate gradient algorithms as method of solution has been tested on two different non conservative models (total enthalpy constant or not).

In the ranges of the Mach numbers and Reynolds numbers of cases A1 to A6, the two models give globally similar results.

The space discretization is based on centered approximation and no artificial viscosity is added, which give satisfactory results for the Reynolds numbers considered in the above computations. However to deal with higher Reynolds numbers (for instance case A7, Re = 10000.), the introduction of an artificial viscosity would be definitely necessary.

Acknowledgements

This work was partly supported by DRET under contract 83/403.

References

[1] C. POULETTY, Génération et Optimisation de maillages en éléments finis, Application à la résolution de quelques équations en Mécanique des Fluides, Thèse de Docteur-Ingénieur, Ecole Centrale, Déc. 85.

[2] R. GLOWINSKI, Numerical Methods for Nonlinear Variational Problems, Springer-Verlag, New-York, 1984.

[3] M.O. BRISTEAU, R. GLOWINSKI, B. MANTEL, J. PERIAUX, P. PERRIER, Numerical Methods for the Simulation of Viscous Flows modelled by the Navier-Stokes Equations, in Finite Element in Fluids, Vol. 6, ed. by R.H. Gallagher, J.T. Oden, O.C. Zienkiewicz, Wiley, Chichester, 1985.

[4] J. GOUSSEBAILE, F. HECHT, G. LABADIE, L. REINHART, Finite Element Solution of the Shallow Water Equations by a Quasi-direct Decomposition Procedure, Int. Journal for Numerical Methods in Fluids, Vol. 4, N° 12, pp. 1117-1136.

[5] M.O. BRISTEAU, R. GLOWINSKI, B. DIMOYAT, J. PERIAUX, P. PERRIER, O. PIRONNEAU, Finite Element Methods for the Compressible Navier-Stokes Equations, in Proceedings of the AIAA Computation of Fluid Dynamics Conference, Danvers, July 13-15, 1983, AIAA Paper 83-1890.

[6] R. GLOWINSKI, O. PIRONNEAU, On a Mixed Finite Element Approximation of the Stokes Problem. Convergence of the Approximate Solution, Numerische Mathematik, 33, 1979, pp. 397-424.

[7] M.O. BRISTEAU, R. GLOWINSKI, B. MANTEL, J. PERIAUX, P. PERRIER, Numerical Methods for the Time Dependent Compressible Navier-Stokes Equations, in Computing Methods in Applied Sciences and Engineering VI, R. Glowinski, J.L. Lions eds., North-Holland, Amsterdam, 1984, pp.565-583.

[8] A. BUCKLEY, A. LENIR, QN-like variable storage conjugate gradients, Math. Programming, 27, (1983), pp. 155-175.

[9] A. PERRONNET, "The Club Modulef : A Library of Subroutines for Finite Element Analysis", Computing Methods in Applied Sciences and Engineering, Lecture Notes in Mathematics, Springer-Verlag, Vol. 704, Berlin, 1979, pp. 127-153.

[10] B. PALMERIO, V. BILLEY, A. DERVIEUX, J. PERIAUX, Self-adaptive mesh refinements and Finite Element Methods for Solving the Euler Equations, Numerical Methods for Fluid Dynamics II, K.W. Morton, M.J. Baines Eds., Clarendon Press, Oxford, 1986, pp. 369-387.

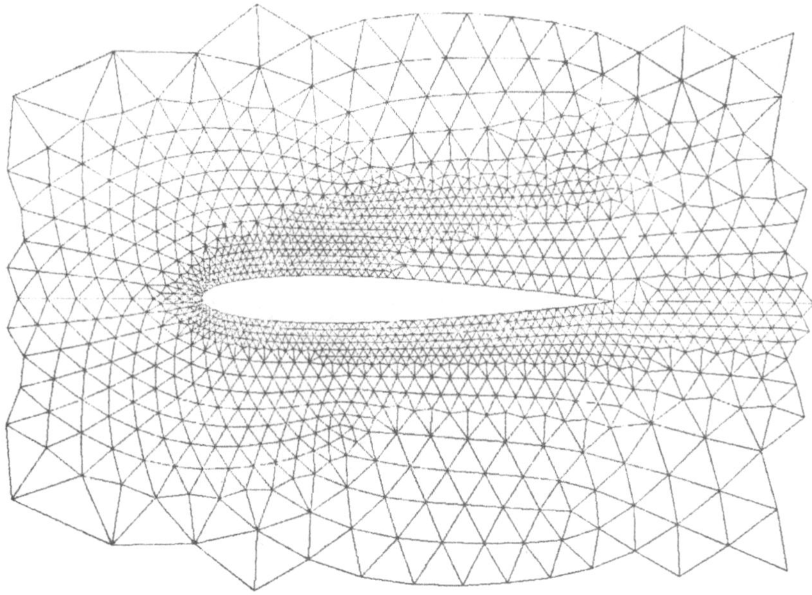

Figure 2 : Enlargement of the mesh for case A2.
Nodes : 3060, Elements : 5936.

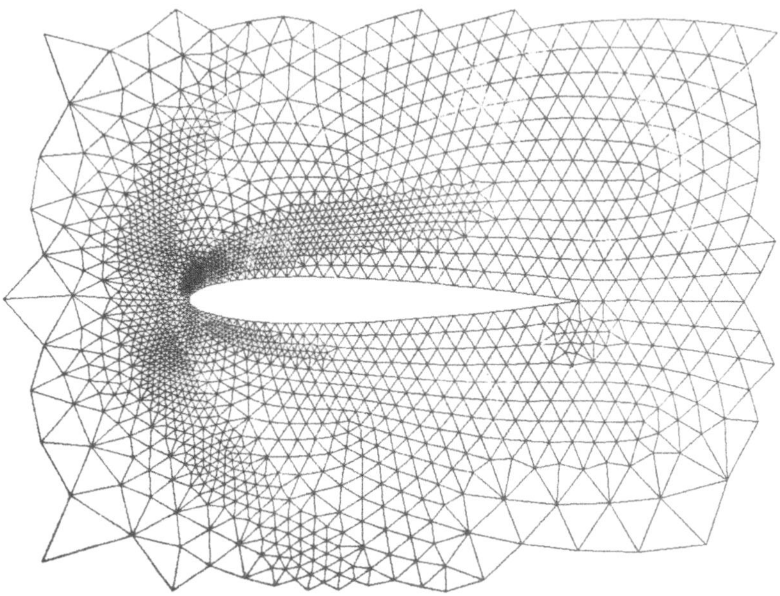

Figure 3 : Enlargement of the mesh for case A3.
Nodes : 2646, Elements : 5172.

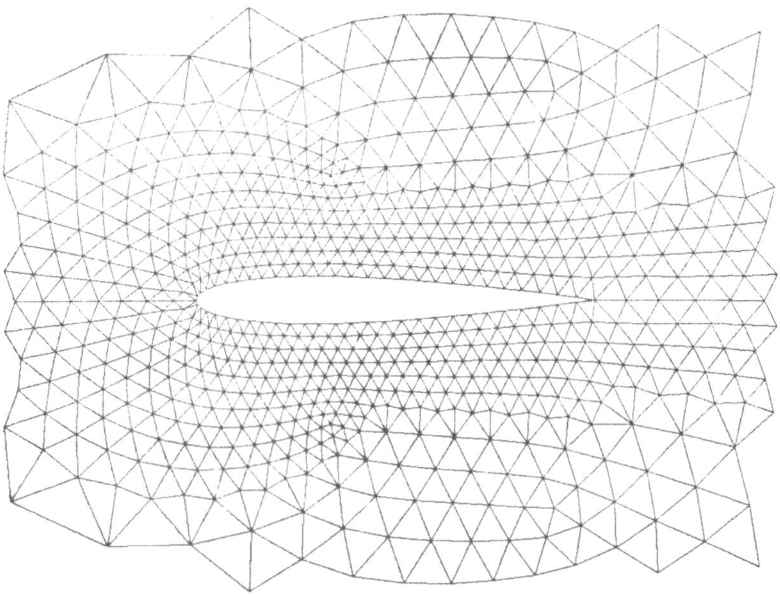

Figure 4 : Enlargement of the mesh for case A5.
Nodes : 2354, Elements : 4568.

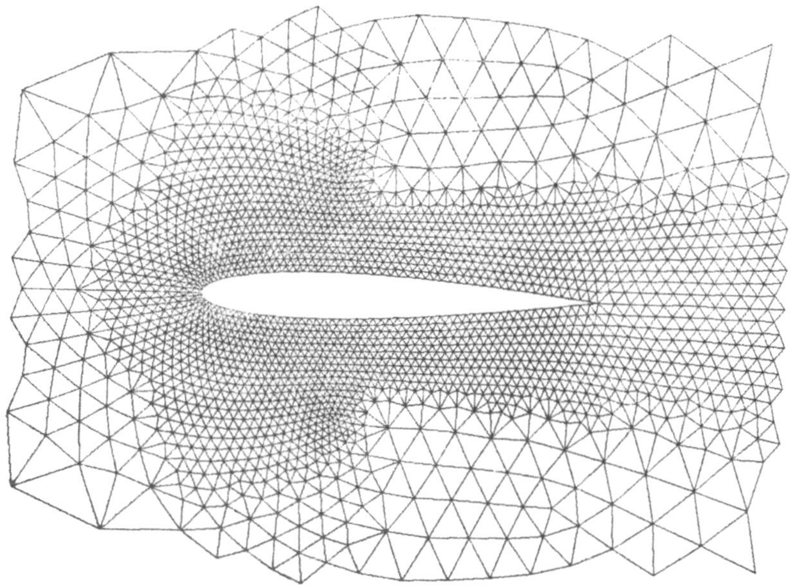

Figure 5 : Enlargement of the mesh for case A6.
Nodes : 4154, Elements : 8112.

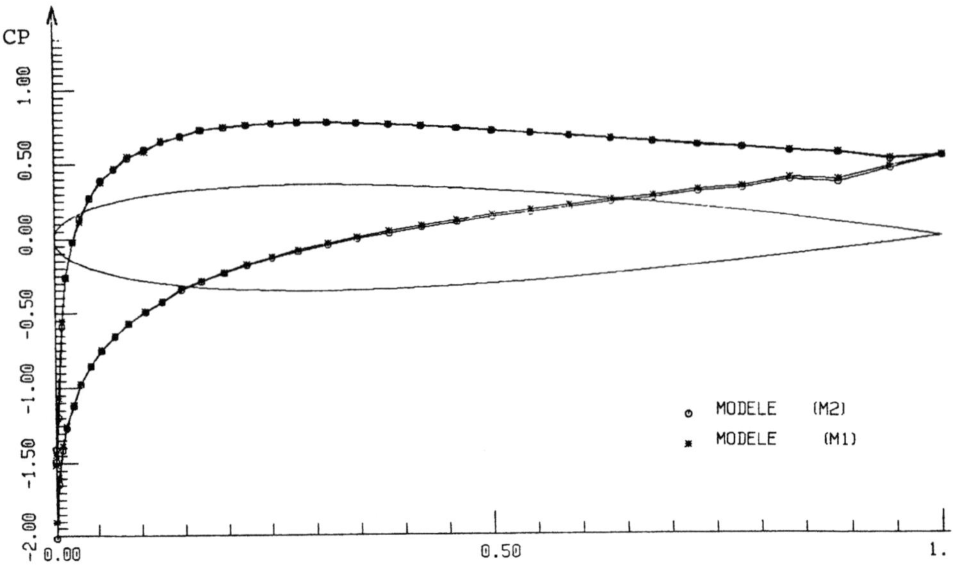

Figure 6 : Pressure coefficient

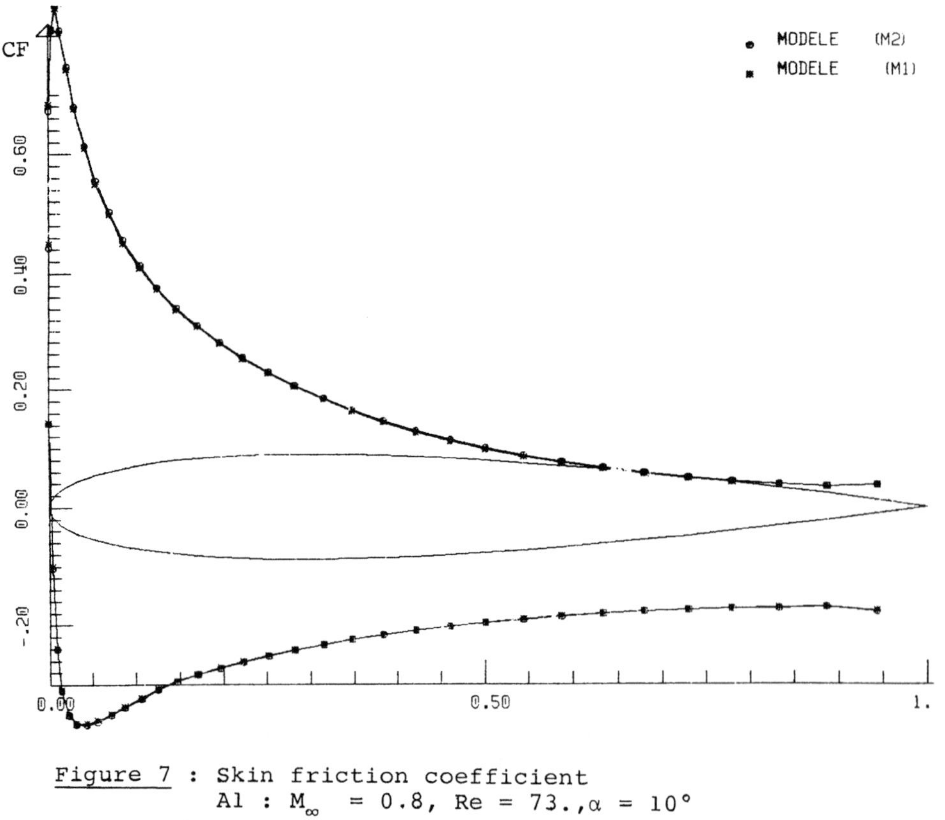

Figure 7 : Skin friction coefficient
A1 : M_∞ = 0.8, Re = 73., α = 10°

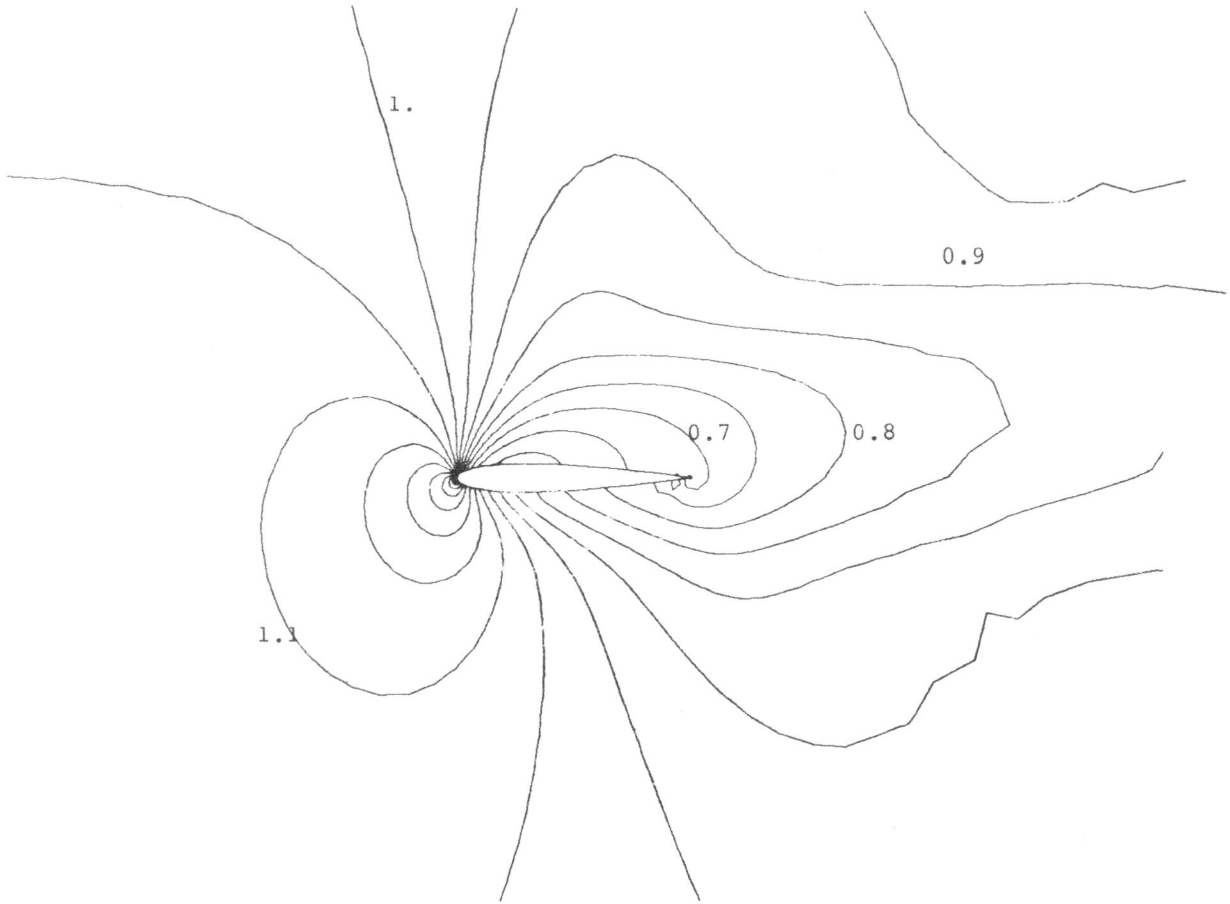

Figure 8 : Iso-density lines, $\Delta\rho/\rho_\infty = 0.05$
A1 : $M_\infty = 0.8$, Re = 73., $\alpha = 10°$

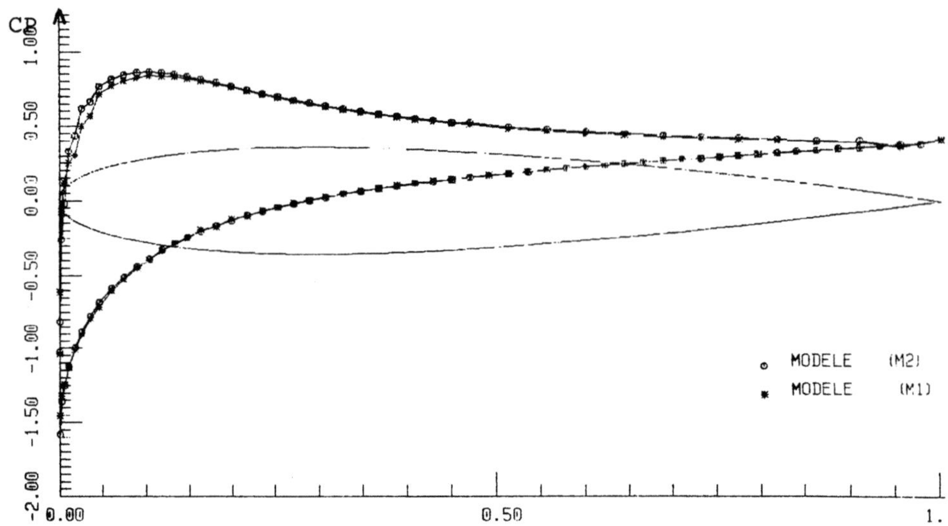

Figure 9 : Pressure coefficient

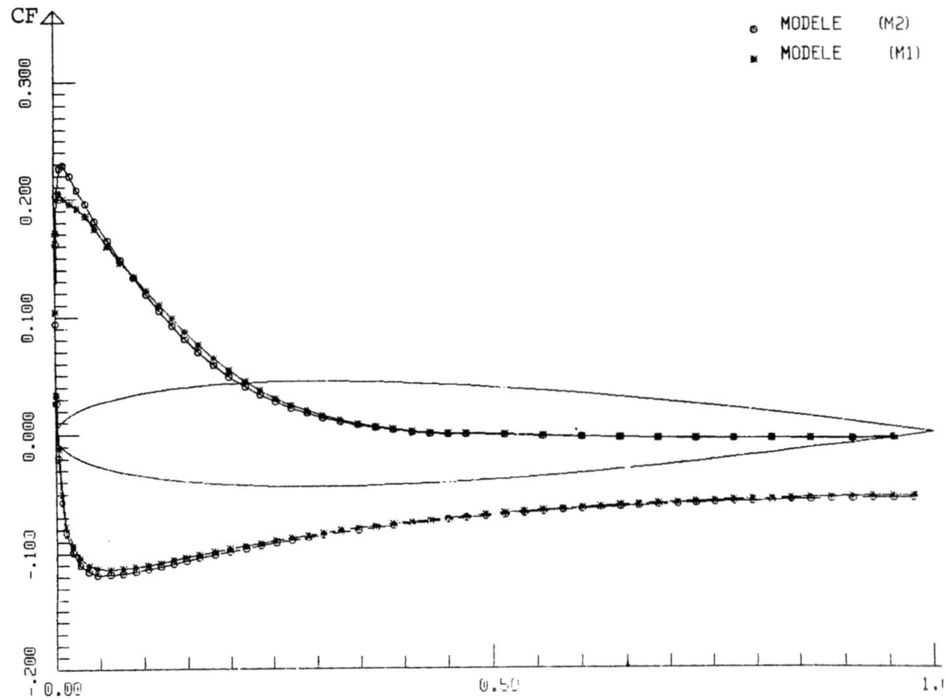

Figure 10 : Skin friction coefficient
A2 : M_∞ = .8, Re = 500., α = 10°.

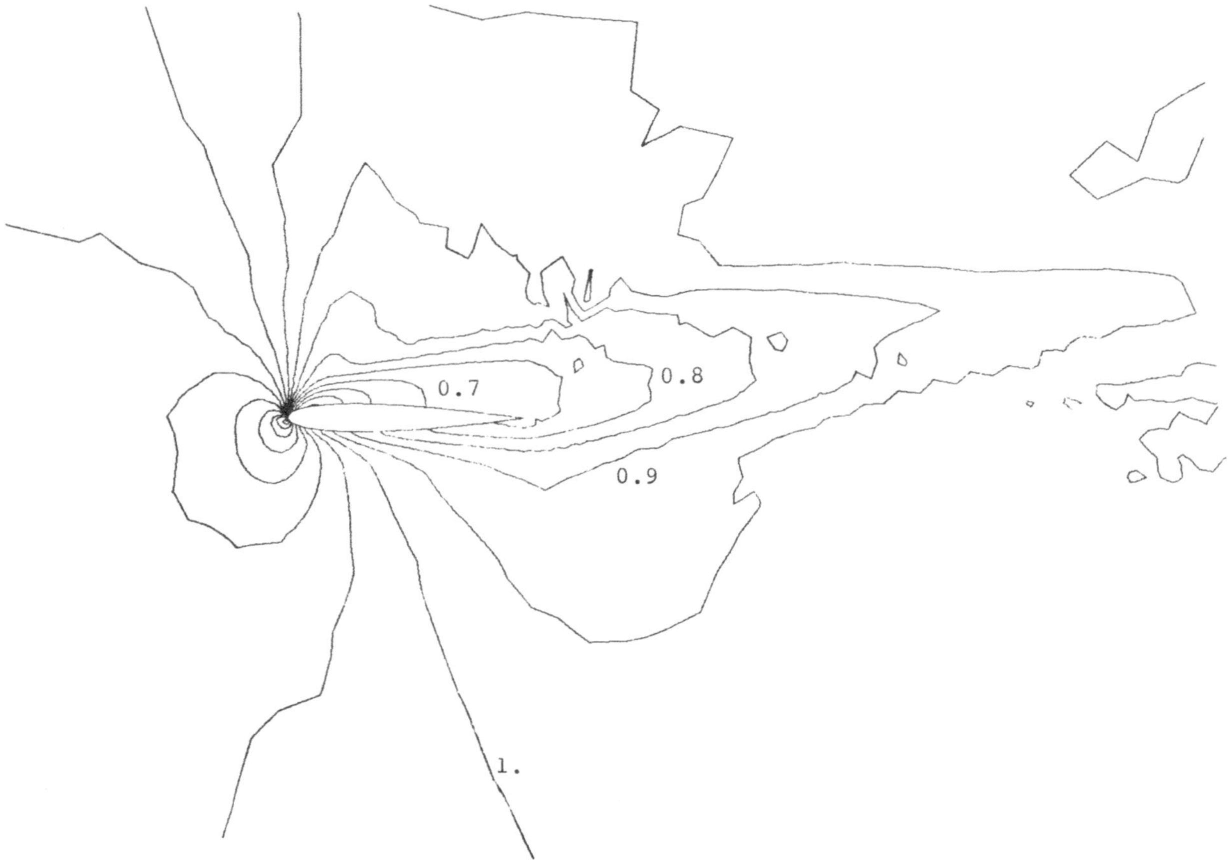

<u>Figure 11</u> : Iso-density lines, $\Delta\rho/\rho_\infty = 0.05$
A2 : $M_\infty = .8$, $Re = 500.$, $\alpha = 10°$.

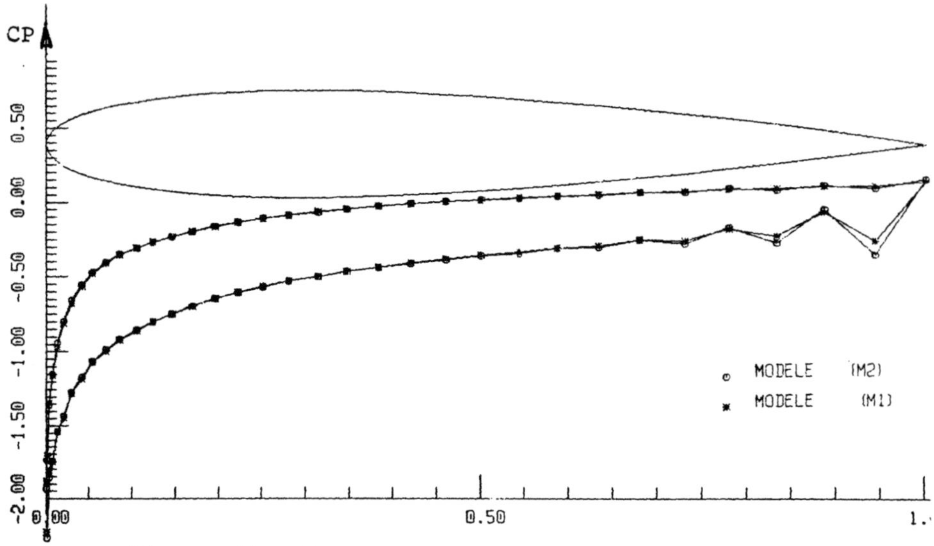

Figure 12 : Pressure coefficient

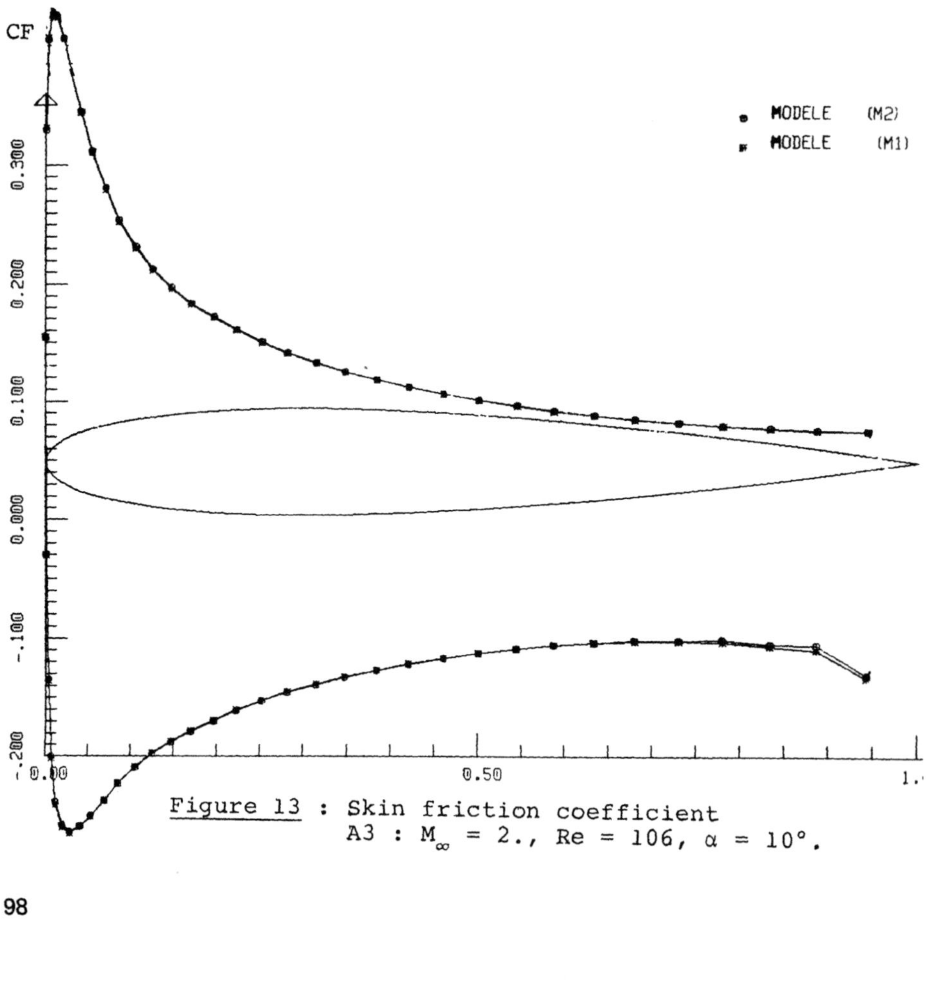

Figure 13 : Skin friction coefficient
A3 : M_∞ = 2., Re = 106, α = 10°.

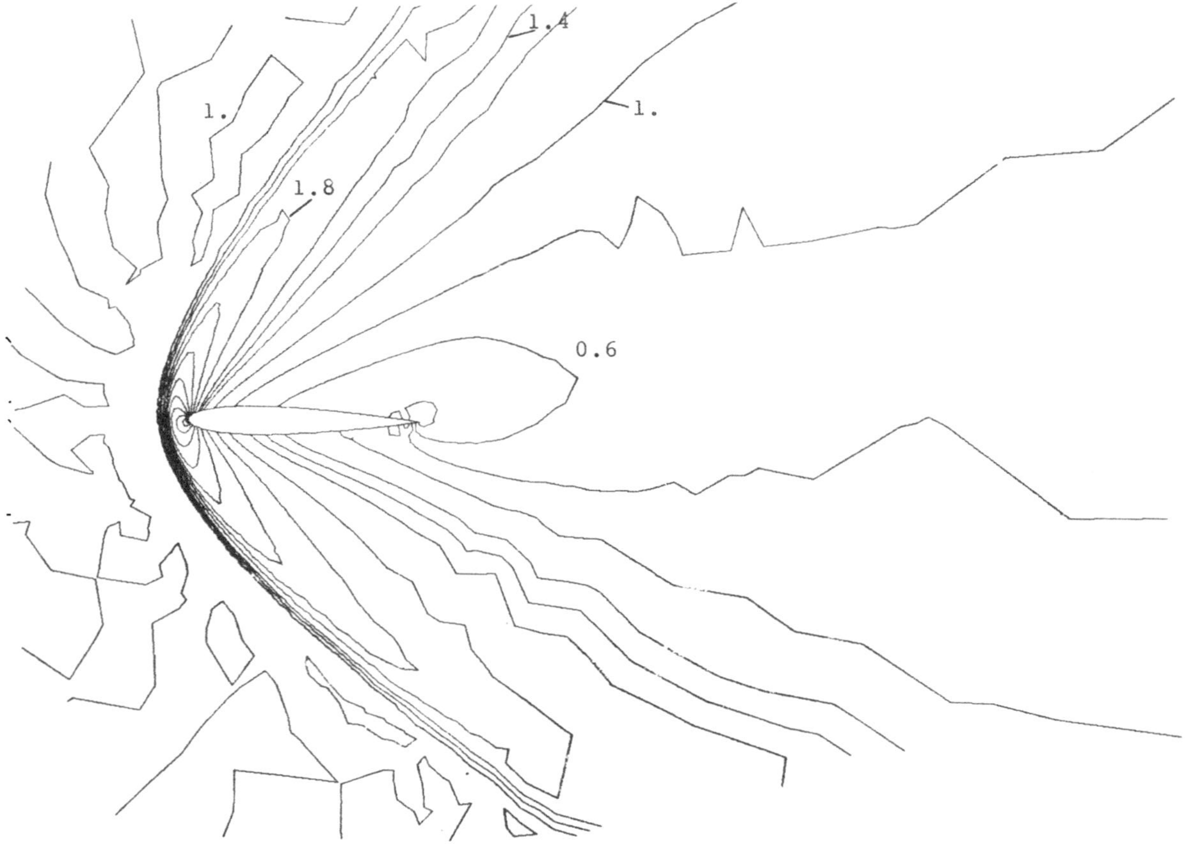

<u>Figure 14</u> : Iso-density lines, $\Delta\rho/\rho_\infty = 0.2$.
A3 : $M_\infty = 2.$, Re = 106., $\alpha = 10°$.

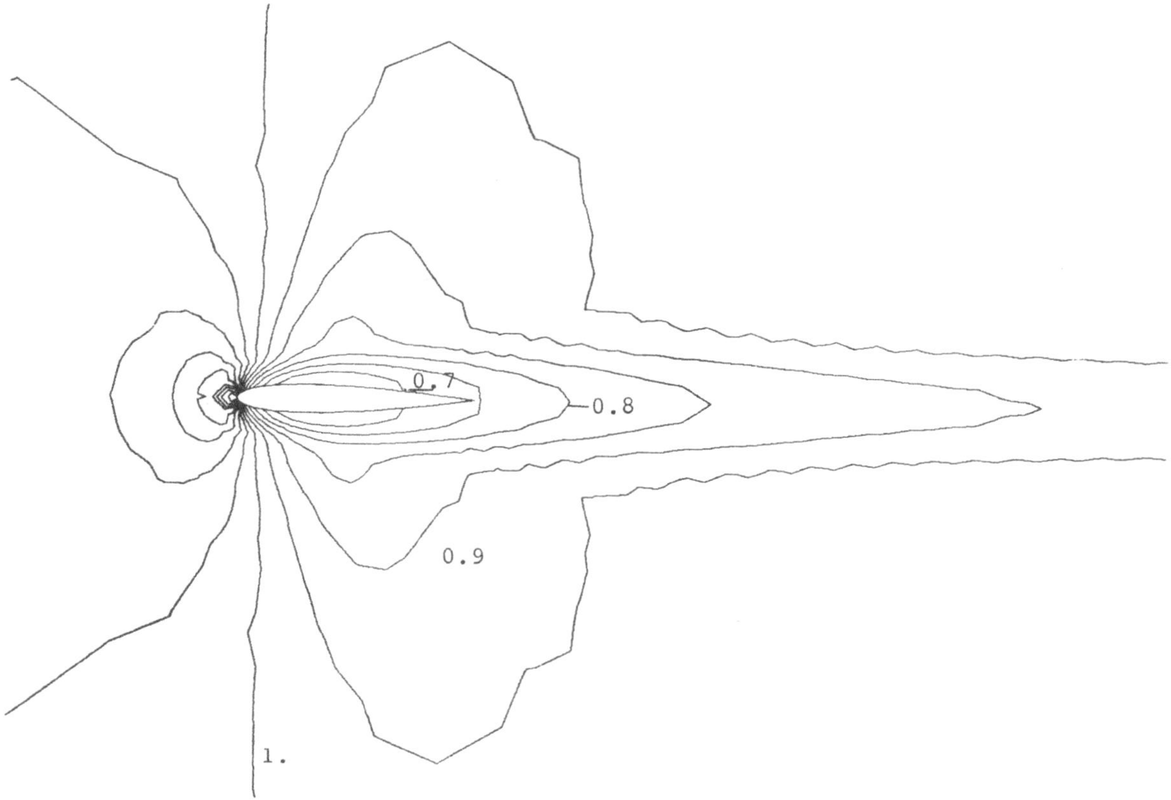

Figure 17 : Iso-density lines, $\Delta\rho/\rho_\infty = 0.05$
A5 : $M_\infty = 0.85$, Re = 500., $\alpha = 0°$.

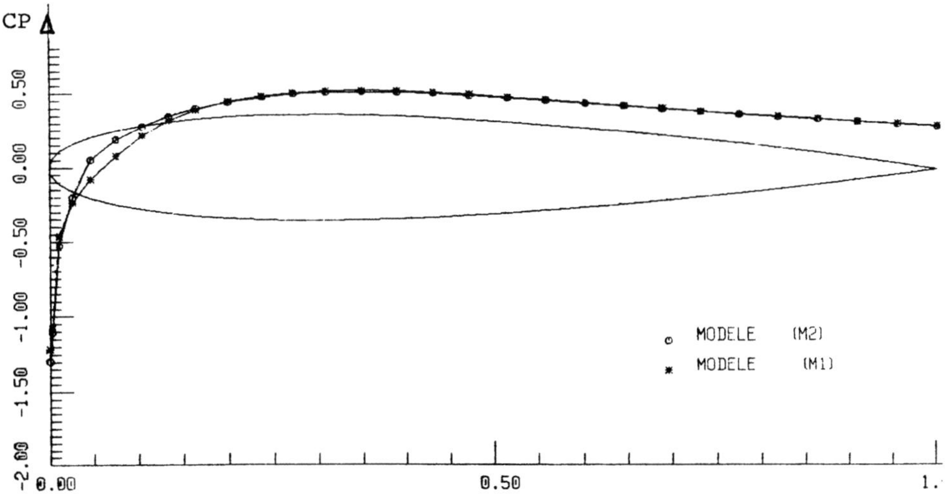

Figure 15 : Pressure coefficient

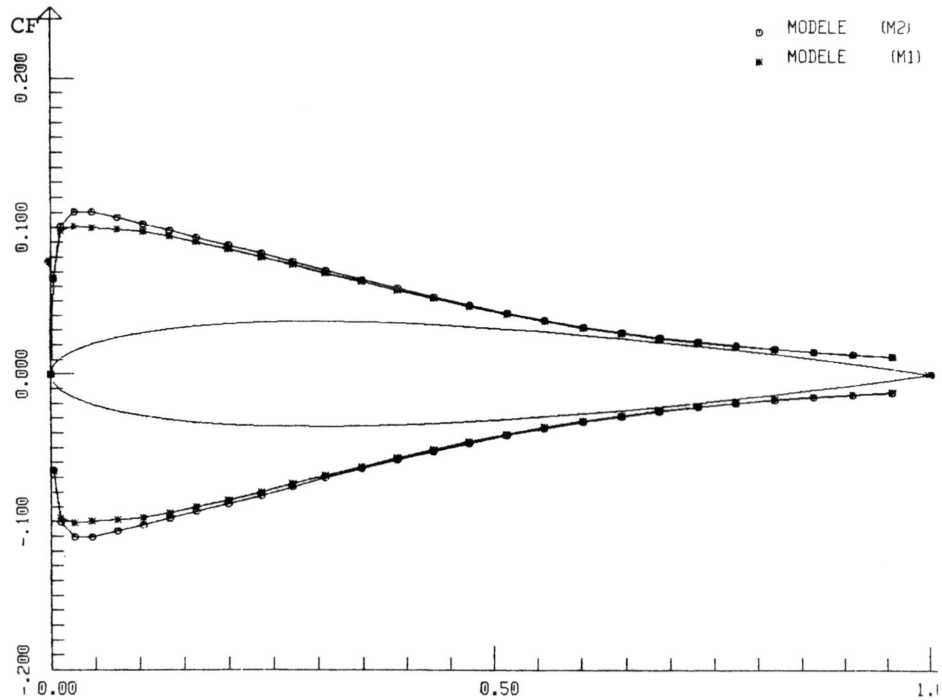

Figure 16 : Skin friction coefficient
A5 : $M_\infty = 0.85$, Re = 500., $\alpha = 0°$.

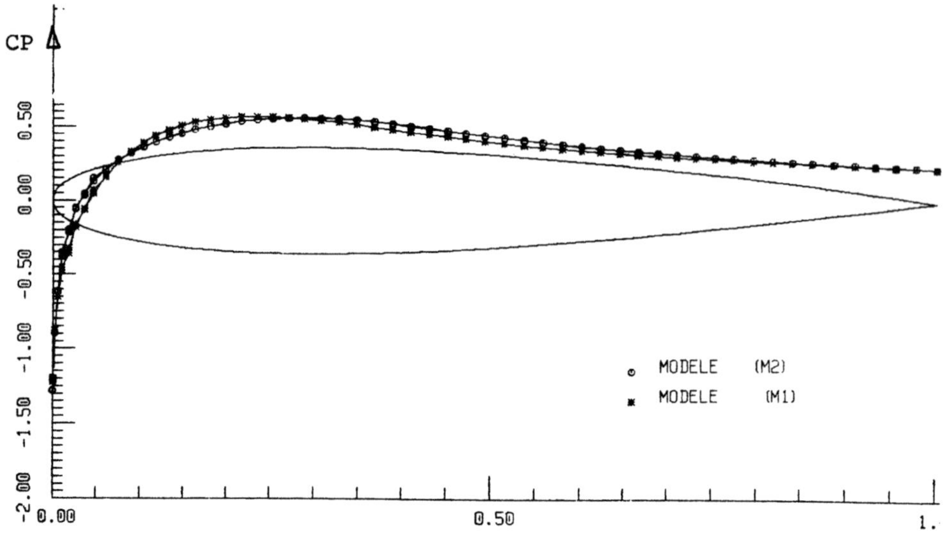

Figure 18 : Pressure coefficient

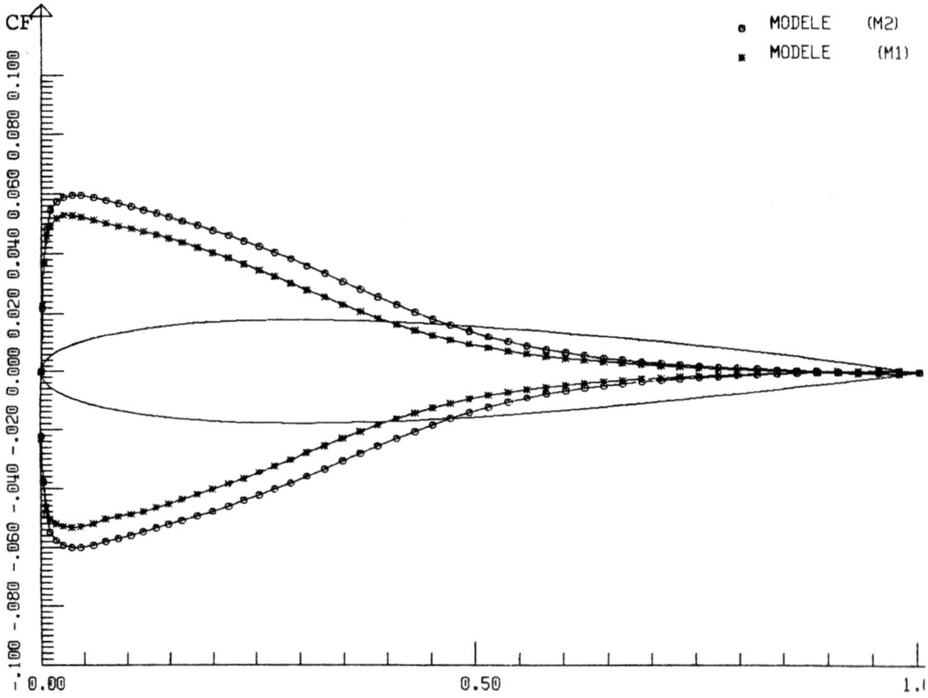

Figure 19 : Skin friction coefficient.
A6 : $M_\infty = 0.85$, Re = 2000., $\alpha = 0°$.

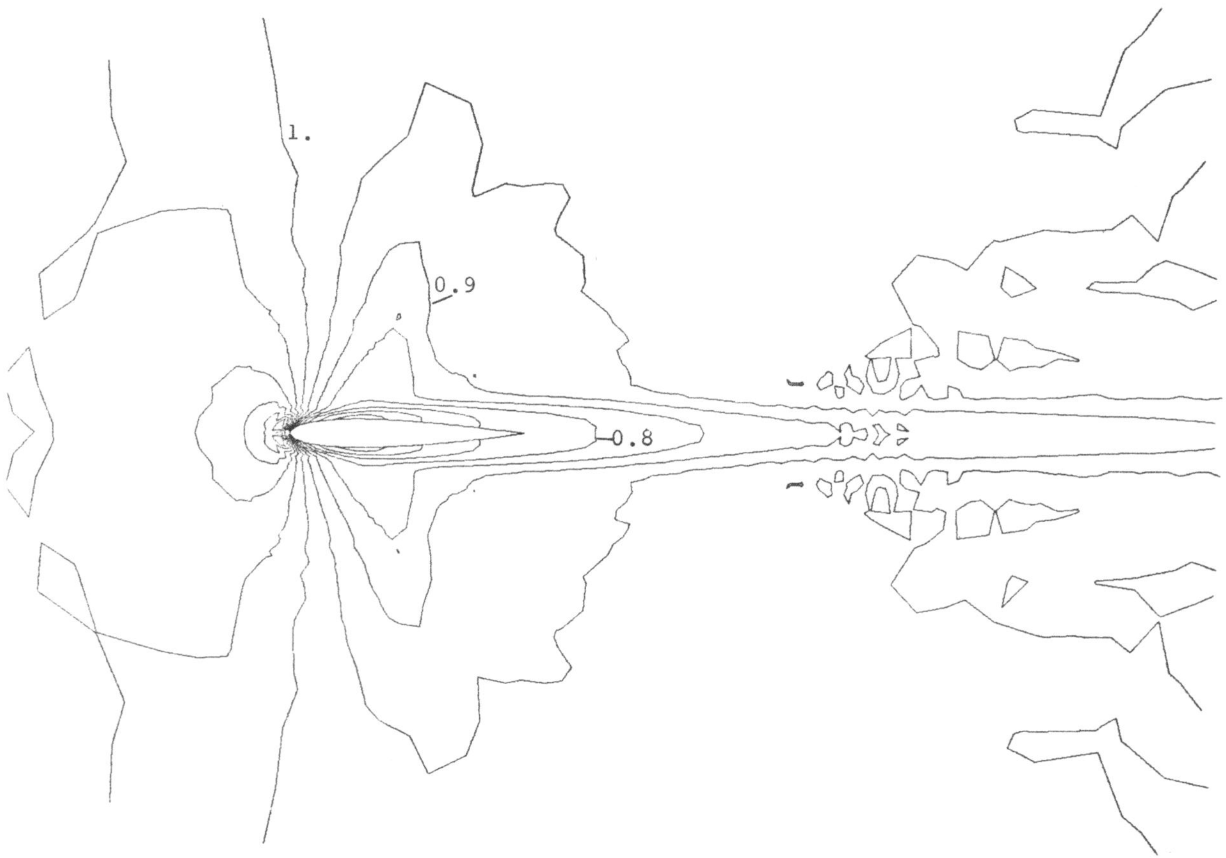

<u>Figure 20</u> : Iso-density lines, $\Delta\rho/\rho_\infty = 0.05$
A6 : $M_\infty = 0.85$, Re = 2000., $\alpha = 0°$.

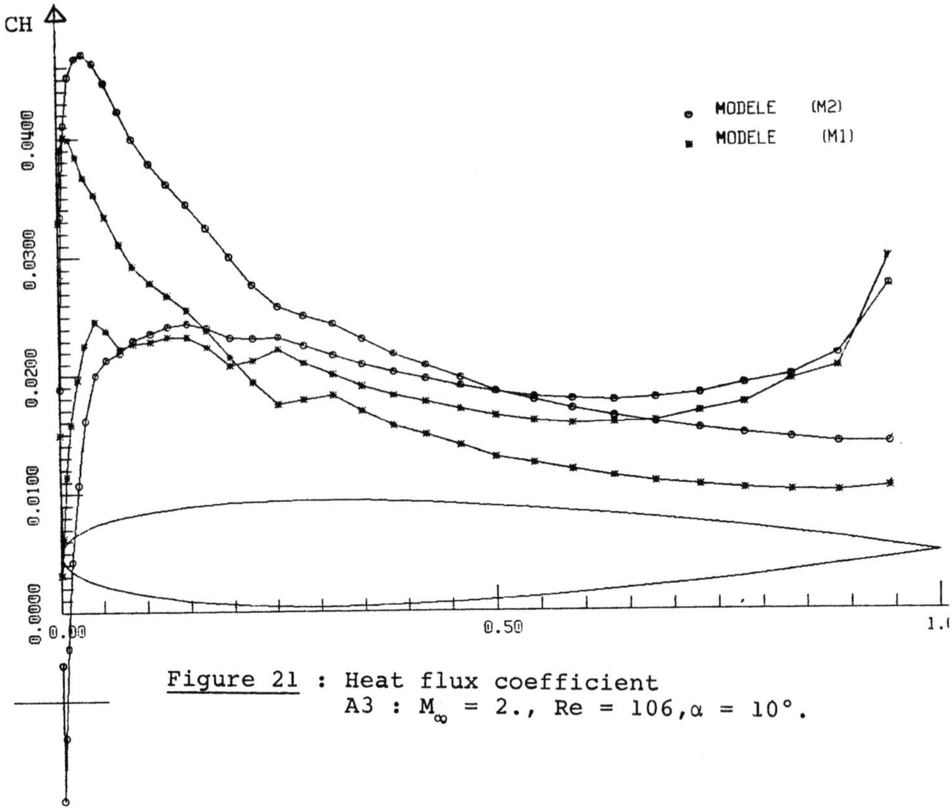

<u>Figure 21</u> : Heat flux coefficient
A3 : M_∞ = 2., Re = 106, α = 10°.

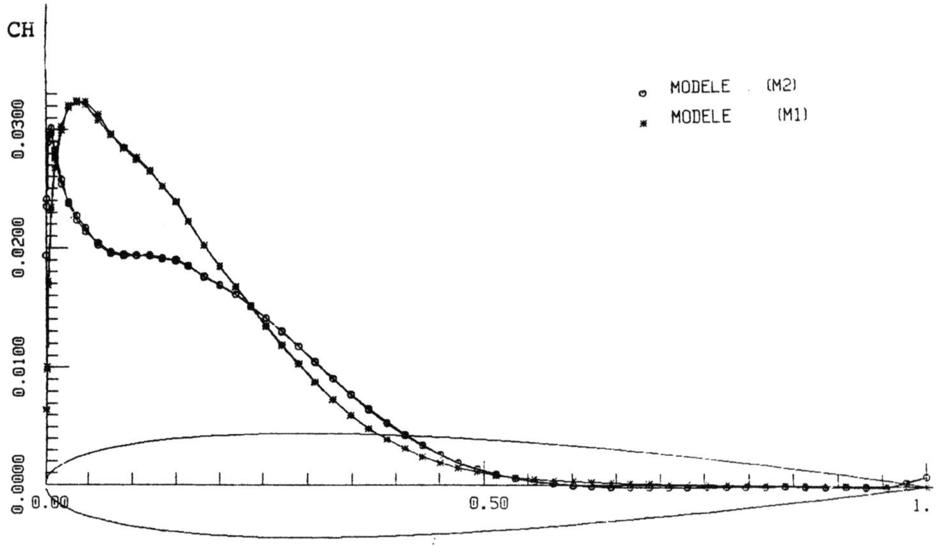

<u>Figure 22</u> : Heat flux coefficient.
A6 : M_∞ = 0.85, Re = 2000, α = 0°.

COMPUTATION OF VISCOUS TRANSONIC FLOWS USING AN UNSTEADY TYPE METHOD AND A ZONAL GRID REFINEMENT TECHNIQUE

L. Cambier
Office National d'Etudes et de Recherches Aérospatiales
BP 72 - 92322 Châtillon Cedex, France

NUMERICAL METHOD

The Navier-Stokes equations are numerically solved by an unsteady type method with an explicit predictor-corrector scheme. The equations are discretized in conservative form using MacCormack's finite difference scheme directly in the physical plane on an arbitrary curvilinear mesh. This discretization method is described in [1, 2].

An artificial viscosity term is added after the computation of the corrector step. The time value being denoted by the superscript n ($t^n = n\ \Delta t$), and the corrector value being denoted by n+1, the final value of a variable f at time n+1 is given by :

$$f^{n+1} = f^{n+1} + D_o(f^n) = f^{n+1} - (Q_{io}\ \delta_i^{(4)} f^n + Q_{jo}\ \delta_j^{(4)} f^n)$$

with $\delta_i^{(4)} f = f_{i+2,j} - 4\ f_{i+1,j} + 6\ f_{i,j} - 4\ f_{i-1,j} + f_{i-2,j}$

$$\delta_j^{(4)} f = f_{i,j+2} - 4\ f_{i,j+1} + 6\ f_{i,j} - 4\ f_{i,j-1} + f_{i,j-2}$$

D_o is a smoothing term added to each one of the equations in order to ensure the stability of the scheme in the perfect fluid region. Q_{io} and Q_{jo} are two constants chosen as low as possible for each numerical application.

Near the wall, the mesh refinement required to represent the rapid variations of the flow properties in the direction normal to the wall is obtained by adding to a regular contraction a zonal grid refinement technique, consisting of successive halvings of the mesh size from one zone to the following one. This technique of dichotomy is described in [3] and [4] and has been successfully applied to high Reynolds number flow calculations. For these calculations, the technique of dichotomy provides a rapid transition from a relatively coarse mesh in the outer quasi-inviscid flow to a very fine mesh in the viscous sublayer with a smoothly varying mesh size in each zone. This technique can also be applied to lower Reynolds number calculations (such as the proposed test cases) considering a small number of zones. Coupling of adjacent zones is done by overlapping. A systematic procedure has been set up to sweep the computational domain and ensure the numerical matching in such a way that the maximum allowable time step can be used in each zone.

The computational domain Ω for problem A is represented on Fig. 1. The global mesh is of C-type and details on its generation are given in [5].

The boundary conditions on the external boundary Γ of the computational domain Ω are the following :

i) On the part Γ_1 of the boundary Γ,
- all the unknowns are given if the flow is entering the domain Ω

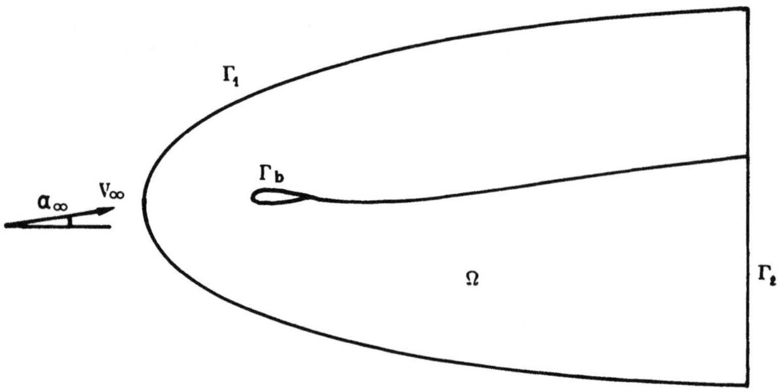

Fig. 1 - Computational domain.

with a supersonic normal velocity,

- the direction of the velocity, the total enthalpy and the stagnation pressure are prescribed if the flow is entering the domain Ω with a subsonic normal velocity,
- only the direction of the velocity is prescribed if the flow is leaving the domain Ω with a subsonic normal velocity.

ii) On the part Γ_2 of the boundary Γ,

- there is no boundary condition if the flow is leaving the domain Ω with a supersonic normal velocity,
- the static pressure is given if the flow is leaving the domain Ω with a subsonic normal velocity.

The treatment of these boundary conditions is of inviscid type and is achieved by a technique based on compatibility relations and described in [6].

On the wall Γ_b, we impose a no-slip condition and we set the temperature value. Wall pressure is derived from the momentum equation at the wall. The detail of the computation at the wall can also be found in [6].

COMMENTS ON CALCULATIONS

We have run our Navier-Stokes code on the seven test cases of the external problem A. The calculations have been carried out on a CRAY 1S/2000 computer. Due to the simplicity of the explicit method, the code presents a good rate of vectorization.

We used a mesh well-adapted to each test case. Firstly, the mesh line issued from the trailing edge makes an angle with the X-axis roughly equal to the angle of attack α_∞. Secondly, the computational domain for test cases with supersonic Mach number at infinity is defined such that the external boundary Γ is never crossed by the flow with a subsonic normal velocity. Lastly, the mesh refinement near the wall (and particularly the number of zones) is chosen according to the Reynolds number value. Figures 2, 3 and 4 show some examples of the meshes used.

The values of the two coefficients Q_{io} and Q_{jo} in the smoothing term D_o are equal to 2.10^{-3} for test cases A1, A2, A5, A6 and to 10^{-3} for test cases A3, A4 and A7. For test case A7, a non-linear second order artificial viscosity term has been added only to the momentum equation.

For the first six test cases, we have obtained a steady solution. There is no automatic convergence test and the presented results correspond to a number of iterations chosen a priori. For test case A7, the computed solution does not reach a steady state. The solution oscillates about a symmetrical steady flow. The presented results show the solution obtained at some arbitrary moment during the iterations.

Table 1 gives some information for each test case about the size of the computational domain, the number of zones, the number of mesh points, the number of cycles, the CPU cost, and the values of the lift coefficient C_L, the drag coefficient C_D and the momentum coefficient C_M. The storage required is lower than 400000 words for all the calculations.

Figure 5 shows the convergence history plots corresponding to the four mandatory test cases. Note that the total energy residuals have been considered instead of the internal energy residuals.

Figures 6 to 20 show some outputs of the seven test cases. For each mandatory test case, we present the pressure coefficient distribution, the skin friction coefficient distribution and the iso-Mach number lines. The iso-density lines corresponding to test cases A1 and A3 are also presented in order to allow a comparison with experiment. For the optional test cases A4 and A6, we only show the pressure coefficient distribution. For test case A7, we present the iso-Mach number lines (Fig. 20) which exhibit a Bénard-Karman vortex street type solution.

REFERENCES

[1] Le Balleur, J.C., Peyret, R., Viviand, H., "Numerical Studies in High Reynolds Number Aerodynamics". Computers and Fluids, Vol. 8, pp. 1-30, 1980.

[2] Hollanders, H., Viviand, H., "The Numerical Treatment of Compressible High Reynolds Number Flows". Computational Fluid Dynamics, Vol. 2, Ed. Kollmann W., Hemisphere Publ. Corp., 1980.

[3] Viviand, H., Ghazzi, W., "Numerical Solution of the Navier-Stokes Equations at High Reynolds Numbers with Applications to the Blunt Body Problem". Lecture Notes in Physics, 59, Springer-Verlag, 1976.

[4] Cambier, L., Ghazzi, W., Veuillot, J.P., Viviand, H., "A Multi-domain Approach for the Computation of Viscous Transonic Flows by Unsteady Methods". Recent Advances in Numerical Methods in Fluids, Vol. 3, Pineridge Press, Swansea, 1984.

[5] Le Balleur, J.C., "Strong Matching Method for Computing Transonic Viscous Flows Including Wakes and Separations. Lifting Airfoils". La Recherche Aérospatiale 1981-3 (English version pp. 21-43), 1981.

[6] Cambier, L., Ghazzi, W., Veuillot, J.P., Viviand, H., "Une approche par domaines pour le calcul d'écoulements compressibles". Computing Methods in Applied Sciences and Engineering V, Ed. Glowinski R. and Lions J.L., North-Holland, 1982.

Table 1 - Some information about the computations and integral values

Case	M_∞,	α_∞,	Re	Size of Ω	Nb mesh points (Nb zones)	Nb cycles	CPU cost per cycle (s)	CPU cost (s)	C_L	C_D	C_M
A1	.8	10°	73	$-9.9<x<10.5$ $-13.9<y<15.4$	193x45 (2)	20000	0.164	3280	0.5465	0.6280	0.2118
A2	.8	10°	500	$-9.6<x<10.4$ $-13.8<y<15.3$	265x55 (3)	10000	0.358	3580	0.4342	0.2656	0.1190
A3	2.	10°	106	$-0.5<x<3.0$ $-4.5<y<4.5$	193x56 (2)	10000	0.157	1570	0.3289	0.4826	0.1698
A4	2.	10°	1000	$-0.5<x<3.0$ $-4.5<y<4.5$	193x72 (3)	6000	0.325	1950	0.3427	0.2535	0.1560
A5	.85	0°	500	$-9.9<x<10.5$ $-13.9<y<13.9$	193x47 (3)	10000	0.223	2230	0.0007	0.2221	-.0001
A6	.85	0°	2000	$-9.9<x<10.5$ $-13.9<y<13.9$	193x47 (3)	10000	0.223	2230	0.0021	0.1164	0.0006
A7	.85	0°	10000	$-9.6<x<10.4$ $-13.8<y<13.8$	265x58 (4)	-	0.575	-	-	-	-

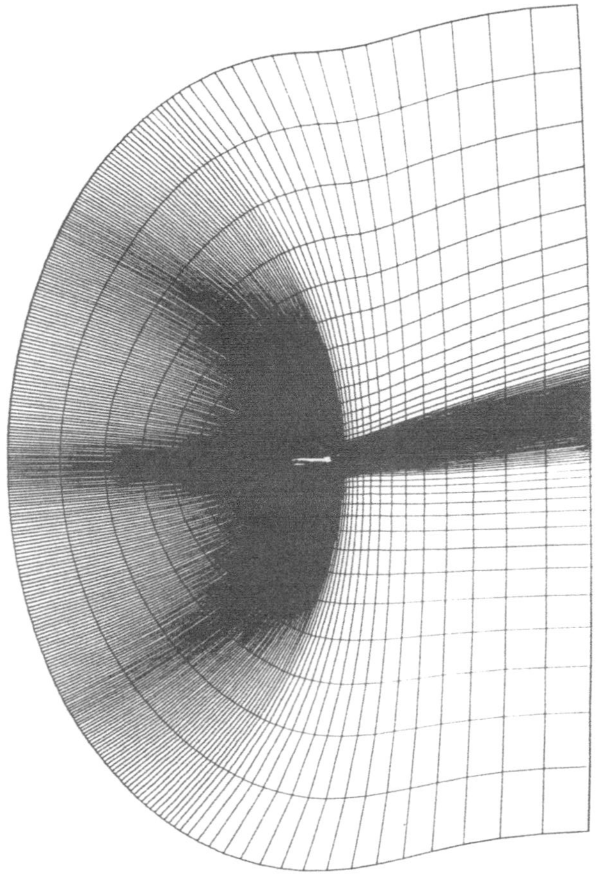

Fig. 2 - Global view of the grid for case A2 (265 x 55 mesh points).

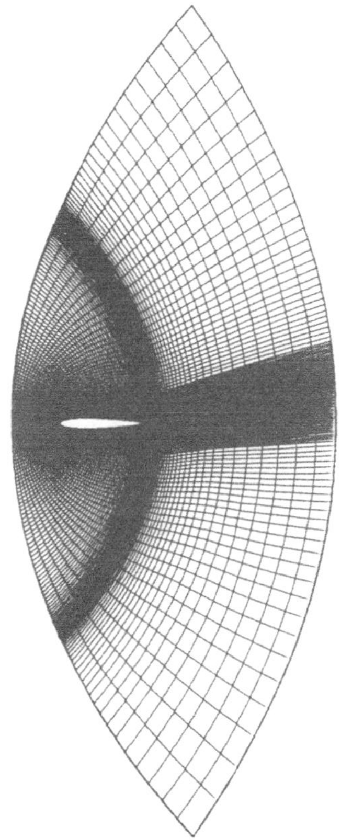

Fig. 3 - Global view of the grid for case A3 (193 x 56 mesh points).

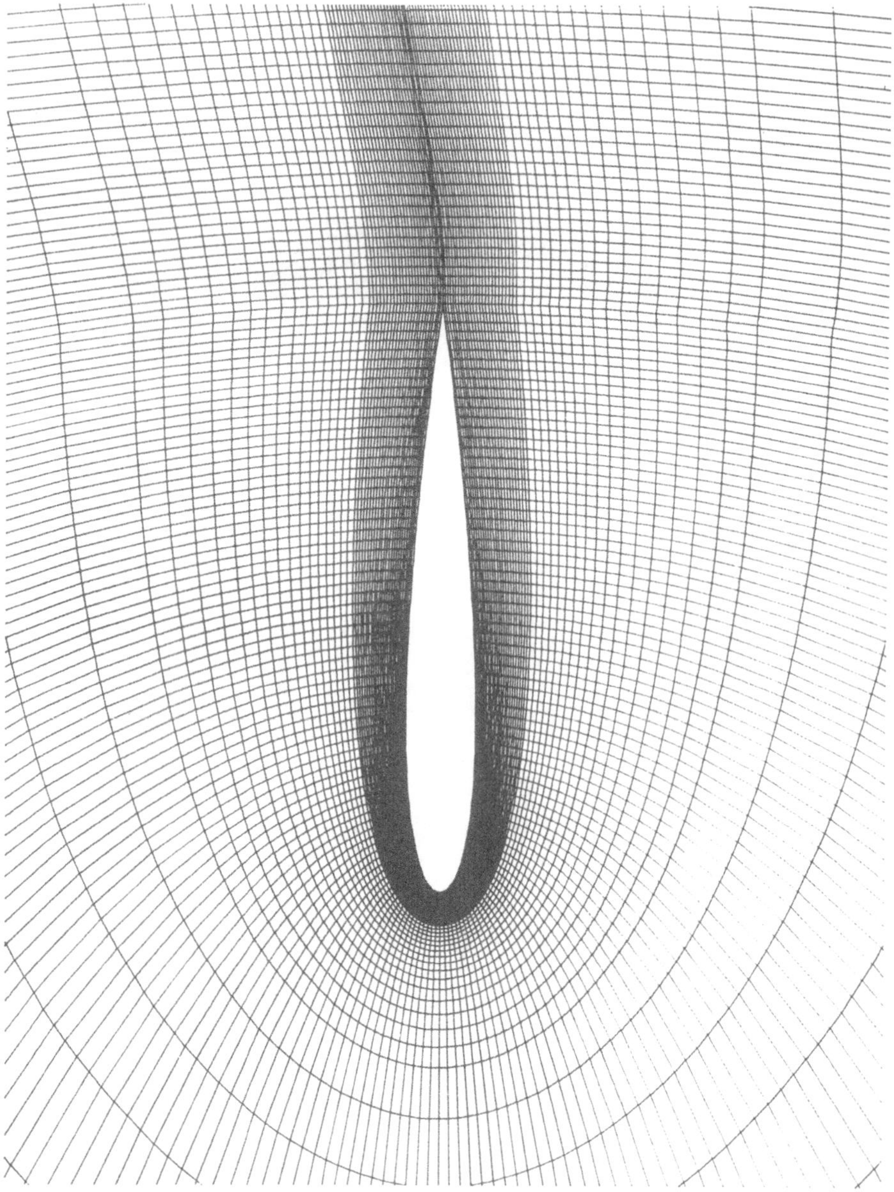

Fig. 4 - Close-up near the airfoil of the grid for case A2 (3 zones).

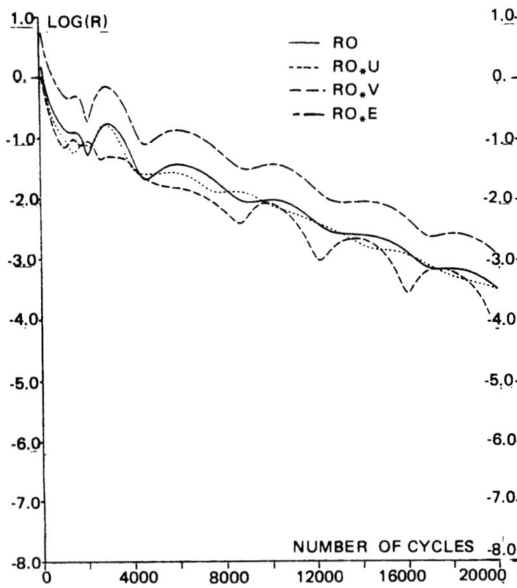

a) Case A1 : $M_\infty=.8$, $\alpha_\infty=10°$, Re=73

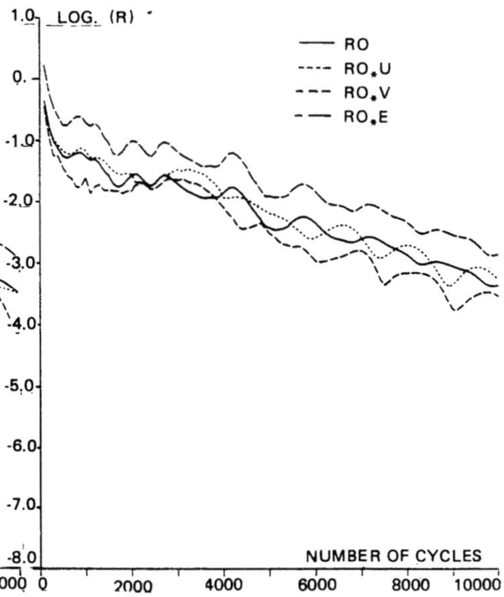

b) Case A2 : $M_\infty=.8$, $\alpha_\infty=10°$, Re=500

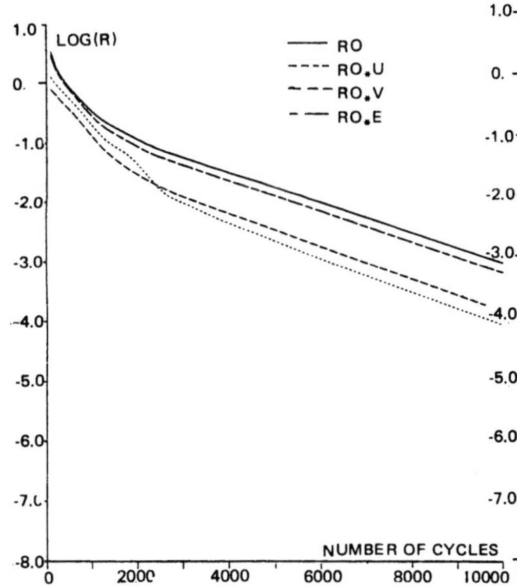

c) Case A3 : $M_\infty=2.$, $\alpha_\infty=10°$, Re=106

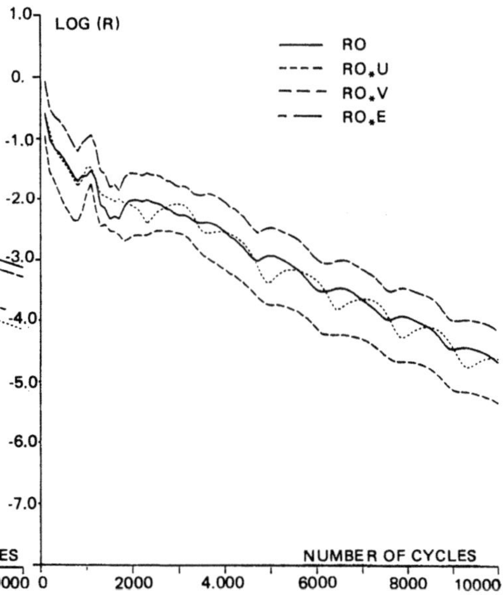

d) Case A5 : $M_\infty=.85$, $\alpha_\infty=0°$, Re=500

Fig. 5 - Convergence history plots (L_2-norm of residual).

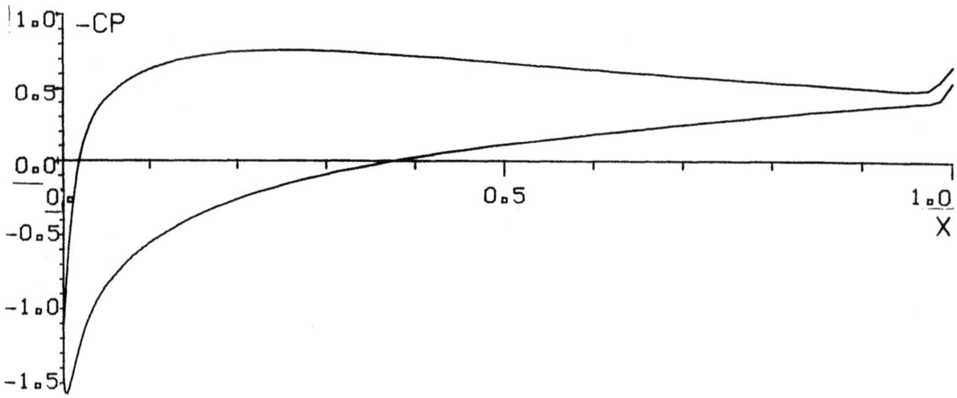

Fig. 6 - Pressure coefficient for case A1 : M_∞=.8, α_∞=10°, Re=73.

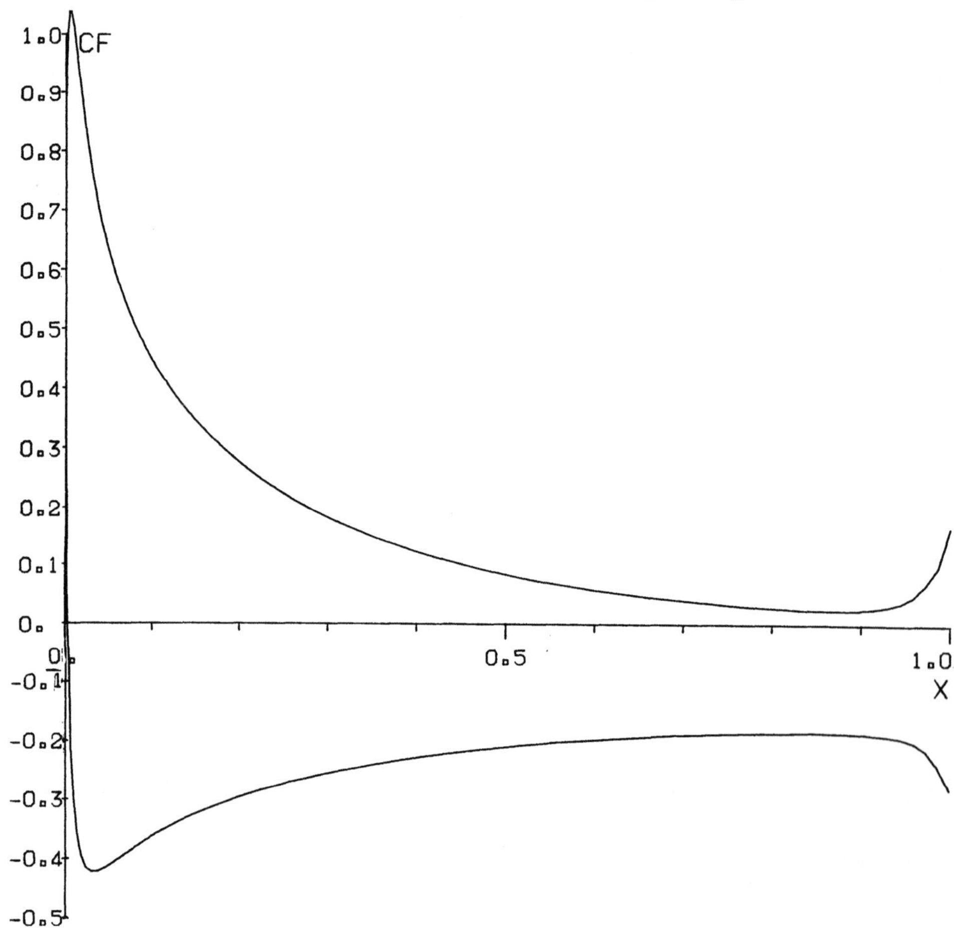

Fig. 7 - Skin friction for case A1 : M_∞=.8, α_∞=10°, Re=73.

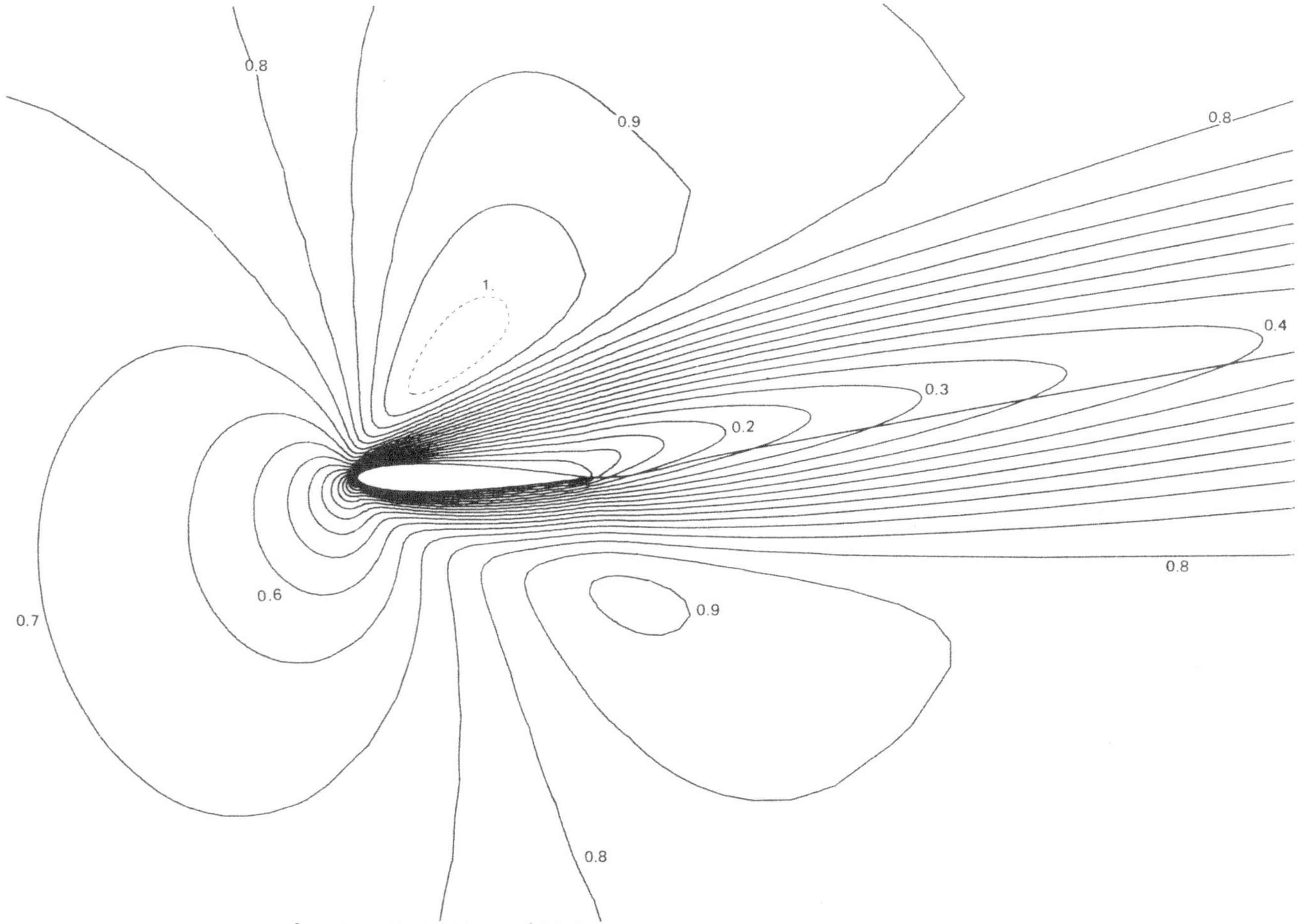

Fig. 8 - Iso-Mach lines ($\Delta M=0.05$) for case A1 : $M_\infty=.8$, $\alpha_\infty=10°$, Re=73.

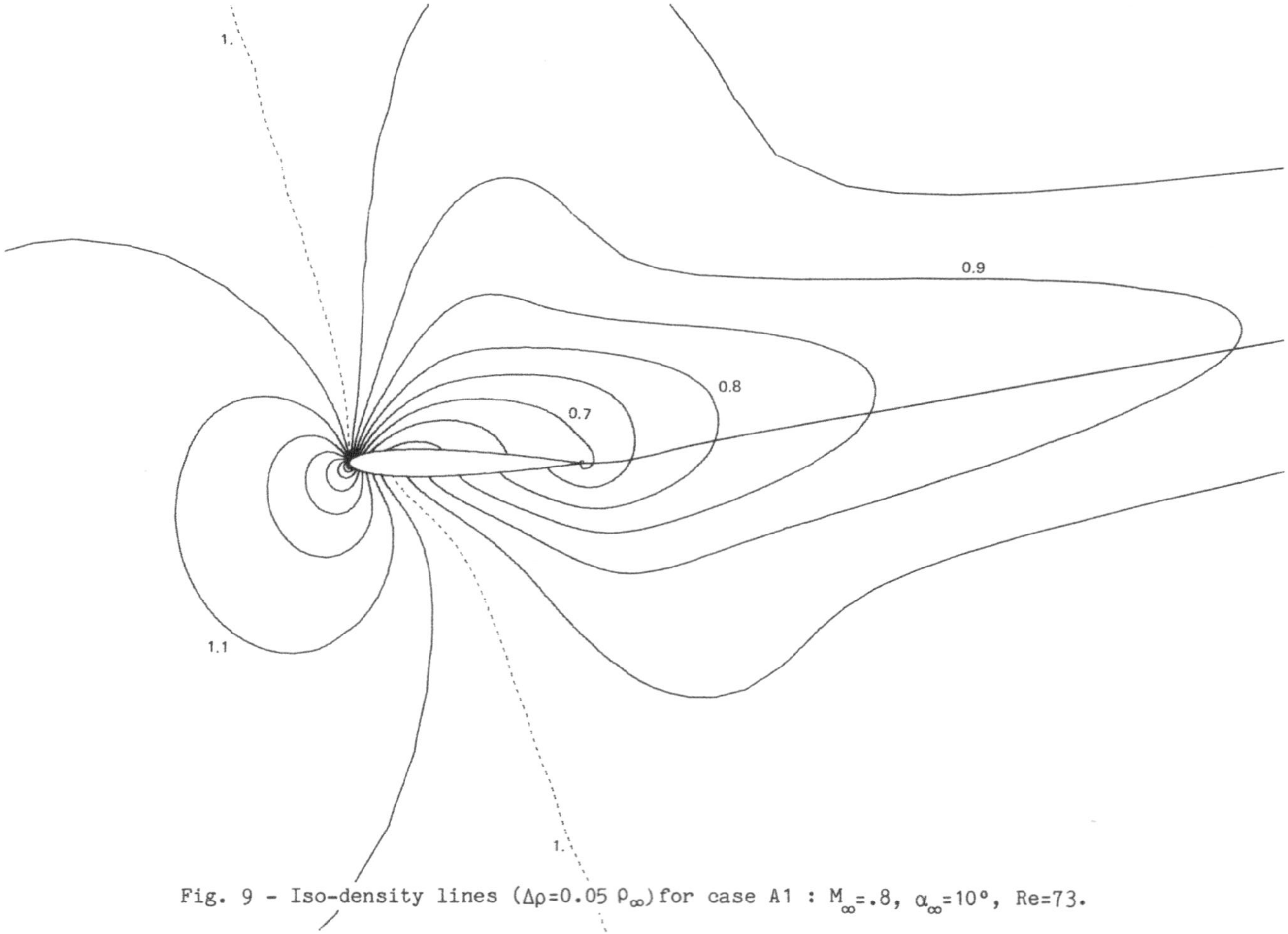

Fig. 9 - Iso-density lines ($\Delta\rho=0.05\,\rho_\infty$) for case A1 : $M_\infty=.8$, $\alpha_\infty=10°$, Re=73.

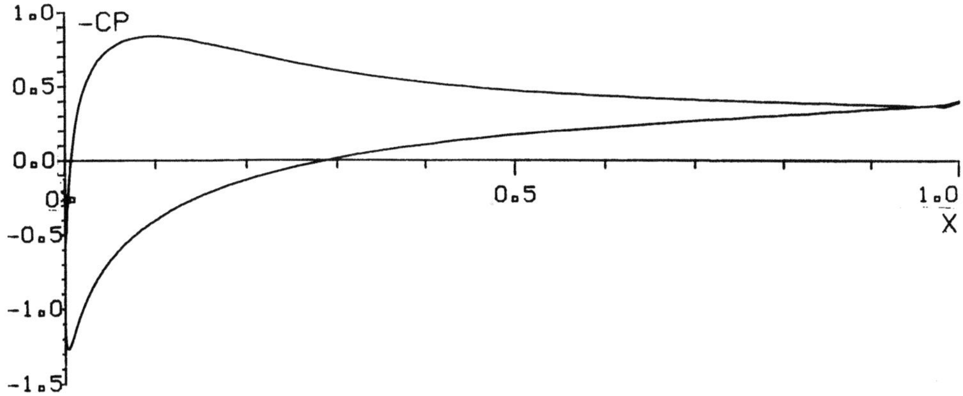

Fig. 10 - Pressure coefficient for case A2 : $M_\infty=.8$, $\alpha_\infty=10°$, Re=500.

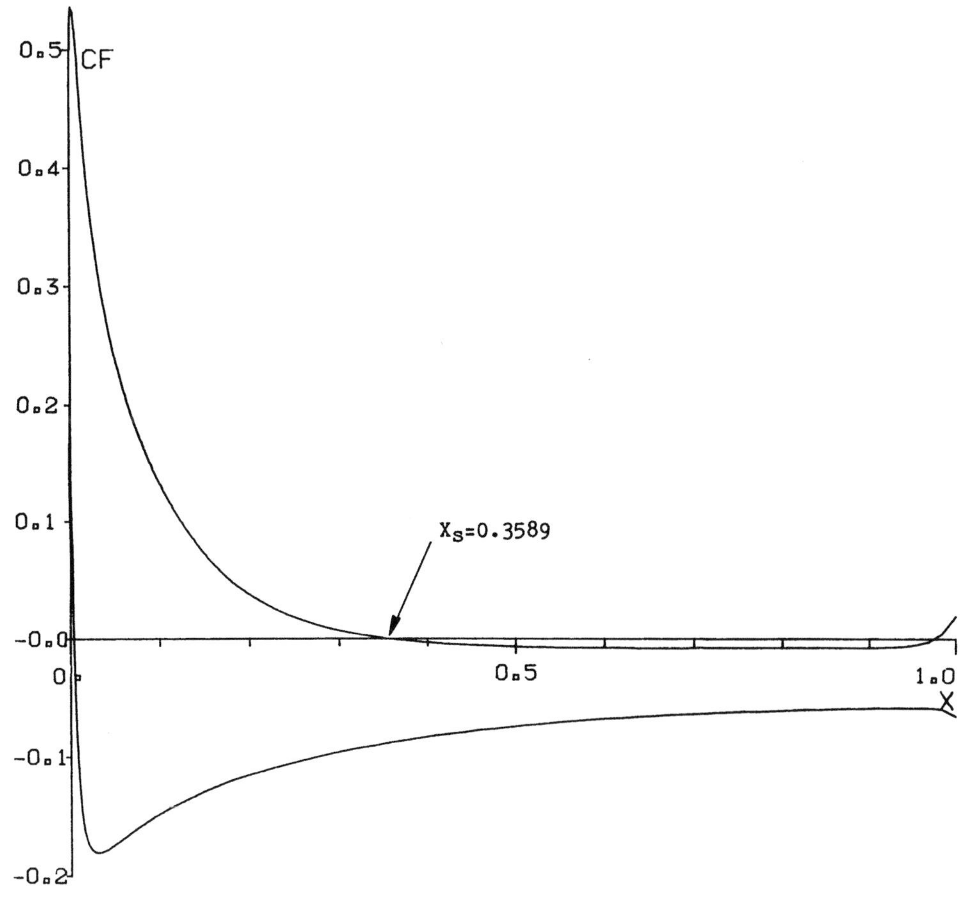

Fig. 11 - Skin friction coefficient for case A2 : $M_\infty=.8$, $\alpha_\infty=10°$, Re=500.

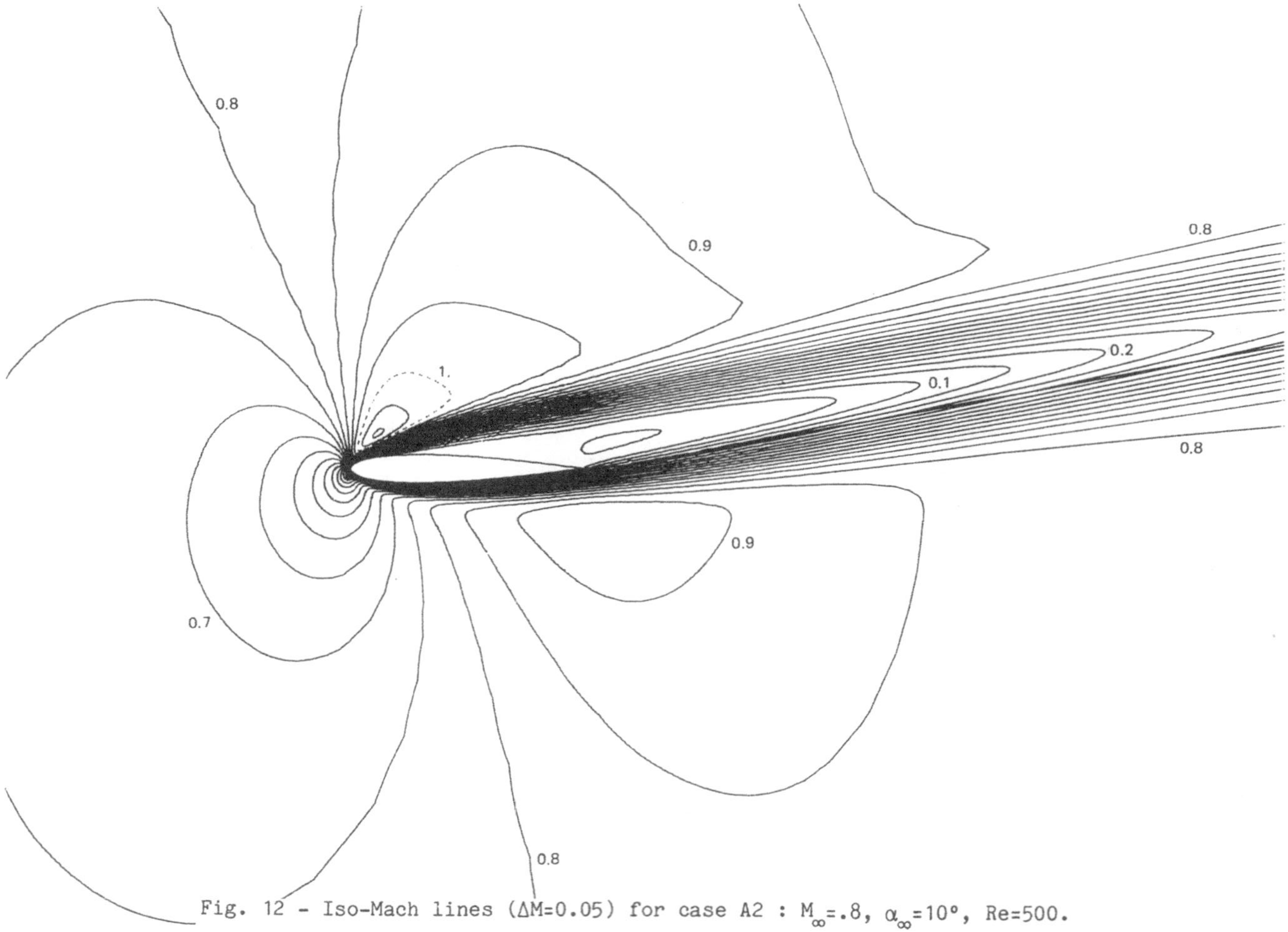

Fig. 12 - Iso-Mach lines ($\Delta M=0.05$) for case A2 : $M_\infty=.8$, $\alpha_\infty=10°$, Re=500.

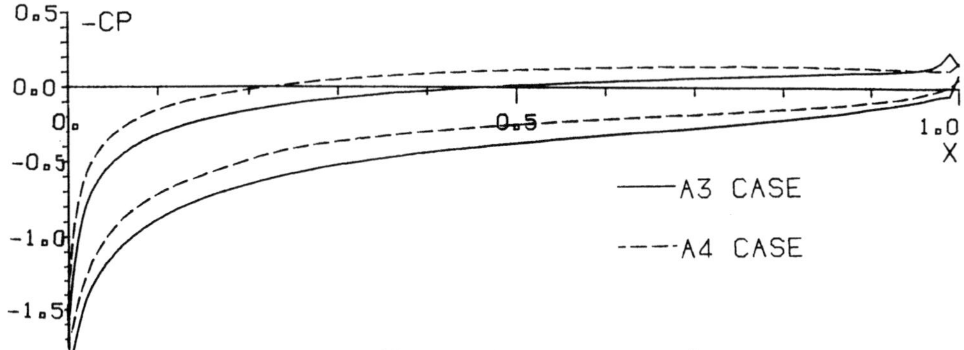

Fig. 13 - Pressure coefficient for case A3 : M_∞=2., α_∞=10°, Re= 10^6,
and case A4 : M_∞=2., α_∞=10°, Re= 1000.

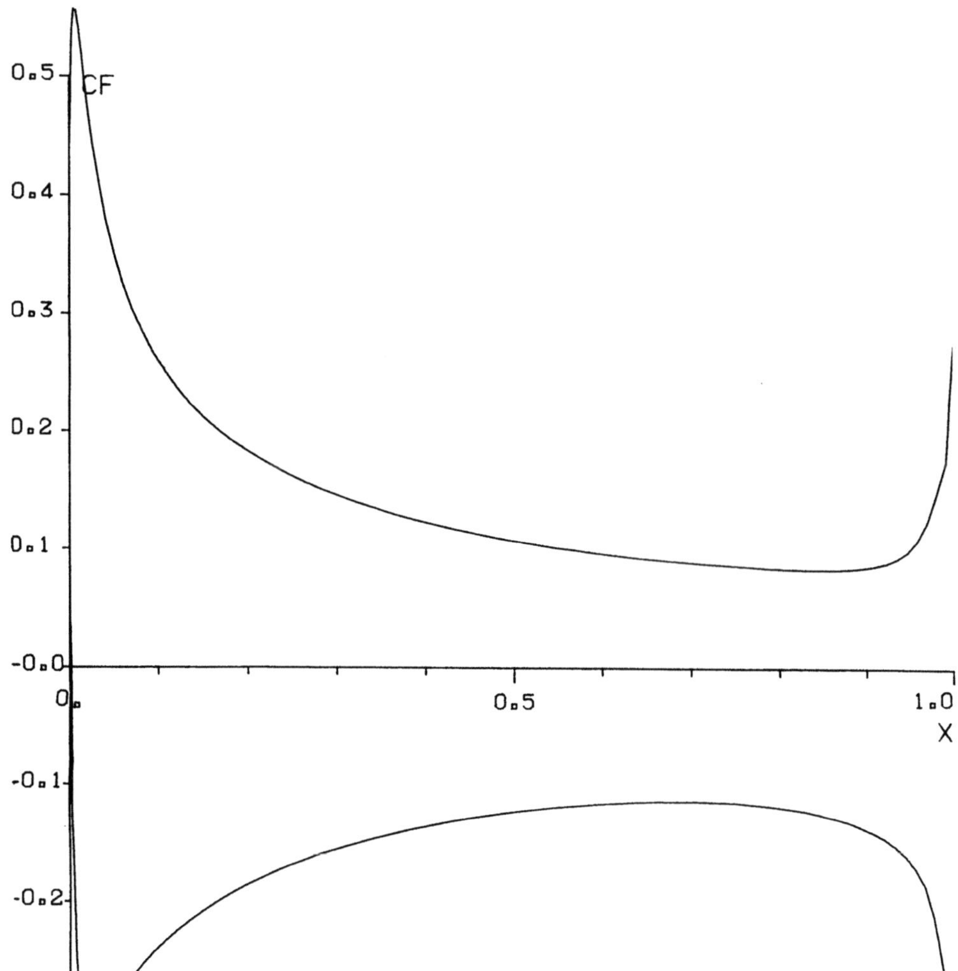

Fig. 14 - Skin friction coefficient
for case A3 : M_∞=2., α_∞=10°, Re=10^6.

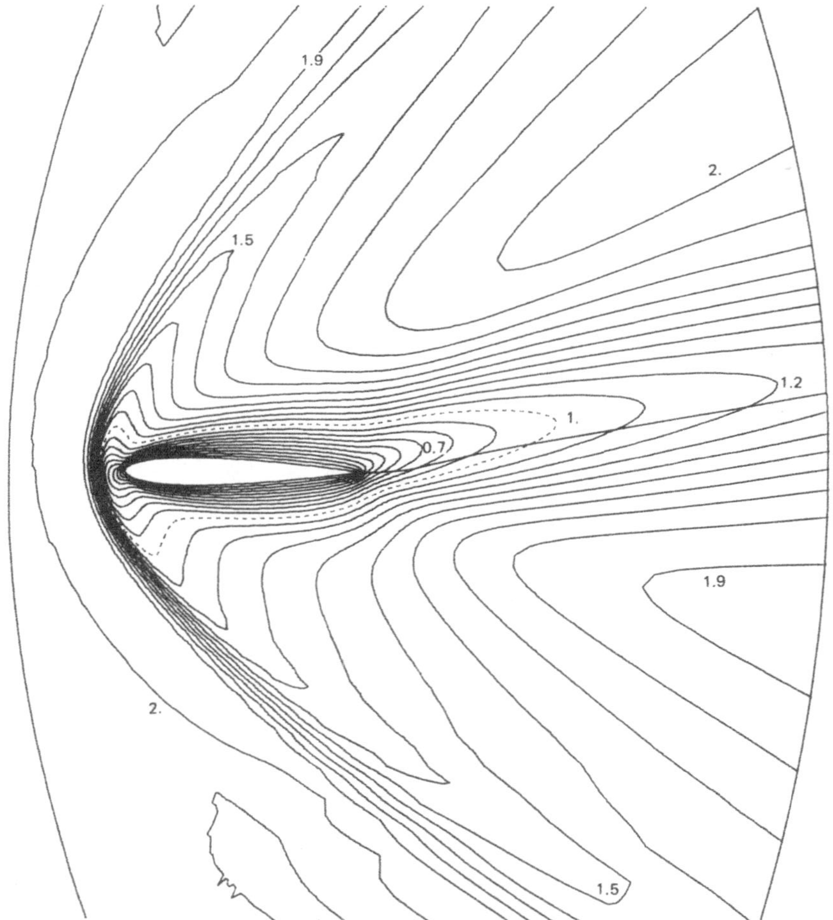

Fig. 15 - Iso-Mach lines ($\Delta M=0.1$) for case A3 : $M_\infty=2.$, $\alpha_\infty=10°$, Re=106.

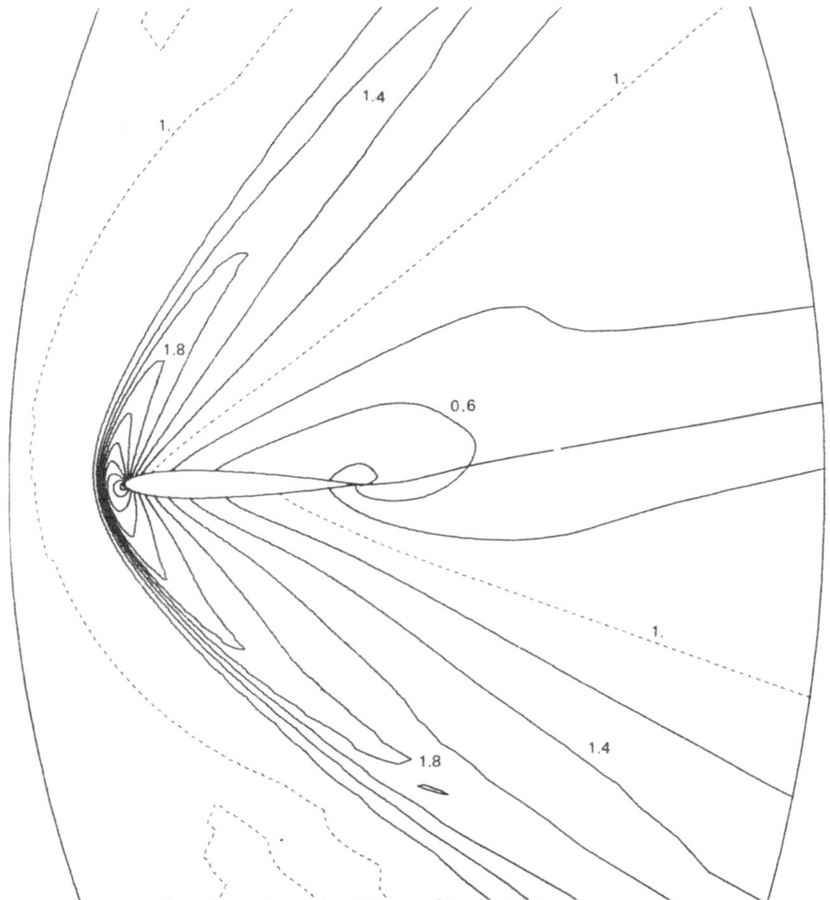

Fig. 16 - Iso-density lines ($\Delta\rho = 0.2\ \rho_\infty$) for case A3 : M_∞=2., α_∞=10°, Re=106.

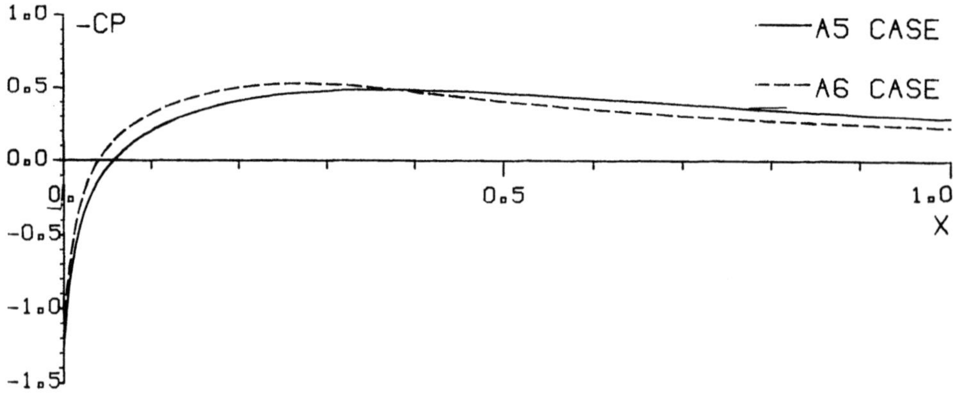

Fig 17 - Pressure coefficient for case A5 : M_∞=0.85, α_∞=0°, Re= 500,
and case A6 : M_∞=0.85, α_∞=0°, Re=2000.

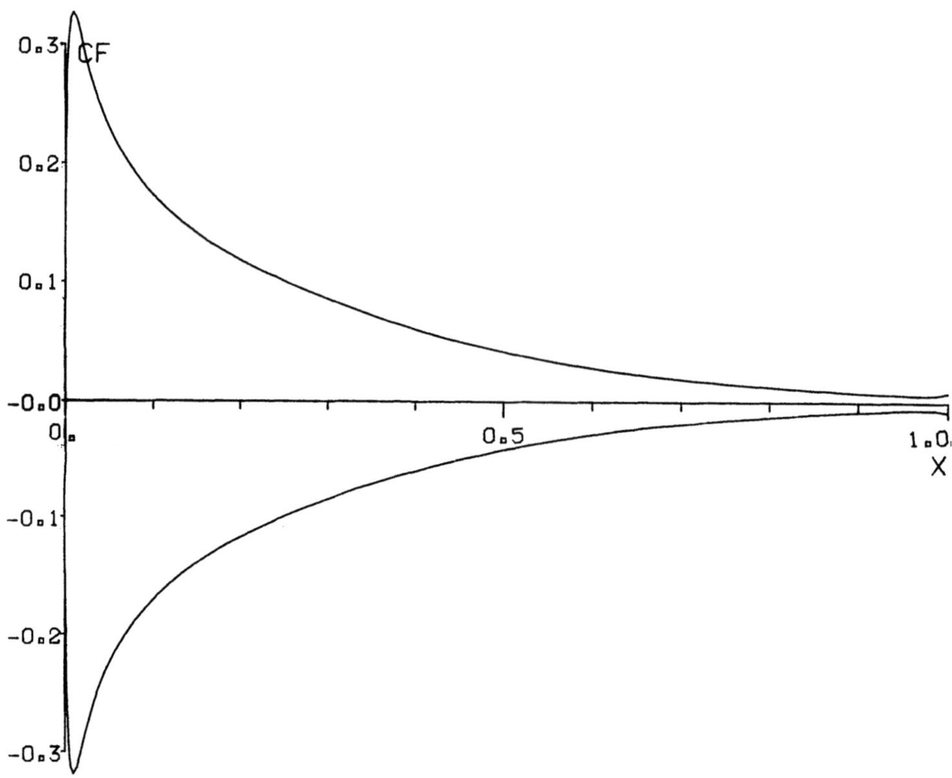

Fig 18 - Skin friction coefficient for case A5 : M_∞=0.85, α_∞=0°, Re= 500.

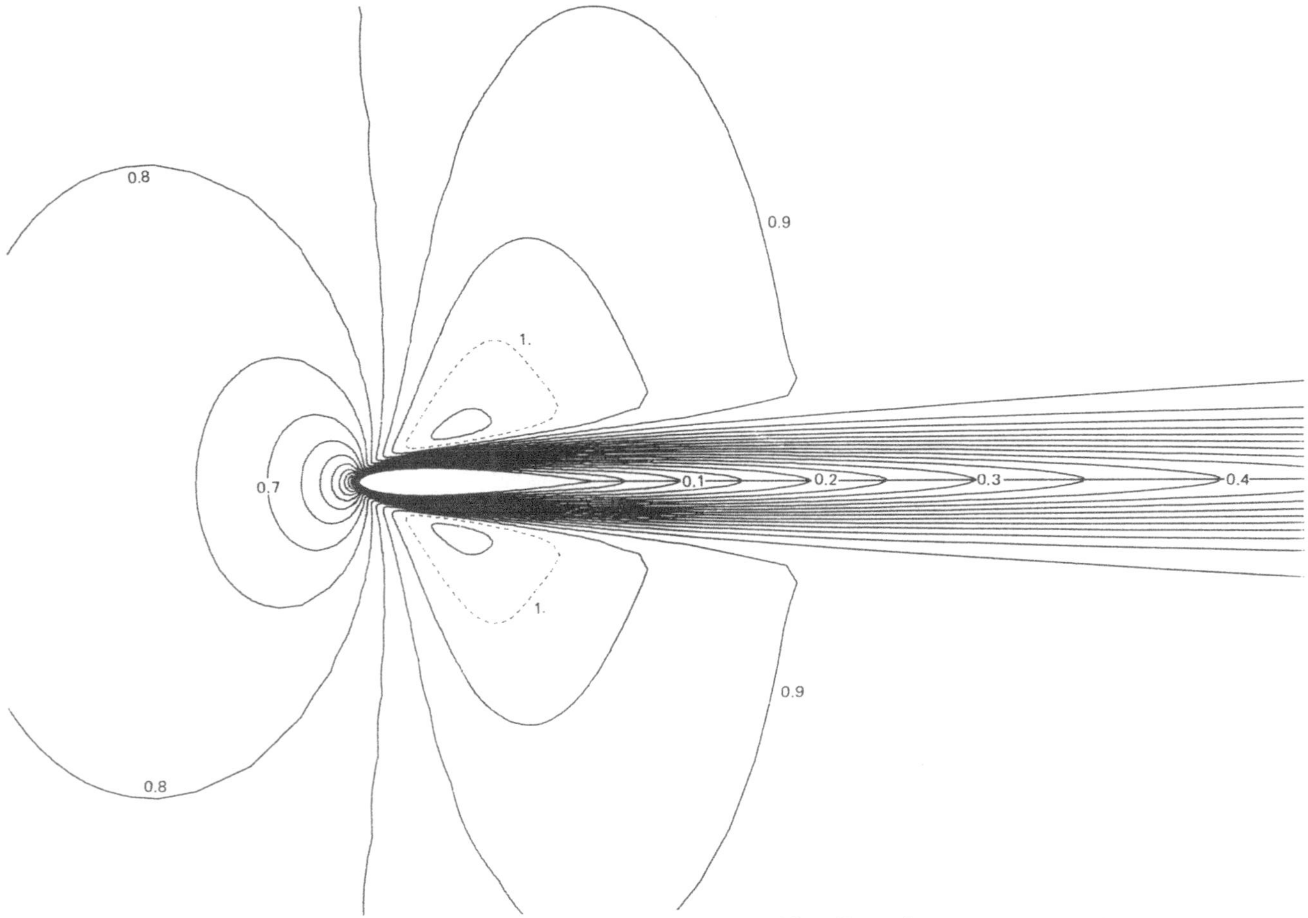

Fig. 19 - Iso-Mach lines ($\Delta M=0.05$) for case A5 : $M_\infty=0.85$, $\alpha_\infty=0°$, Re=500.

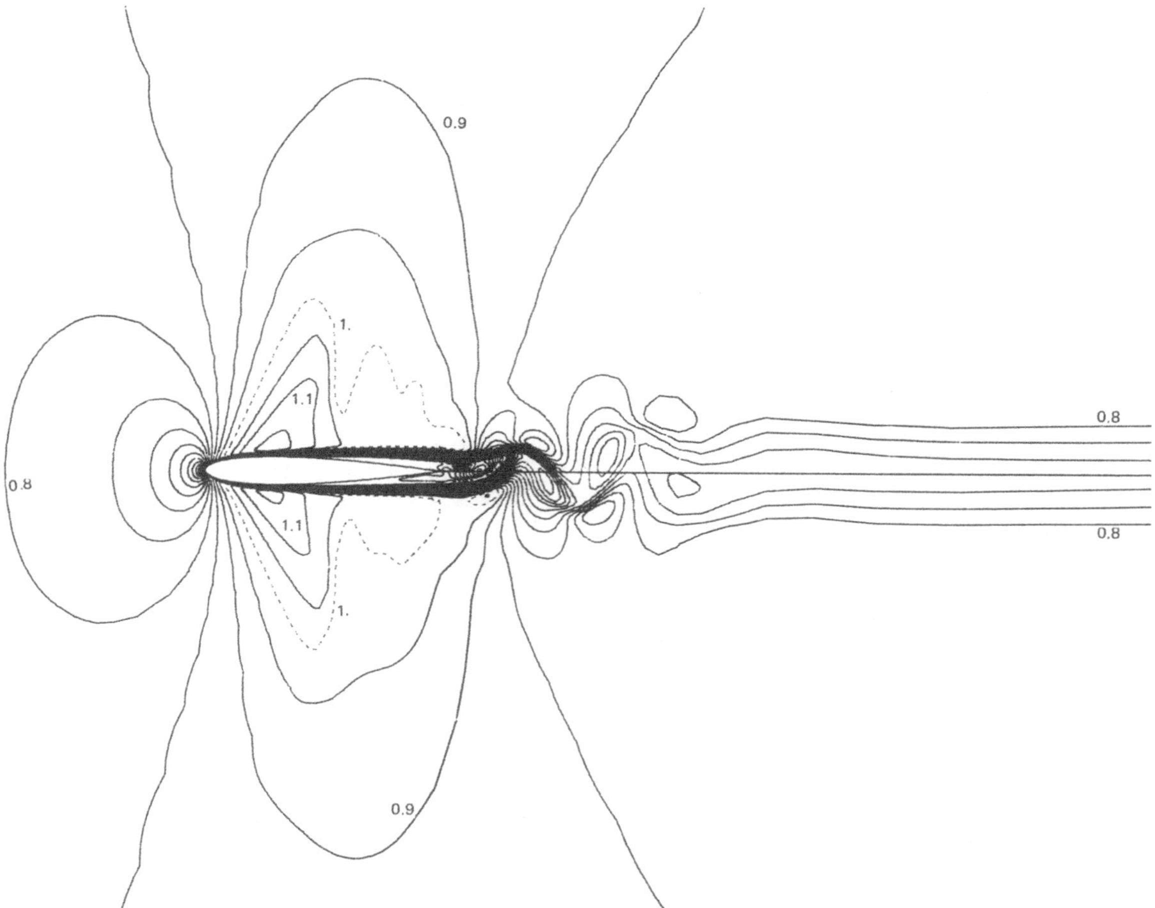

Fig. 20 - Iso-Mach lines (ΔM=0.05) for case A7 : M_∞=0.85, α_∞=0°, Re= 10000.

A MULTISTAGE MULTIGRID METHOD FOR THE COMPRESSIBLE NAVIER STOKES EQUATIONS

F. Grasso
Ist. di Gasdinamica, P.le Tecchio 80, Napoli, Italy 80125

A. Jameson, L. Martinelli
Dept. of Mech. and Aero. Eng., Princeton N.J., 08544

SUMMARY

Transonic and supersonic laminar flows around a NACA0012 airfoil are computed for different values of the Reynolds number and angle of attack by means of a three-stage explicit finite volume method. Convergence is enhanced by use of multigrid and local time stepping. However the latter is not employed when the flow presents a transient character as is the case of the transonic test case at Reynolds number equal to 10000. Computations have been performed on (256x64) grid. For some cases both (256x64) and (128x32) meshes are employd and the need of high grid resolution in the presence of disparate characteristic length scales is shown. The method is found to be very accurate and robust by comparison of the computed results with experimental data and other numerical results.

INTRODUCTION

The present paper deals with the development of an accurate and robust explicit finite volume method for the solution of viscous transonic flows around airfoils.

Several implicit or semi-implicit algorithms for the solution of the compressible Navier Stokes equations [1]-[3] have been presented within the last decade. In the recent years there has been a growing interest in explicit methods on account of the rapid progress in high speed computers, as well as in the development of efficient Euler solvers [4]-[5]. Several explicit Navier Stokes solvers have appeared [6]-[8]. The present paper is a companion paper of Ref. [8]. In the latter the method has been tested for a variety of transonic flow situations, either laminar or turbulent, and for very large Reynolds numbers. In the present work the method has been tested and validated for a variety of Reynolds numbers, ranging from 73 to 10000, for values of the Mach number in the range M =.8-2, and for different values of the angle of attack (α).

The governing equations are discretized by a finite volume technique.

Time integration is performed by using a three-stage Runge Kutta algorithm, together with multigrid to accelerate the evolution of the solution. The computational efficiency of the method is increased by an implicit smoothing of the residuals together with local time stepping. The latter is not employed in the high Reynolds number case that shows a transient character.

In Ref. [8] it is shown that a cell centered formulation (i.e. variables defined at cell centers) is best suited for Neumann type boundary conditions, while a corner point formulation (i.e. variables defined at grid nodes) is more adequate with Dirichlet-type b.c.. The test cases here computed require that the temperature be specified at the wall, and consequently a corner point formulation is used.

The computations have been performed on a scalar computer. However the algorithm is fully vectorizable and should be able to exploit the characteristics of parallel machines.

In the next sections the governing equations are presented and the numerical algorithm is discussed. Finally the results of laminar flows over a NACA0012 airfoil are presented and some concluding remarks given.

GOVERNING EQUATIONS

A dimensionless conservation form of the Navier Stokes equations is used with the following definition of the non-dimensional variables

$(x^*,y^*) = (x/c;\ y/c)$, where c is the chord

$$t^* = t(p_\infty/\rho_\infty)^{1/2}/c$$

$$\rho^* = \rho/\rho_\infty$$

$$u^* = u/\,(p_\infty/\rho_\infty)^{1/2}$$

$$v^* = v/\,(p_\infty/\rho_\infty)^{1/2}$$

$$E^* = E/(p_\infty/\rho_\infty)$$

$$p^* = p/p_\infty$$

$$\mu^* = \gamma^{1/2} M_\infty/Re\ \mu/\mu_\infty$$

$$\lambda^* = c_p \mu^*/Pr\ .$$

Dropping the * for the sake of brevity, the integral form of the equations is

$$\frac{\partial}{\partial t}\int_V W dV = -\oint_{\partial V}(f n_x + g n_y)dS + \int_V\left(\frac{\partial P}{\partial x} + \frac{\partial Q}{\partial y}\right)dV \qquad (1)$$

where W, f(g) and P(Q) are respectively the vector unknown, the non dissipative and the viscous flux components in x(y), and they are given by:

$$W = [\rho, \rho u, \rho v, \rho E]^T$$

$$f = [\rho u, \rho u^2+p, \rho uv, \rho u(E+p/\rho)]^T$$

$$g = [\rho v, \rho vu, \rho v^2+p, \rho v(E+p/\rho)]^T$$

$$P = [0, \sigma_{xx}, \sigma_{xy}, u\sigma_{xx}+v\sigma_{xy}-q_x]^T$$

$$Q = [0, \sigma_{xy}, \sigma_{yy}, u\sigma_{xy}+v\sigma_{yy}-q_y]^T$$

where

$$\underline{\underline{\sigma}} = \mu[(\nabla\underline{u}+\nabla\underline{u}^T) - \frac{2}{3}\nabla\cdot\underline{u}\underline{\underline{U}}]$$

$$\underline{q} = -\frac{\gamma}{Pr}\mu\nabla e .$$

The pressure p is obtained from the equation of state:

$$p = \rho e = (\gamma-1)[E-(u^2+v^2)/2] .$$

The Chapman-Rubesin approximation is employed to evaluate the viscosity coefficient.

NUMERICAL SOLUTION

The computational domain is divided into arbitrary quadrilateral cells. The system of governing equations is reduced to a system of ordinary differential equations (ODE) by using the method of lines to decouple the spatial and temporal terms. Use of mean value theorem and mid-point rule (to evaluate the volume and surface integrals) yield the following ODE:

$$\frac{d}{dt}(WV) = -C_w+D_wV+AD_w \tag{2}$$

where V, C_w, D_w and AD_w are respectively the sum of the volumes of the four cells surrounding node ij, the net inviscid, dissipative and adaptive dissipation contributions.

A corner point formulation is employed to accurately satisfy the boundary conditions of Dirichlet-type required by the test cases. The net inviscid contribution at each node ij is obtained as the sum of the net contributions over the four surrounding cells so as to increase the computational efficiency and to guarantee conservation. The following expression is obtained:

$$C_w = \sum_{c=1,4} \left(\sum_{k_c=1,4} [(\hat{f}n_x+\hat{g}n_y)\Delta S]_{k_c} \right) \tag{3}$$

where c is the index of the (four) cells surrounding node ij, and k_c is the face index of cell c.

The numerical flux $(\hat{f},\hat{g})$ at cell face k_c is computed as the average of the corresponding values at the two adjacent cells. Such a procedure is computationally efficient and it yields second order accuracy on a smooth grid.

Gauss theorem is used first to evaluate the numerical diffusion flux components $(\hat{P},\hat{Q})$ at cell centers and then to compute their divergence at node ij. For example, the net contribution for the x-momentum component is [8]:

$$D_{\rho u} = \Big([(\hat{\sigma}_{xx_1}-\hat{\sigma}_{xx_3})(yc_2-yc_4)-(\hat{\sigma}_{xx_2}-\hat{\sigma}_{xx_4})(yc_1-yc_3)] + \\ -[(\hat{\sigma}_{xy_1}-\hat{\sigma}_{xy_3})(xc_2-xc_4)-(\hat{\sigma}_{xy_2}-\hat{\sigma}_{xy_4})(xc_1-xc_3)]\Big)/ \\ [(xc_1-xc_3)(yc_2-yc_4)-(xc_2-xc_4)(yc_1-yc_3)] \tag{4}$$

where $\hat{}$ and xc (yc) are respectively the discretized stress tensor, and x(y) coordinate of the center of cell c.

The adaptive dissipation (AD_w), as designed in Ref. [9], is still added to the scheme to inhibit even/odd mesh points decoupling, and to prevent non linear instabilities. However the amount of artificial dissipation added is 10^{-3} times smaller than the physical one.

ODE is integrated in time by using a three-stage Runge Kutta scheme that guarantees damping of high frequency modes. The physical and adaptive dissipative contributions are evaluated at the first stage in terms of the solution at time level n, and are not reevaluated at each stage.

The solution is advanced in time as follows

$$w^{(0)} = w^n$$

$$w^{(k)} = w^{(0)} - \alpha_k \frac{\Delta t}{V} [C_w^{(k-1)} - D_w^{(0)} V - AD_w^{(0)}]$$

$$w^{n+1} = w^{(3)}$$

where $\alpha_1 = \alpha_2 = .6$; $\alpha_3 = 1$.

The efficiency of the method is increased by advancing the solution at each computational cell at its own Δt, so that any disturbance propagates out of the field at a rate proportional to the number of points in the outward direction. Use of such local time stepping fails when the solution presents a transient character, as in case 6 of Table 1 corresponding to $M_\infty = .85$, Re = 10000, $\alpha = 0°$. The vortices that form at the trailing edge are responsible for the unsteady behaviour, and the solution is advanced in time with a time step corresponding to the most critical stability restriction. Implicit smoothing of the residuals is also used to increase the region of stability, yielding a higher CFL limit [9].

The multigrid strategy devised by Jameson for the solution of the Euler equations, and extended to the Navier Stokes equations by these authors [8], has also been used in the present work. Auxiliary meshes are introduced by doubling the mesh spacing. The coarser grid variables are initialized so as to conserve mass, momentum and energy. The multistage scheme is reformulated with the results that the solution on a coarse grid is driven by the residuals collected on the next finer grid [8], [9]. The process is repeated on successively coarser grids. After reaching the coarsest grid, the corrections are passed back to the next finer grid by bilinear interpolation.

Boundary conditions

At the wall the following boundary conditions are imposed:

$$u = v = 0$$

$$T_w = T_\infty \left(1 + \frac{\gamma - 1}{2} M_\infty^2\right) .$$

The pressure is obtained from the equation of state:

$$p_w = \rho_w T_w$$

where ρ_w is computed from mass conservation.

The far field boundary is located 18 chords away from the airfoil, and conditions, based on the introduction of Riemann invariants for a one dimensional flow normal to the boundary, are employed.

Mesh description

The mesh used in the computations is a body fitted C-mesh obtained by using a hyperbolic grid generator, kindly provided by L. Wigton of the Boeing Co. The computational domain is discretized in 256x64 cells on the finer mesh, however for some test cases computations have also been performed on a coarser grid (128x32). The first point in the boundary layer is located at .002 chords away from the wall, and 5/8 of the total number of points in the streamwise direction are fitted on the airfoil.

RESULTS AND DISCUSSION

Transonic and supersonic laminar flows around NACA0012 airfoils have been computed for different values of the Reynolds number and angle of attack. All the calculations have been carried out on a IBM 3081 scalar machine; the CPU time for a typical run on a 256x64 grid (see Fig. 1) with three levels of multigrid (and 500 multigrid cycles) is 149 minutes.

For most of the cases convergence is quite satisfactory. The steady state is reached within 300-500 (multigrid) cycles and the RMS of density residual reduces to $0(10^{-3}-10^{-4})$.

When assessing the validity of a computational tool, in the absence of analytical results, it is extremely important to compare the computed results with available experimental data and/or computed results. Figs. 2-3 show a comparison of the computed density with the experimental one of Ref. [10], corresponding to cases 1 and 3 of Table 1. One observes that overall the agreement is rather satisfactory. In both cases there are some disagreements at the trailing edge. However the results shown correspond to rather large values of the RMS of density residuals. Note that in the supersonic case the location and thickness of the detached bow shock is very well predicted.

The robustness of the method and its accuracy is also shown by the comparison of the results for the other cases with those obtained by the others. The computed results show the capability of the method to accurately predict transonic and supersonic flows at different Reynolds numbers ranging from very low to high values (Figs. 4-23). Some discrepancies in the level of the predicted skin friction coefficient are observed, most likely on account of the Chapman-Rubesin approximation employed to evaluate the viscosity coefficient.

Case 6 of Table 1 is the most interesting and critical test case. At $M = .85$, $Re=10000$ and $\alpha = 0°$ an unsteady behaviour of the flow is observed and vortices are shed from the trailing edge as shown in Figs. 26-27. On account of the transient character, local time stepping is not employed and a time step corresponding to the most critical stability restriction is used. A non symmetric field is obtained ($c_L=.1386$, $c_D=.0723$) and no convergence is achieved. Figs. 28-31 show the results for the same case obtained on an intermediate grid (128x32). When computing on such a grid no vortices form. Moreover a pseudo-convergence is reached, and a symmetric solution is obtained ($c_L=.0$, $c_D=.046$) with a value of the RMS of density residual equal to -2.59.

The two completely different results, obtained when using two different mesh refinements, indicate that higher temporal and spatial accuracies are required in the presence of disparate length and time scales. Under these circumstancies the local time stepping technique cannot be used and highly refined grids must be employed or else adaptive mesh enrichment should be used [11]-[13] to accurately detect small scale structures.

REFERENCES

[1] Steger J.L., "Implicit Finite Difference Simulation of Flow About Arbitrary Geometries with Applications to Airfoils," AIAA paper no. 77-665, 1977.

[2] Beam, R. and Warming, R., "An implicit Factored Scheme for the Compressible Navier-Stokes Equations," AIAA J. Vol. 16, 1978.

[3] Barth, J., Pulliam, T. and Bunning, P., "Navier Stokes Computations for Exotic Airfoils," AIAA paper no. 85-0109, 1985.

[4] Jameson, A., Schmidt, W. and Turkel, E., "Numerical Solutions of the Euler Equations by Finite Volume Methods Using Runge-Kutta Time-Stepping Schemes," AIAA paper no. 81-1259, 1981.

[5] Jameson, A., Baker, J. and Weatherhill, N., "Calculation of the Inviscid Transonic Flow over a Complete Aircraft," AIAA paper no. 86-0103, 1986.

[6] Agarwaal, R. and Deese, L., "Computations of Transonic Viscous Airfoil, Inlet and Wing Flowfields," AIAA paper no. 84-1551, 1984.

[7] Swanson, R.C. and Turkel, E., "A Multistage Time-Stepping Scheme for the Navier Stokes Equations," AIAA paper no. 85-0035, 1985.

[8] Martinelli, L., Jameson, A. and Grasso, F., "A Multigrid Method for the Navier Stokes-Equations," AIAA paper no. 86-0208, 1986.

[9] Jameson, A., "Transonic Flow Calculations," MAE Report no. 1651, Princeton University, 1983.

[10] Allegre, J., Lengrand, J.C. and Raffin, M., "Champs d'Ecoulements Rarefiée Compressibles Autour d'un Profil," SESSIA Rep. 894/82.380, 1982.

[11] Berger, M. and Jameson, A., 9th ICNMFD, Saclay, France, 1985.

[12] Dannenhoffer, J.F. and Baron, J.R., "Grid Adaptation for the 2-D Euler Equations," AIAA paper no. 85-0484, 1985.

[13] Bassi, F., Grasso, F. and Savini, M., 10th ICNMFD, Beijing, China, 1986.

ACKNOWLEDGEMENT

The authors wish to thank Drs. B. Mantel and J. Periaux of AMD-BA for their help in plotting the results.

Table 1 - Integral Values of Computed Test Cases

Case	Grid Points	M_∞	Re	α	c_L	c_D	log(dρ/dt)
1	256x64	.8	73	10°	.6657	.2522	-2.34
2	256x64	.8	500	10°	.4797	.1597	-3.50
3	256x64	2.0	106	10°	.3716	.1851	-1.74
4	256x64	.85	500	0°	.0	.0964	-3.75
5	256x64	.85	2000	0°	.0	.068	-3.62
6.1	128x32	.85	10000	0°	.0	.046	-2.59
6.2	256x64				.1386	.0723	-0.10

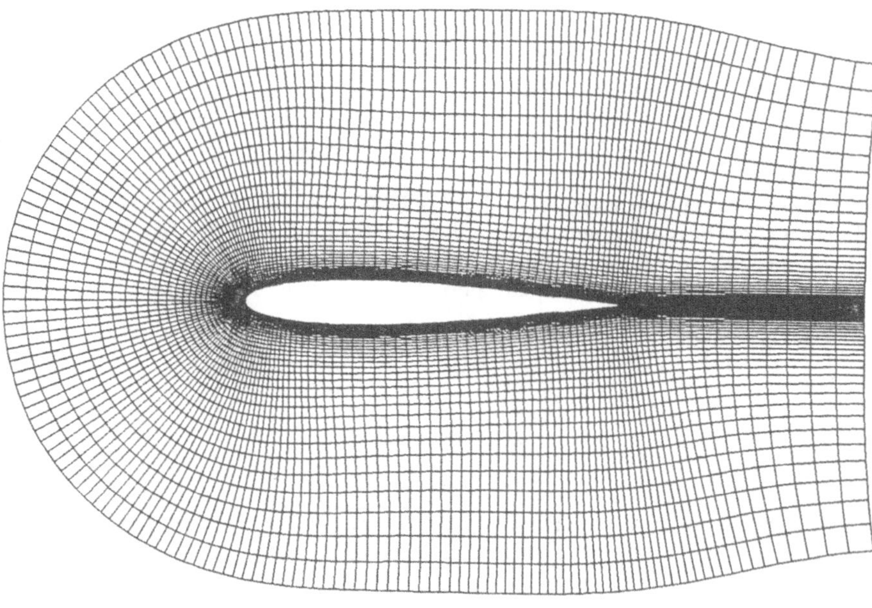

Fig. 1 Computational Grid

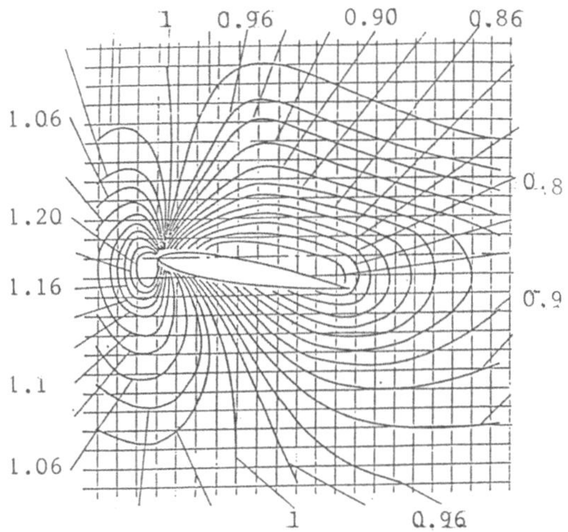

Fig. 2.a - Experimental Density Contours

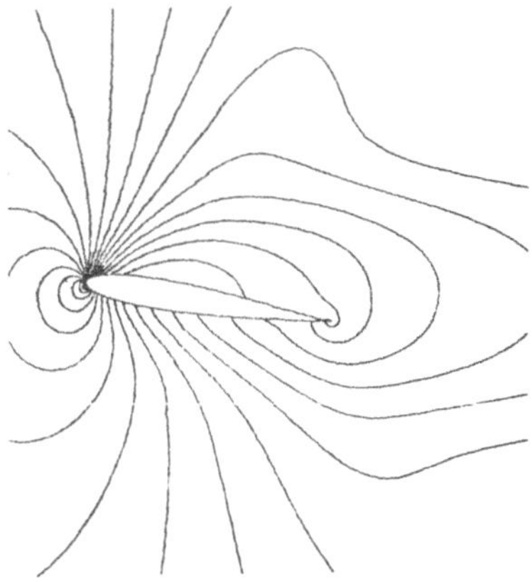

Fig. 2.b - Computed Density Contours (DR = .05)

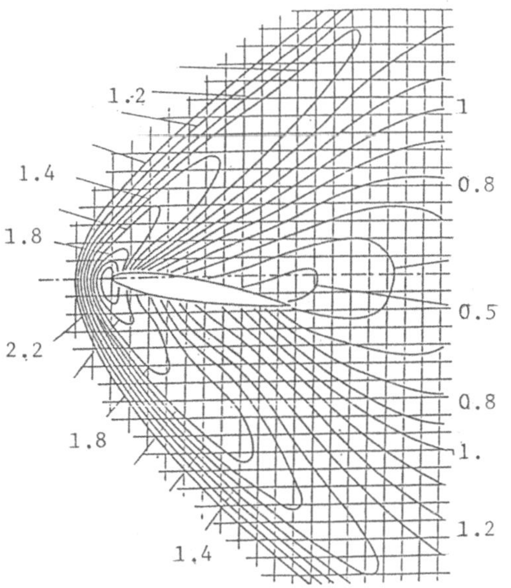

Fig. 3.a - Experimental Density Contours

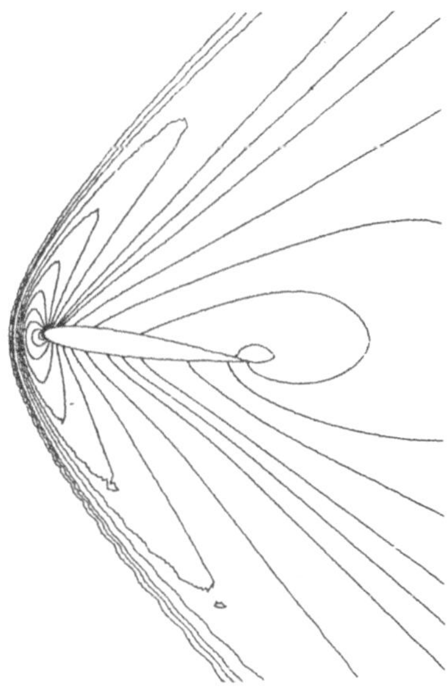

Fig. 3.b - Computed Density Contours (DR = .2)

CASE 1 (256x64)

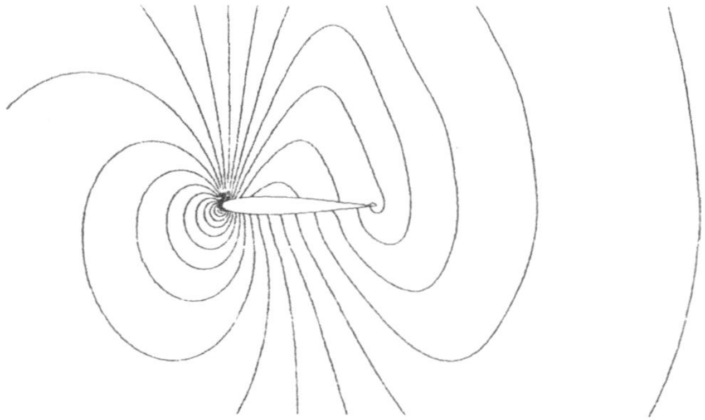

Fig. 4 - Pressure Contours (DP = .05)

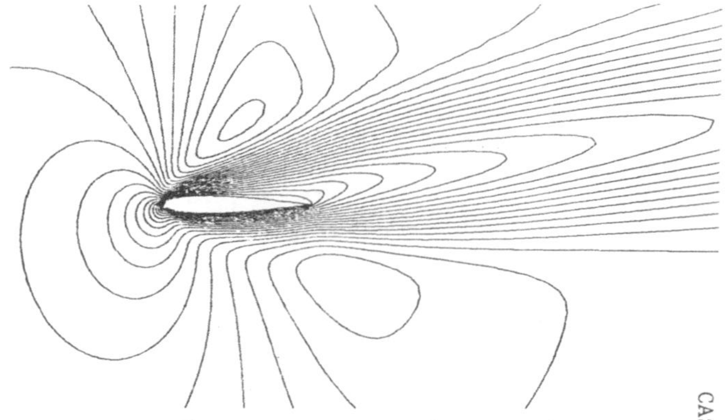

Fig. 5 - Mach Contours (DM = .05)

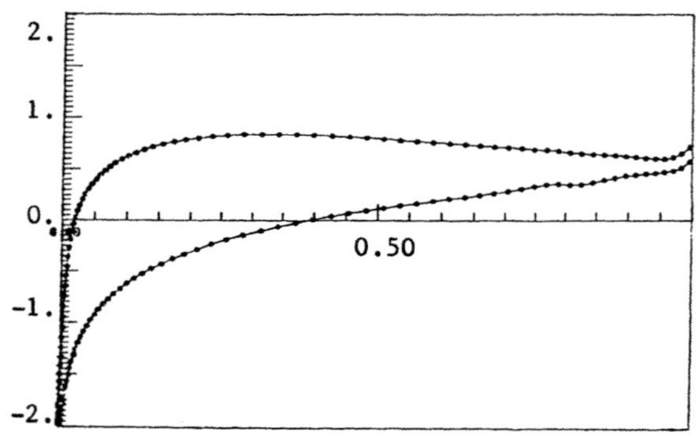

Fig. 6 - Pressure Coefficient vs x

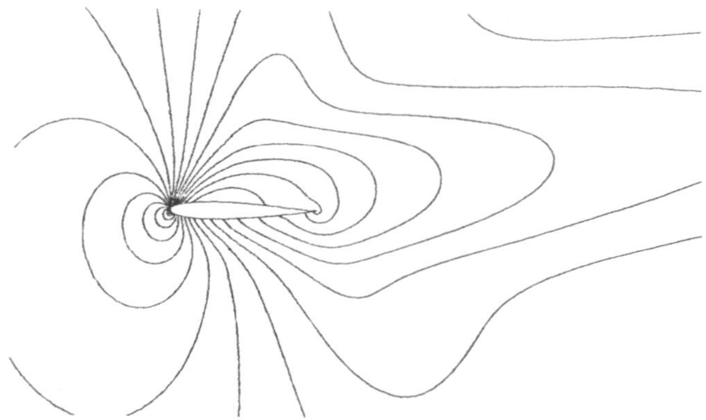

Fig. 7 - Density Contours (DR=.05)

CASE 2 (256x64)

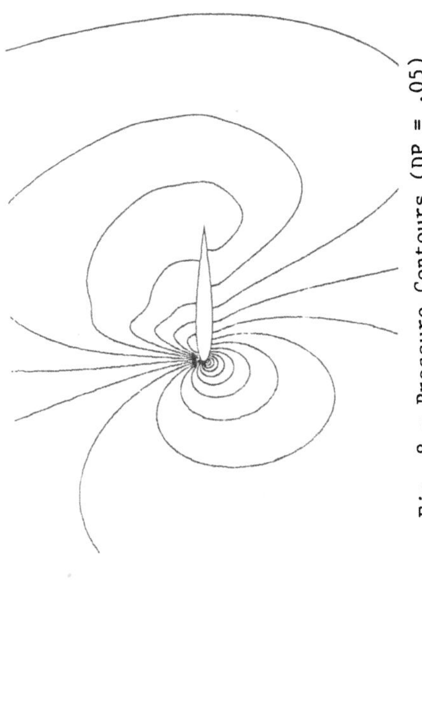

Fig. 8 - Pressure Contours (DP = .05)

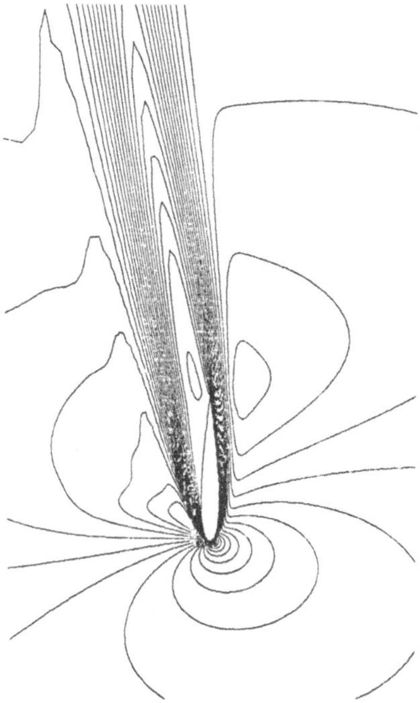

Fig. 9 - Mach Contours (DM = .05)

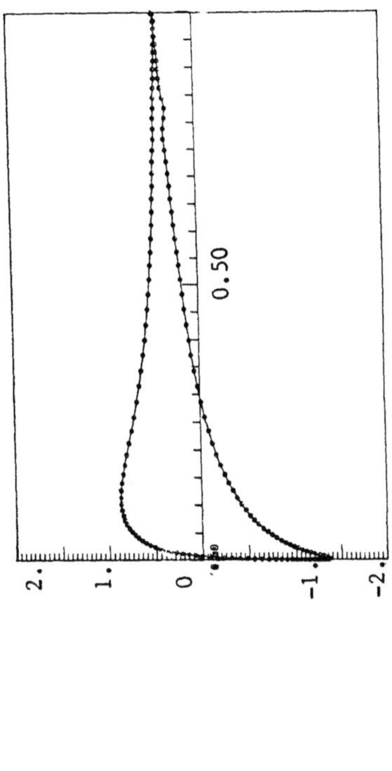

Fig. 10 - Pressure Coefficient vs x

Fig. 11 - Density Contours (DR=.05)

CASE 3 (256x64)

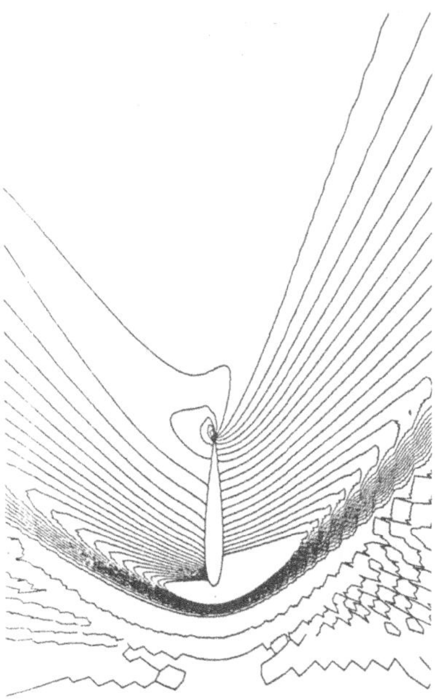

Fig. 12 - Pressure Contours (DP = .10)

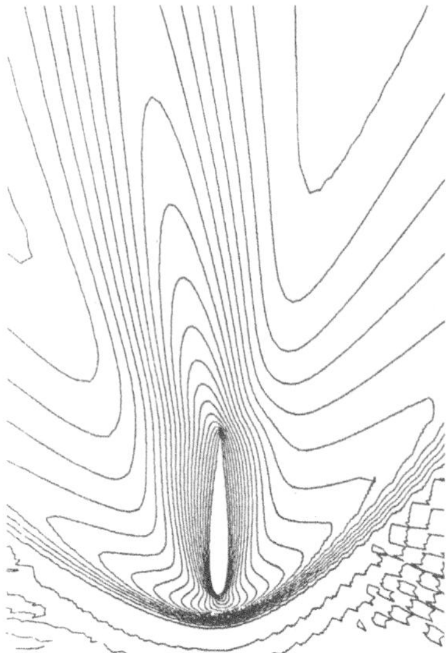

Fig. 13 - Mach Contours (DM = .10)

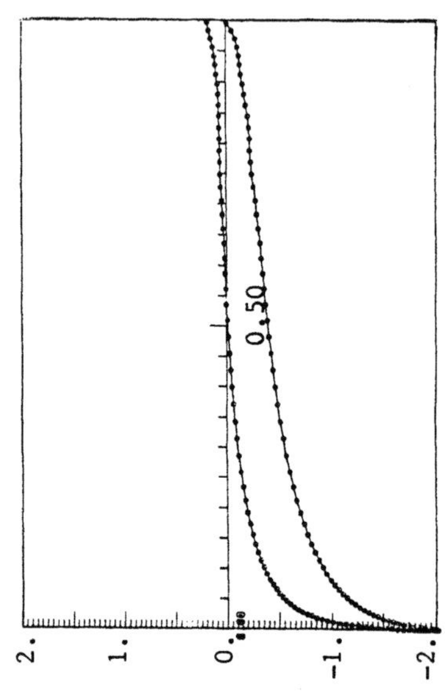

Fig. 14 - Pressure Coefficient vs x

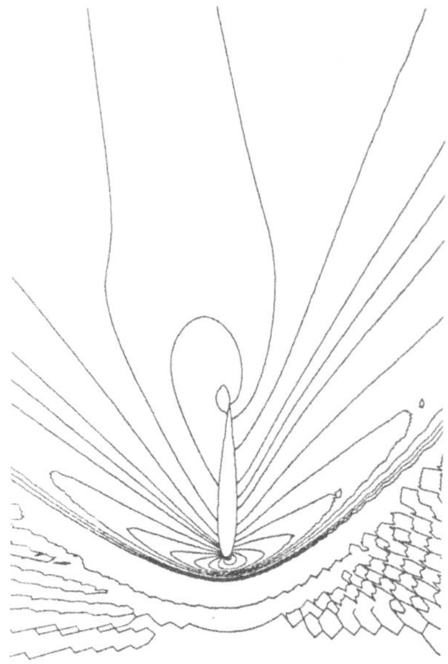

Fig. 15 - Density Contours (DR=.10)

CASE 4 (256x64)

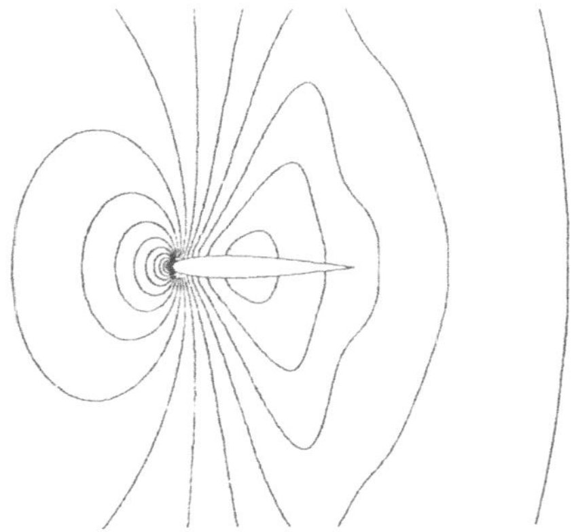

Fig. 16 - Pressure Contours (DP = .05)

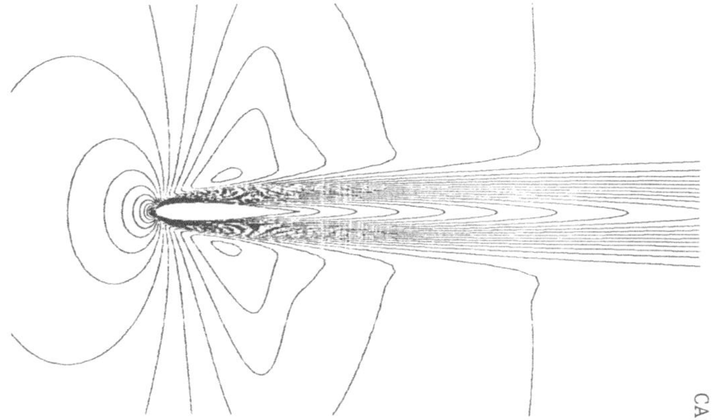

Fig. 17 - Mach Contours (DM = .05)

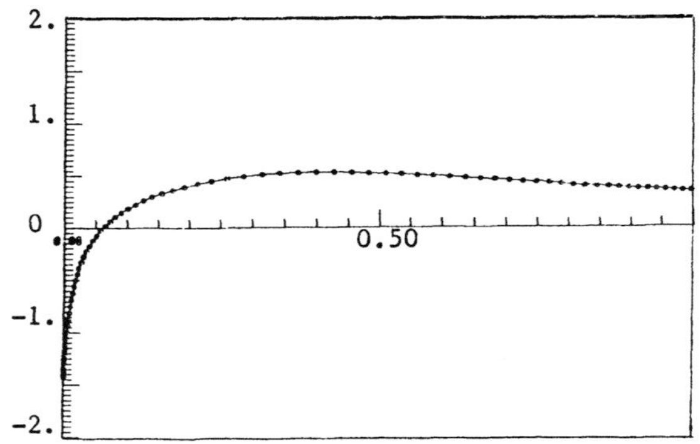

Fig. 18 - Pressure Coefficient vs x

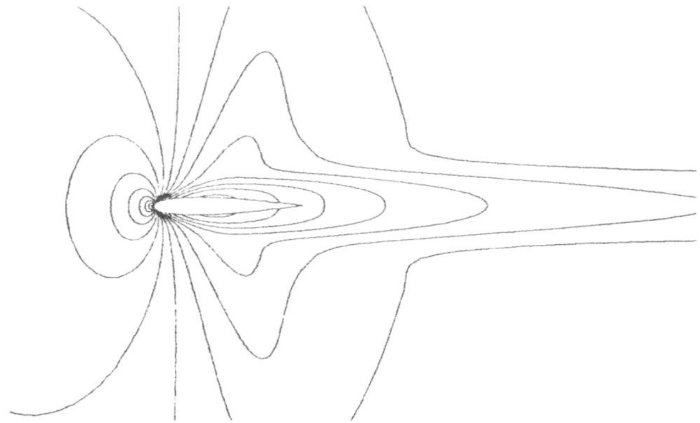

Fig. 19 - Density Contours (DR=.05)

CASE 5 (256x64)

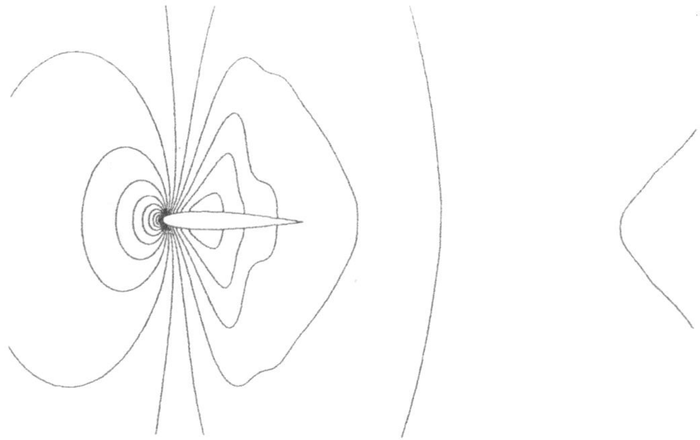

Fig. 20 - Pressure Contours (DP = .05)

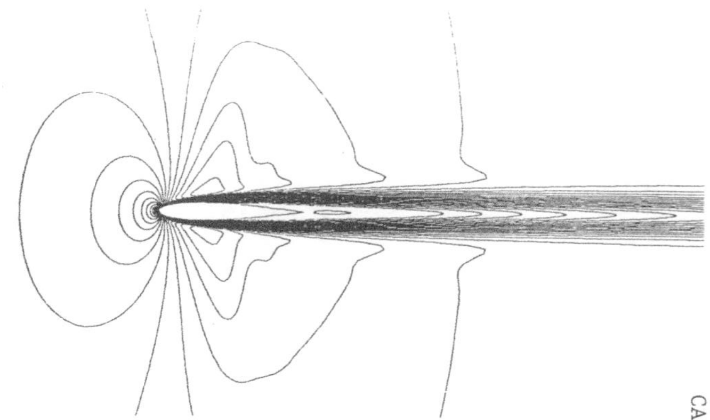

Fig. 21 - Mach Contours (DM = .05)

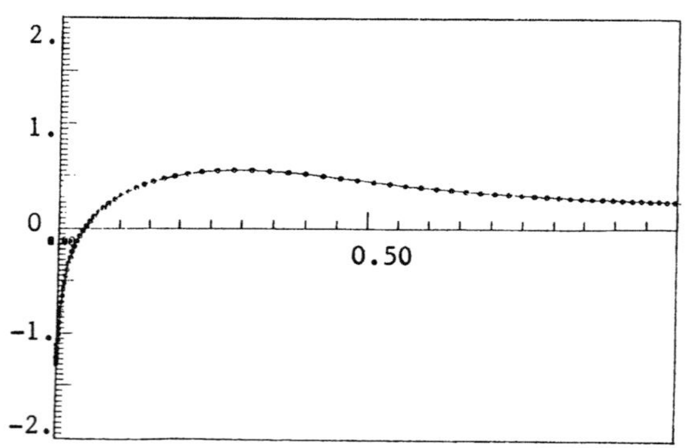

Fig. 22 - Pressure Coefficient vs x

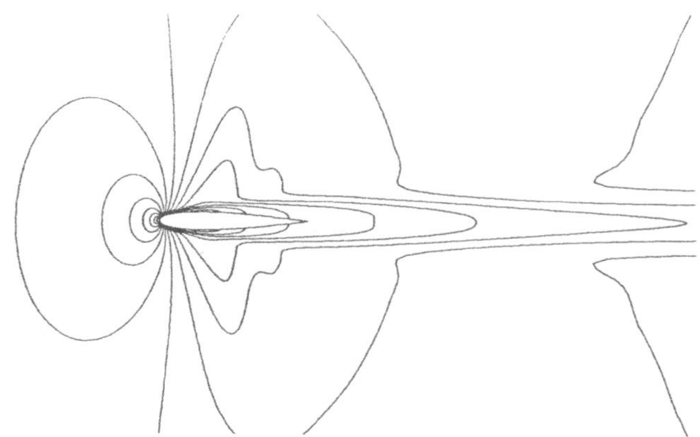

Fig. 23 - Density Contours (DR=.05)

CASE 6 (256x64)

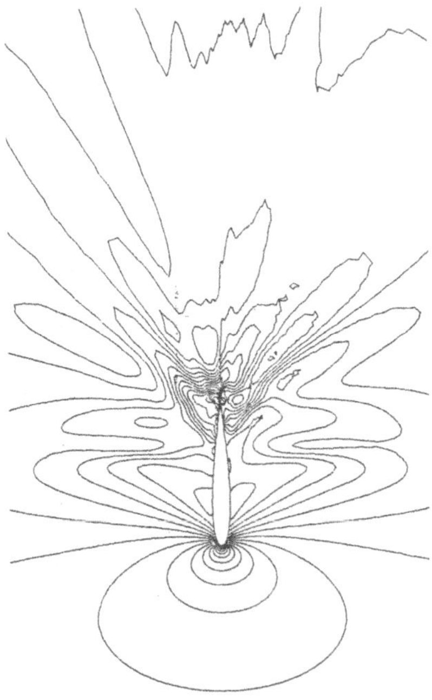

Fig. 24 - Pressure Contours (DP = .05)

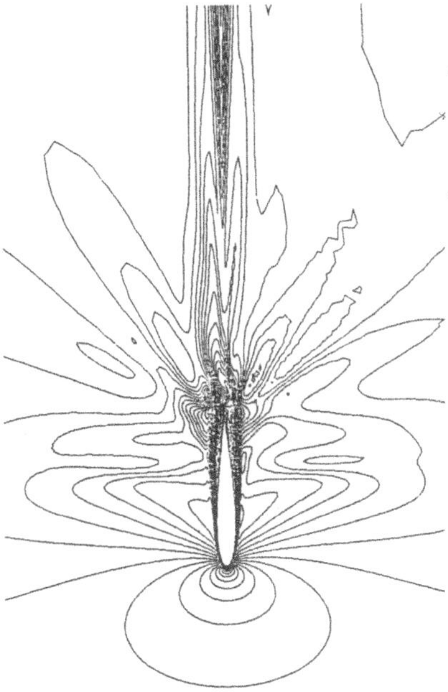

Fig. 25 - Mach Contours (DM = .05)

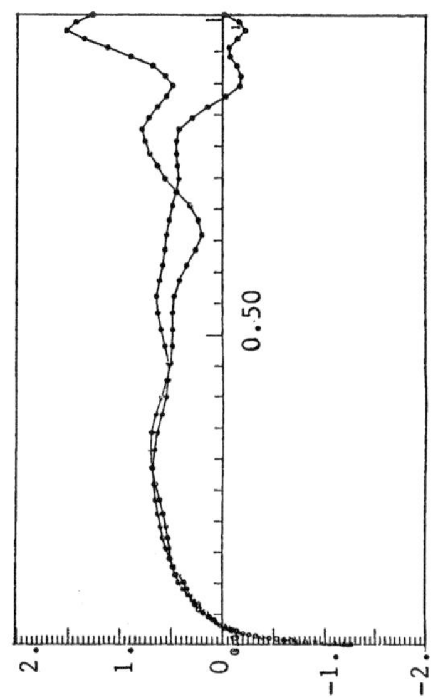

Fig. 26 - Pressure Coefficient vs x

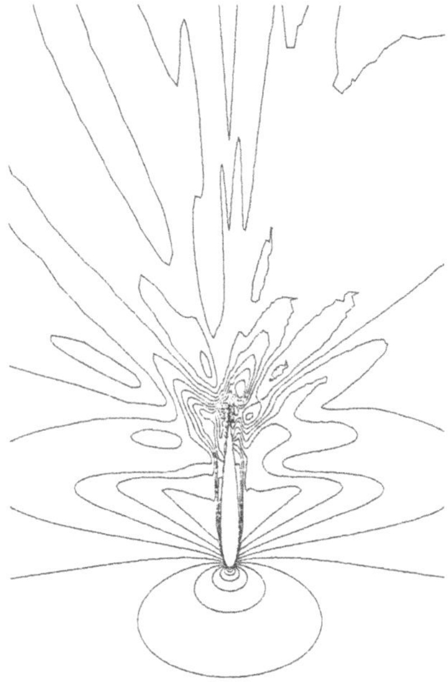

Fig. 27 - Density Contours (DR=.05)

CASE 6 (128x32)

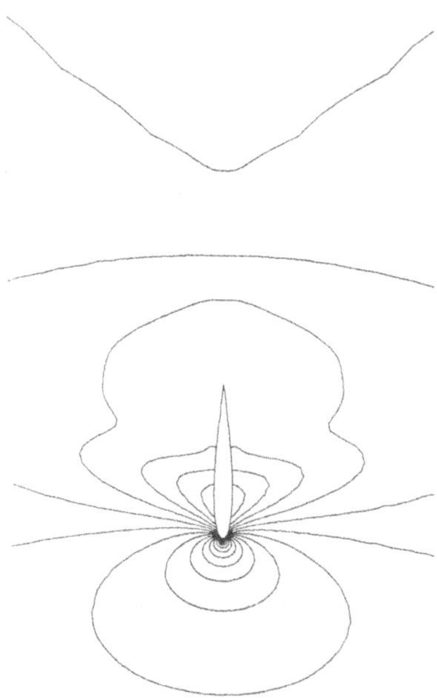

Fig. 28 - Pressure Contours (DP = .05)

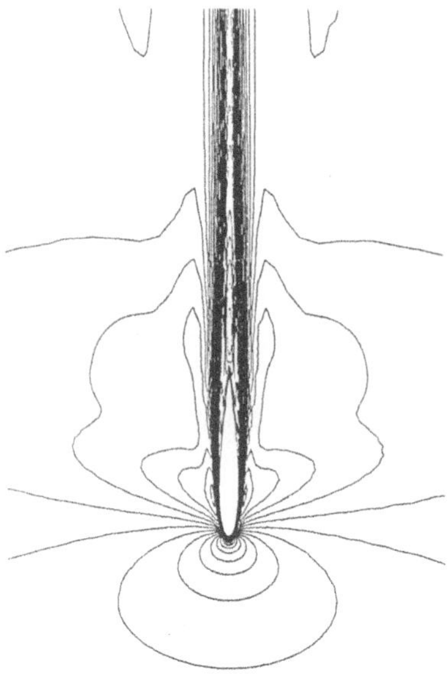

Fig. 29 - Mach Contours (DM = .05)

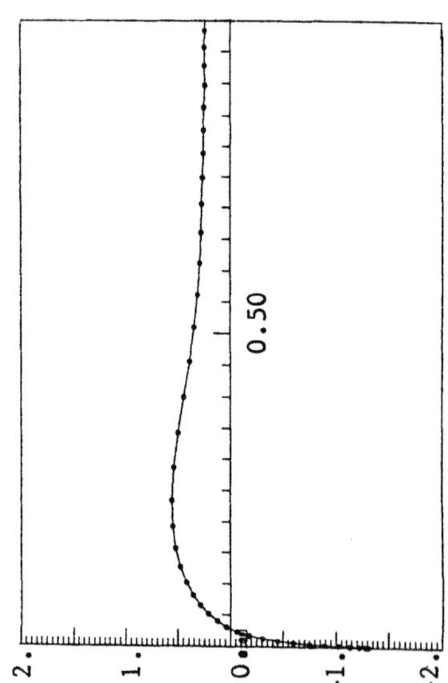

Fig. 30 - Pressure Coefficient vs x

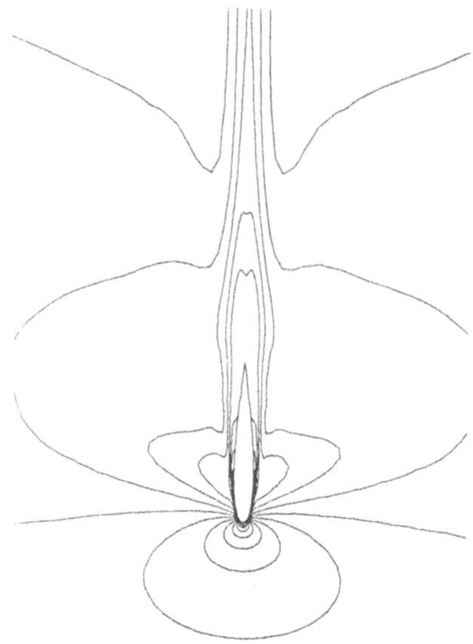

Fig. 31 - Density Contours (DR=.05)

Solutions of the Navier-Stokes Equations for Sub- and Supersonic Flows in Rarefied Gases

Werner Haase
Dornier GmbH, 7990 Friedrichshafen, FRG

Abstract

For sub- and supersonic flows in rarefied gases solutions of the Navier-Stokes equations are presented. The governing equations are solved by means of a finite volume technique combined with a Runge-Kutta time stepping scheme. The method is applied to two-dimensional, isothermal flows around a NACA0012 airfoil. Available measurements will be used for a verification of the computed results.

Numerical Method

Governing Equations

The Navier-Stokes equations describing two-dimensional, unsteady and compressible flows in conservation form read

$$\frac{\partial U}{\partial t} + \frac{\partial F}{\partial x} + \frac{\partial G}{\partial y} = 0 \tag{1}$$

where:

$$U = \begin{bmatrix} \rho \\ \rho u \\ \rho v \\ E \end{bmatrix}, \quad F = \begin{bmatrix} \rho u \\ \rho u^2 + p + \tau_{xx} \\ \rho uv + \tau_{xy} \\ (E + p + \tau_{xx})u + \tau_{xy}v + Q_x \end{bmatrix}, \quad G = \begin{bmatrix} \rho v \\ \rho uv + \tau_{xy} \\ \rho v^2 + p + \tau_{yy} \\ (E + p + \tau_{yy})v + \tau_{xy}u + Q_y \end{bmatrix} \tag{2}$$

and

$$\begin{aligned} \tau_{xx} &= -\lambda\left(\frac{\partial u}{\partial x} + \frac{\partial v}{\partial y}\right) - 2\mu\frac{\partial u}{\partial x}, \qquad \tau_{yy} = -\lambda\left(\frac{\partial u}{\partial x} + \frac{\partial v}{\partial y}\right) - 2\mu\frac{\partial v}{\partial y}, \\ \tau_{xy} &= -\mu\left(\frac{\partial u}{\partial y} + \frac{\partial v}{\partial x}\right), \\ Q_x &= -\gamma\frac{\mu}{Pr}\frac{\partial e}{\partial x}, \qquad Q_y = -\gamma\frac{\mu}{Pr}\frac{\partial e}{\partial y} \end{aligned} \tag{3}$$

with density ρ and mean total energy per unit volume E:

$$E = \rho e + 0.5\rho(u^2 + v^2) \tag{4}$$

The perfect gas equation of state is used to define the mean pressure p via the internal energy e:

$$p = (\gamma - 1)\rho e \tag{5}$$

Furthermore, the bulk viscosity λ is defined as $\lambda = -2/3\ \mu$, the Prandtl number is fixed at 0.72 and the ratio of specific heat, γ, is maintained constant at 1.4.

Finite Volume Method

Applying the integral form of equation (1) to each cell of the computational domain separately where all physical properties are defined to be constant the resulting system of ordinary differential equations in time are solved by the following 3-stage Runge-Kutta method [1]

$$
\begin{aligned}
u^{(0)} &= u^{(n)} \\
u^{(1)} &= u^{(0)} - \alpha_1 P u^{(0)} \\
u^{(2)} &= u^{(0)} - \alpha_2 P u^{(1)} \\
u^{(3)} &= u^{(0)} - \alpha_3 P u^{(2)} \\
u^{(n+1)} &= u^{(3)}
\end{aligned}
\qquad (6)
$$

where n denotes the previous time-level and P represents a spatial difference operator. The viscous terms are treated using central differences throughout the domain and one-sided formulas in the wall normal direction at solid surfaces.

Due to its effectiveness as well as to robustness and in contrast to [2] the 3-stage scheme (6) has been applied. The 4-stage scheme, although ran in different modes, turned out not to be successful in all testcases. Therefore, in the present investigations the 3-stage scheme with the coefficients $\alpha_1 = \alpha_2 = 0.6$, $\alpha_3 = 1.0$ being stable up to a Courant number of 1.8 is used for all calculations.

Filtering Techniques

To prevent an odd-even decoupling, blended second and fourth order artificial dissipation [1] is used. If the filtering technique is applied only once, stability analysis indicates the best damping property as well as the largest extension of the stability region to the left of the real axis giving latitude in the introduction of dissipative terms. In practice, the fourth order filter is active throughout the computational domain except in areas with larger pressure gradients where the second order filter takes over. To avoid filter fluxes through walls, the filter is switched off in the last two cells adjacent to walls.

Introducing the residual averaging approach [1], i.e. collecting the information from residuals implicitly, permits stable calculations beyond the ordinary Courant number limit of the explicit scheme. Therefore, the Courant number has been chosen to 3.5 for all calculations. Furthermore, since only the steady state is of interest, a variable timestep approach has been used accelerating convergence drastically.

Steady State Definition

The steady state is defined to be reached if the force coefficients (drag and lift) as well as the absolute sum of pressure along surface and wake do not vary more than 0.05% and the total number of supersonic points remains constant within a monitoring sequence of 50 iterations. Typically, an error norm reduction (L2-norm) of approximately three decades is reached and lift and drag coefficients as well as the number of supersonic points approach an asymptotic value.

Boundary Conditions

The boundary conditions read as follows (see also fig.1):
At the ***solid wall boundary*** $\Gamma 1$ no-slip conditions are used, the temperature is set constant to the total temperature at infinity and the pressure is calculated by linear extrapolation to the wall.
A ***wake boundary*** $\Gamma 2$ is defined by overlapping lower and upper wake volumes. A symmetry condition is used for symmetrical flow cases.
The ***farfield conditions*** $\Gamma 3$ are based on the introduction of Riemann invariants [1] for a one-dimensional flow normal through the boundary. For supersonic flow however, all values are fixed for incoming flow and linear extrapolation is used for outgoing flow.

At the ***outflow boundary*** $\Gamma 4$ linear extrapolation is used for density and mass fluxes, the static pressure however, is fixed to the static pressure at infinity.

Mesh Type

For discretization of the computational domain a C-type mesh is used with 110 mesh points for surface discretization and 31 for lower and upper wake representation, i.e. 172 mesh points in total. The mesh spacing normal to the surface is geometrically stretched. A constant number of 20 mesh points is distributed across a distance of 0.05 chordlengths normal to the wall with a first meshline at 0.00037 chordlenghts for cases A1, A2, A3 and 0.00012 for case A5. A total number of 60 grid points are used in the normal direction of the grid. In fig. 2 the mesh used for testcases A1, A2 and A3 is given in the vicinity of the NACA0012 airfoil. In all computations the farfield and outflow boundaries are located at a distance of 20 chordlengths away from the body.

Results

In the following, for the flow about a NACA0012 airfoil at sub- and supersonic speeds, results are presented and discussed for the mandatory testcases A1, A2, A3 and A5. The values for Mach number (Mach), Reynolds number (Re-no), angle of attack (α), force coefficients (c_D, c_L, c_M), location of separation (SEP) as well as computation times (CPU) and total numbers of iterations (ITER) needed to reach the steady state are summarized in table 1.

Testcase A1

For testcase A1 (Ma = 0.8, Re = 73, α = 10°) the pressure coefficient on lower and upper surface is given in fig. 3. The thick straight line is related to the critical pressure coefficient value. In fig. 4 the skin friction distribution is shown. On top of this figure the corresponding streamline pattern is inserted. No separation occurs, although the flow is nearly separated. In fig. 5 density contours are presented and compared with available measurements [3]. The thick line corresponds to the location where ρ/ρ_∞ equals unity, the increment between two contour lines is 0.02 for measurement as well as for computation. Therefore, the computed innermost line along the upper airfoil surface denotes $\rho/\rho_\infty = 0.76$. Comparison of measurement and present computation shows clearly that this line - and consequently all other iso-density lines too - is located at a larger distance away from the airfoil. However, computations performed by [5] and by other workshop participants show the same trend, i.e. in opposite to the measurements, calculated density fields seem - with respect to the airfoil - somewhat more "inflated". A reason for that is given in [3] assuming that the present flow conditions already correspond to the slip region which is situated at the lower boundary of the continuum flow regime [3]. Therefore, is proves necessary to recalculate all low Reynolds number cases with an additional surface slip condition. First investigations [6], quantifying slip effects, present a much better comparison with the available measurements. The Mach contours for flowcase A1 are given in fig. 6. With its center approximately one chordlenght above the airfoil, a small supersonic region can be examined. Referring to the definition of the steady state the L2 norms for $d\rho/dt$, dE/dt and $d(\rho q)/dt$ - fig. 7 - indicate a three decade reduction for density and energy residuals and 2.5 decades for the mass flux residuals. 9900 iterations were needed to reach the steady state as it has been defined above. This is surprisingly high compared to approximately 900 iterations for turbulent flow computations (with nearly the same number of grid points and highly resolved wall layers [4]) and might be related to low Reynolds number effects; see also table 1 and compare testcase A1 and A2 with respect to ITER.

Testcase A2

For the second testcase A2 ($Ma = 0.8$, $Re = 500$, $\alpha = 10°$) the pressure coefficient and the skin friction distributions are given in fig. 8 and fig. 9, respectively. In opposite to case A1, the pressure distribution shows a slight suction "peak" at 10% chordlength followed by a somewhat steeper positive gradient leading directly to separation at 36% chordlength. Fig. 10 presents density contours again normalized with the density at infinity. The increment between two contour lines is 0.02. Comparing the location of the 0.76-line for case A1 and A2 cleary shows the Reynolds number influence, i.e. for case A2 the 0.76-line lies closer to the airfoil. In addition, Mach contours are shown in fig. 11. In comparison to case A1, the supersonic region (within the thick solid line) is now directed to the airfoil with a shape being more familiar to flows having higher Reynolds numbers. The increment between two contour lines is $Ma = 0.05$.

Testcase A3

For the supersonic testcase with the flow parameters $Ma = 2$, $Re = 106$ and $\alpha = 10°$, the pressure coefficient distribution is given in fig. 12, the corresponding skin friction in fig. 13. For this testcase, a comparison between measured [3] and computed density contours is again possible, fig. 14. The thick line indicates the location where ρ/ρ_∞ equals unity and the increments have been chosen to 0.1. Note, that the contour lines 1.7, 1.9 and 2.1 are not present in the measured contour plot. Following [6] again, it might be argued, that differences to the measurements are related to the no-slip conditions used in the present computations. The corresponding Mach number contours are shown in fig. 15 (thick line: $Ma = 1.0$) and the pressure contours (p/p_∞) in fig. 16 where the thick solid line indicates $p/p_\infty = 1.0$, identical to a zero pressure coefficient, see also fig. 12.

Testcase A5

The last mandatory testcase has been defined by $Ma = 0.85$, $Re = 500$ at zero incidence. As it can be seen from figs. 17 and 18, presenting smooth pressure coefficient (without a suction peak) and skin friction distributions, respectively, no separation occurs. In fig. 19 density contour lines are presented with an increment of 0.02 and the corresponding Mach number contours are given in fig. 20 with an increment of 0.05. The thick lines in fig. 20 again define the supersonic regions, being quite far from the body due the low free-stream Reynolds number.

Trailing Edge Skin Friction Problems

In the testcases A1, A2 and A3 a more or less drastic increase in skin friction can be observed at the trailing edge of thê airfoil. Following [4] (although only turbulent flows were discussed), a reason for this physically not plausible behaviour might be an unproper alignement of the trailing edge mesh line with respect to the direction of the trailing edge streamline. For testcase A1 this can be seen comparing the streamline pattern, inserted on top of fig. 4, and the trailing edge meshline in figs. 1 or 2. One can easily recognize that the streamline does not emanate with the (NACA0012 trailing edge) bisector angle, moreover it shows an emanation angle being slightly higher than the lower surface trailing edge angle.

For testcase A5 however, it must be assumed that the symmetrical flow (zero degrees incidence) together with a symmetrical mesh leads to a skin friction distribution without such drastic variation at the trailing edge. In fact, fig. 18 presents a smooth skin friction distribution due to the "automatically" aligned mesh. The trailing edge flow situation ($0.95 < x/c < 1.05$) is given on top of fig. 18 as a velocity vector plot.

References

[1] Jameson, A., Schmidt, W., Turkel, E., "Numerical Solutions of the Euler Equations by Finite Volume Methods using Runge-Kutta Time-Stepping Schemes" ***AIAA Paper*** 81-1259, June. 1981.

[2] Haase, W., Wagner, B., Jameson, A., "Development of a Navier- Stokes Method Based on a Finite Volume Technique for the Unsteady Euler Equations," ***Notes on Numerical Fluid Mechanics***, Vol. 7, Vieweg Verlag, 1983, pp. 99-107.

[3] Allegre, J., Raffin, M., Lengrand, J.C., "Experimental Flowfields Around NACA0012 Airfoils Located in Subsonic and Supersonic Rarefied Air Streams," ***Gamm Workshop on "Numerical Simulation of Compressible Navier-Stokes Flows"***, Nice Dec. 4-6, 1985 - see this volume.

[4] Haase, W., "Influence of Trailing Edge Meshes on Skin Friction in Navier-Stokes Calculations", to appear in: ***AIAA-Journal***, August 1986.

[5] Peyret, R., Taylor, T.D., "Computational Methods for Fluid Flow", ***Springer Series in Computational Physics***, Springer Verlag, 1983.

[6] Allegre, J., "Private Communications" (with respect to the 15th Rarefied Gas Dynamics Symposium, Italy, June 1986).

Table 1

Case	Mach	Re-no	α [°]	c_D	c_L	c_M	SEP [x/c]	CPU [sec]	ITER
A1	0.80	73.0	10	0.6377	0.5459	-0.05119	no	4237	9900
A2	0.80	500.0	10	0.2721	0.4468	-0.00324	0.360	1002	2350
A3	2.00	106.0	10	0.4772	0.3063	-0.05061	no	2808	6600
A5	0.85	500.0	0	0.2176	0.0000	0.00000	no	1599	3750

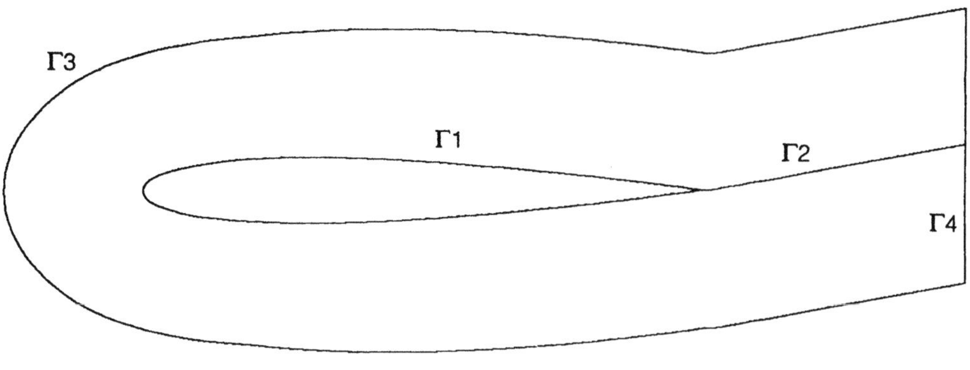

Fig. 1: Definition of Boundaries

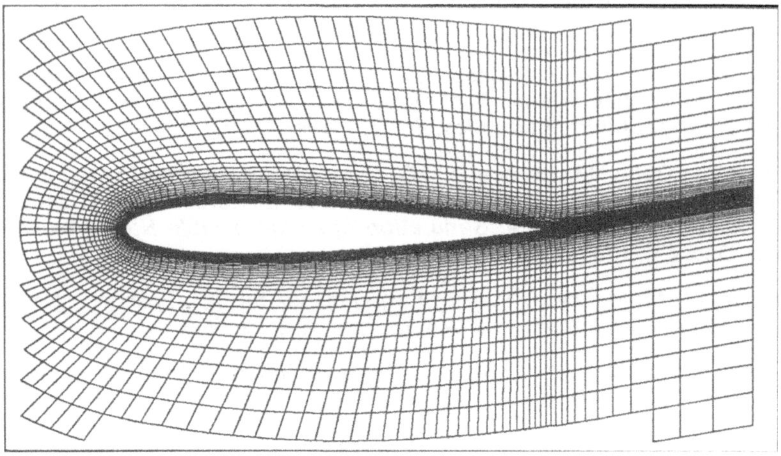

Fig. 2: Computational Mesh in the Vicinity of the NACA0012 Airfoil for Case A1, A2, A3

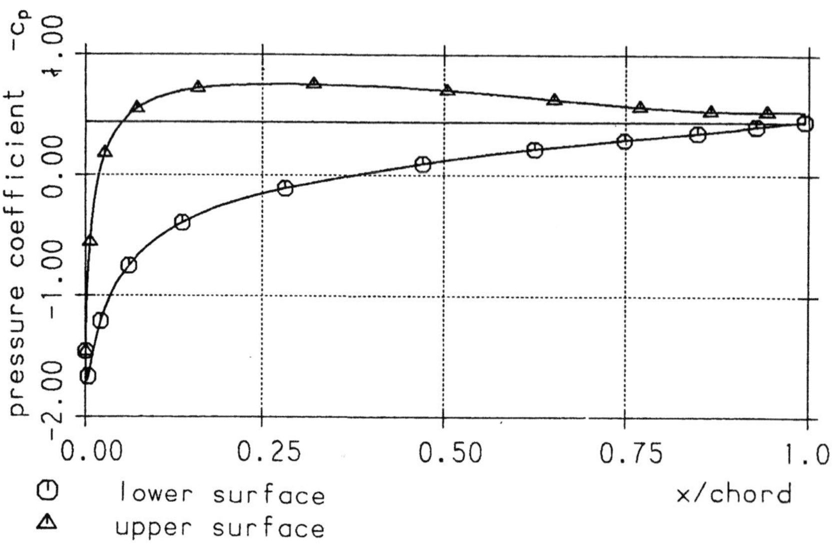

Fig. 3: Pressure Coefficient for Case A1
Ma=0.80, Re=73.0, Alpha=10.0°

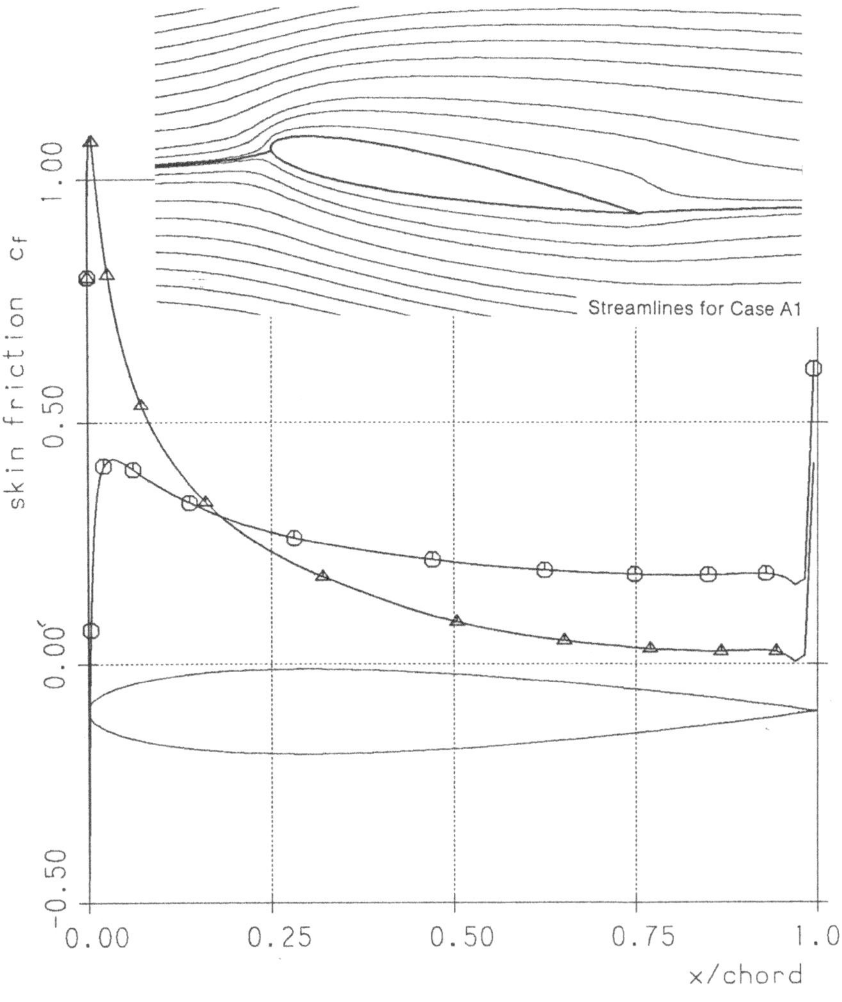

Ⓘ lower surface
△ upper surface

Fig. 4: Skin Friction Distribution for Case A1
Ma=0.80, Re=73.0, Alpha=10.0°

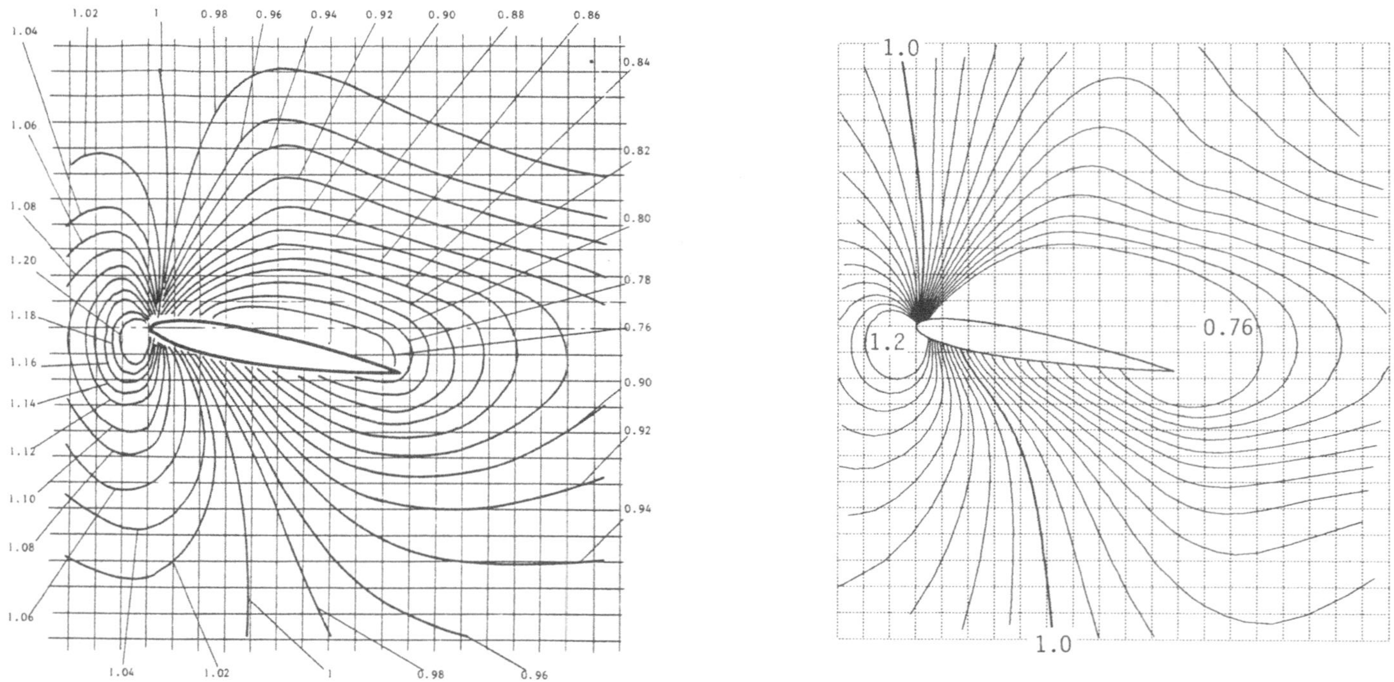

Measurement [3]

Present Calculation

Fig. 5: Density (ρ/ρ_∞) Contours ($\Delta = 0.02$) for Case A1 ($Ma = 0.80$, $Re = 73$, $\alpha = 10°$)

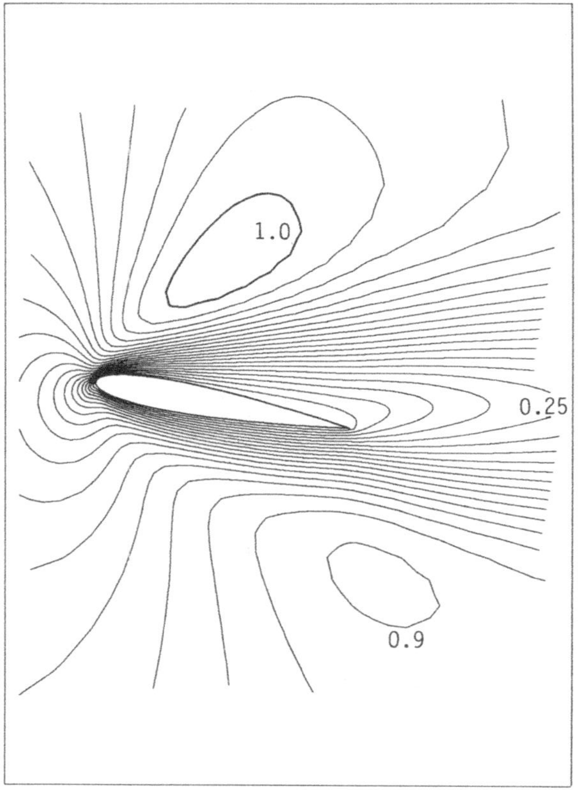

Fig. 6: Mach Number Contours ($\Delta = 0.05$) for Case A1 ($Ma = 0.80$, $Re = 73$, $\alpha = 10°$)

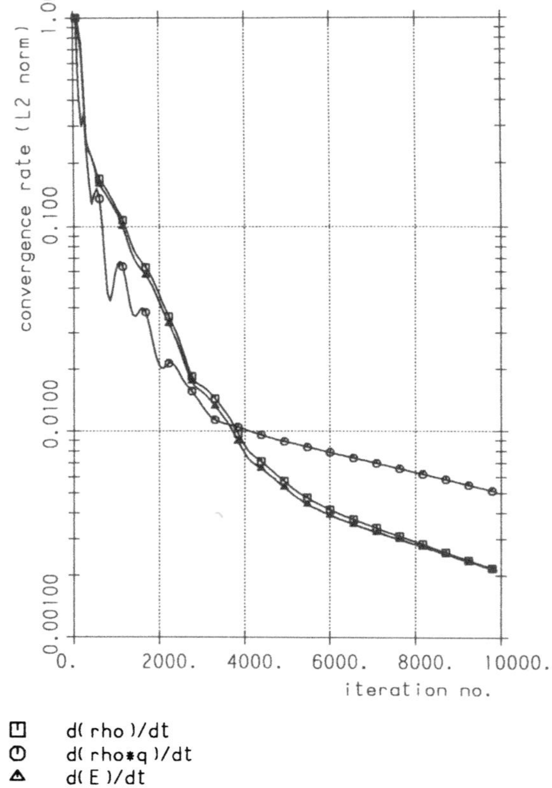

Fig. 7: Convergence Rate for Case A1 Ma=0.80, Re=73.0, Alpha=10.0°

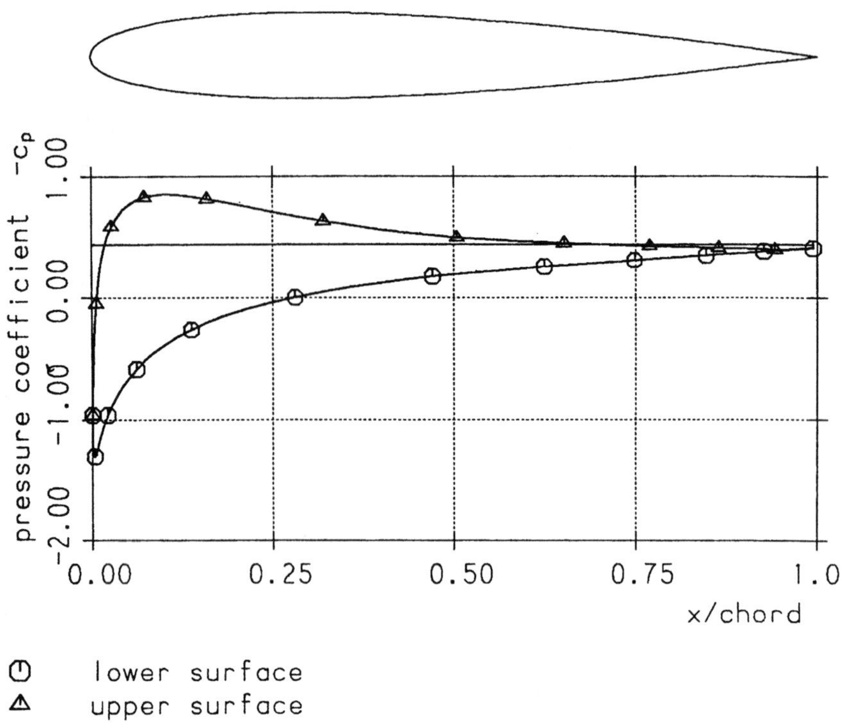

Fig. 8: Pressure Coefficient for Case A2
Ma=0.80, Re=500.0, Alpha=10.0°

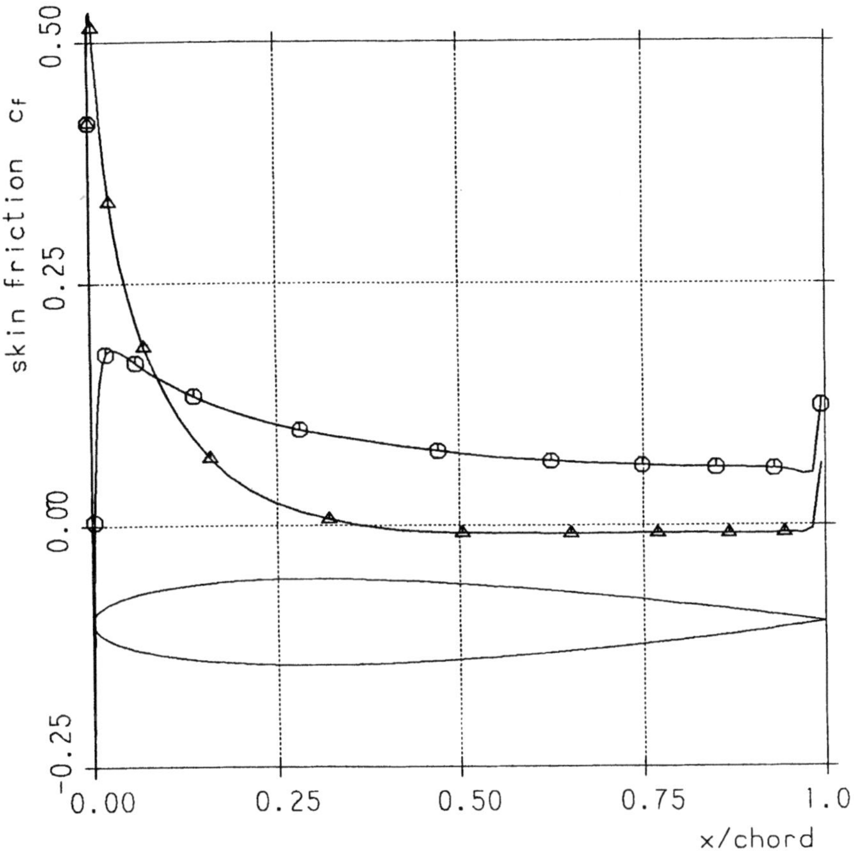

lower surface
upper surface

Fig. 9: Skin Friction Distribution for Case A2
Ma=0.80, Re=500.0, Alpha=10.0°

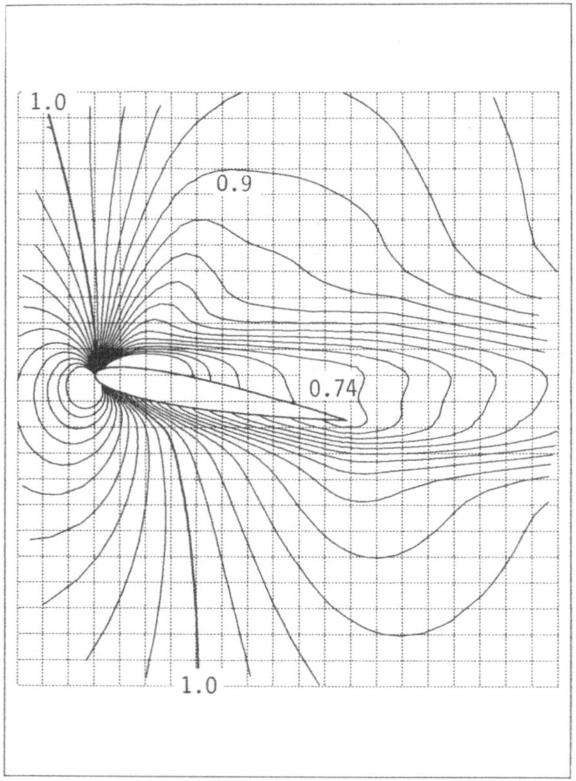

Fig. 10: Density (ρ/ρ_∞) Contours ($\Delta = 0.02$) for Case A2 ($Ma = 0.80$, $Re = 500$, $\alpha = 10°$)

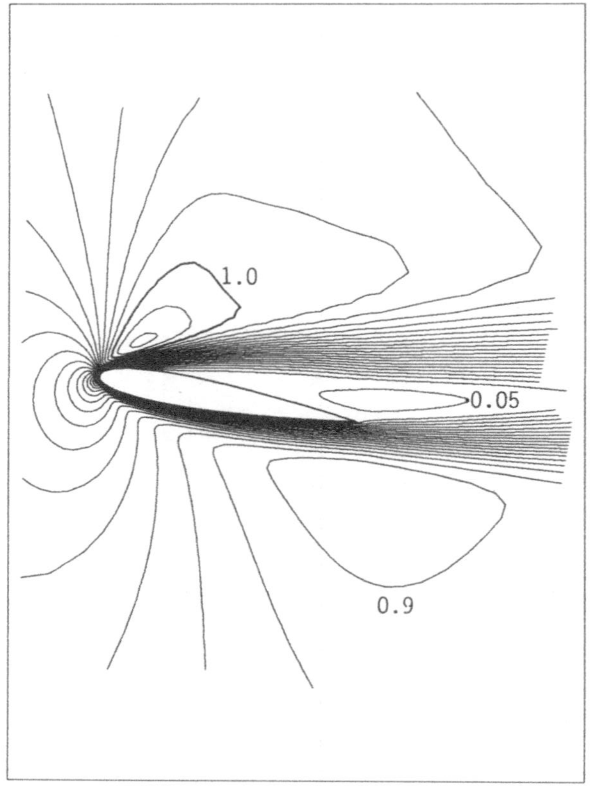

Fig. 11: Mach Number Contours ($\Delta = 0.05$) for Case A2 ($Ma = 0.80$, $Re = 500$, $\alpha = 10°$)

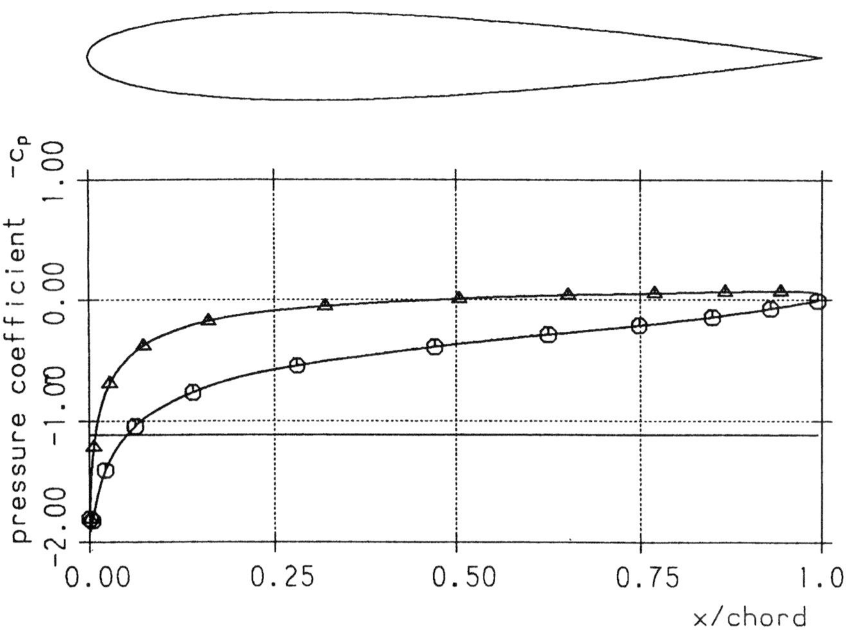

⊕ lower surface
△ upper surface

Fig. 12: Pressure Coefficient for Case A3
Ma=2.00, Re=106.0, Alpha=10.0°

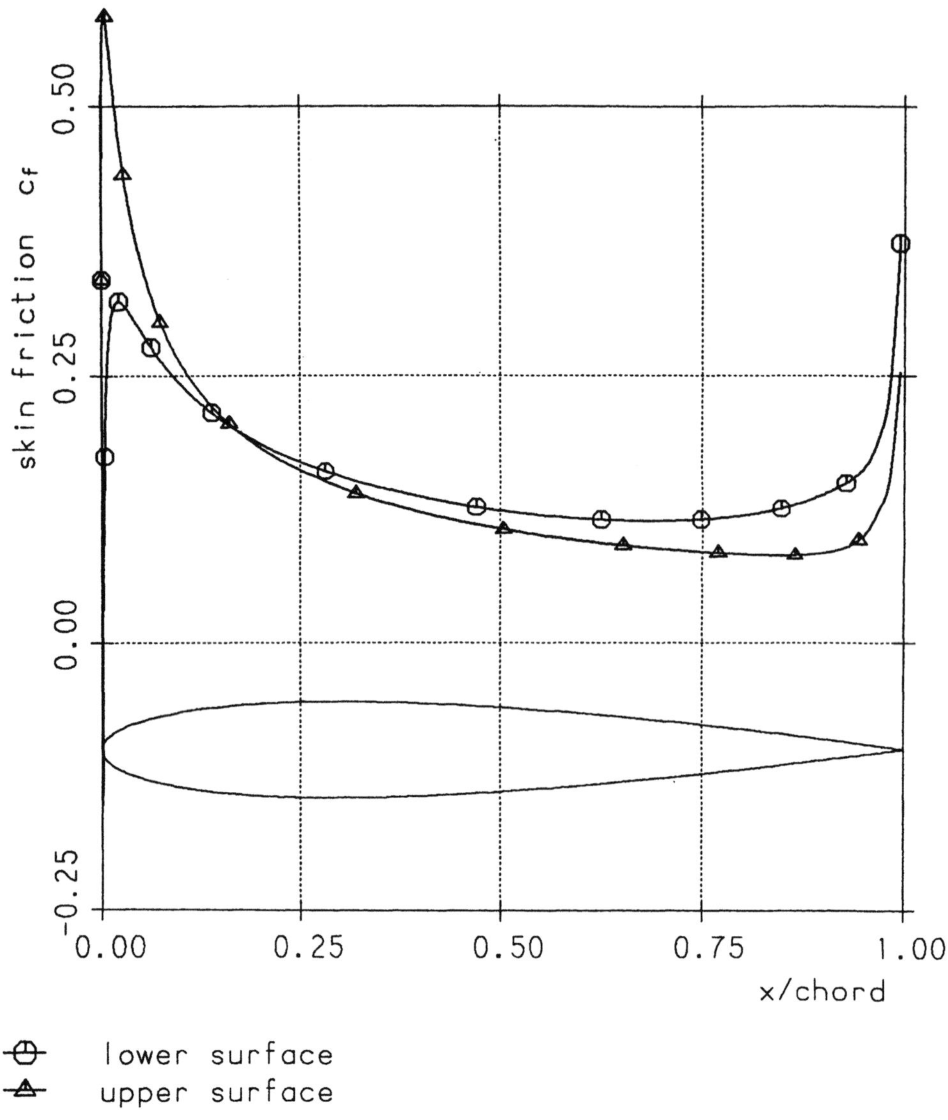

Fig. 13: Skin Friction Distribution for Case A3
Ma=2.00 Re=106.0 Alpha=10.0^0

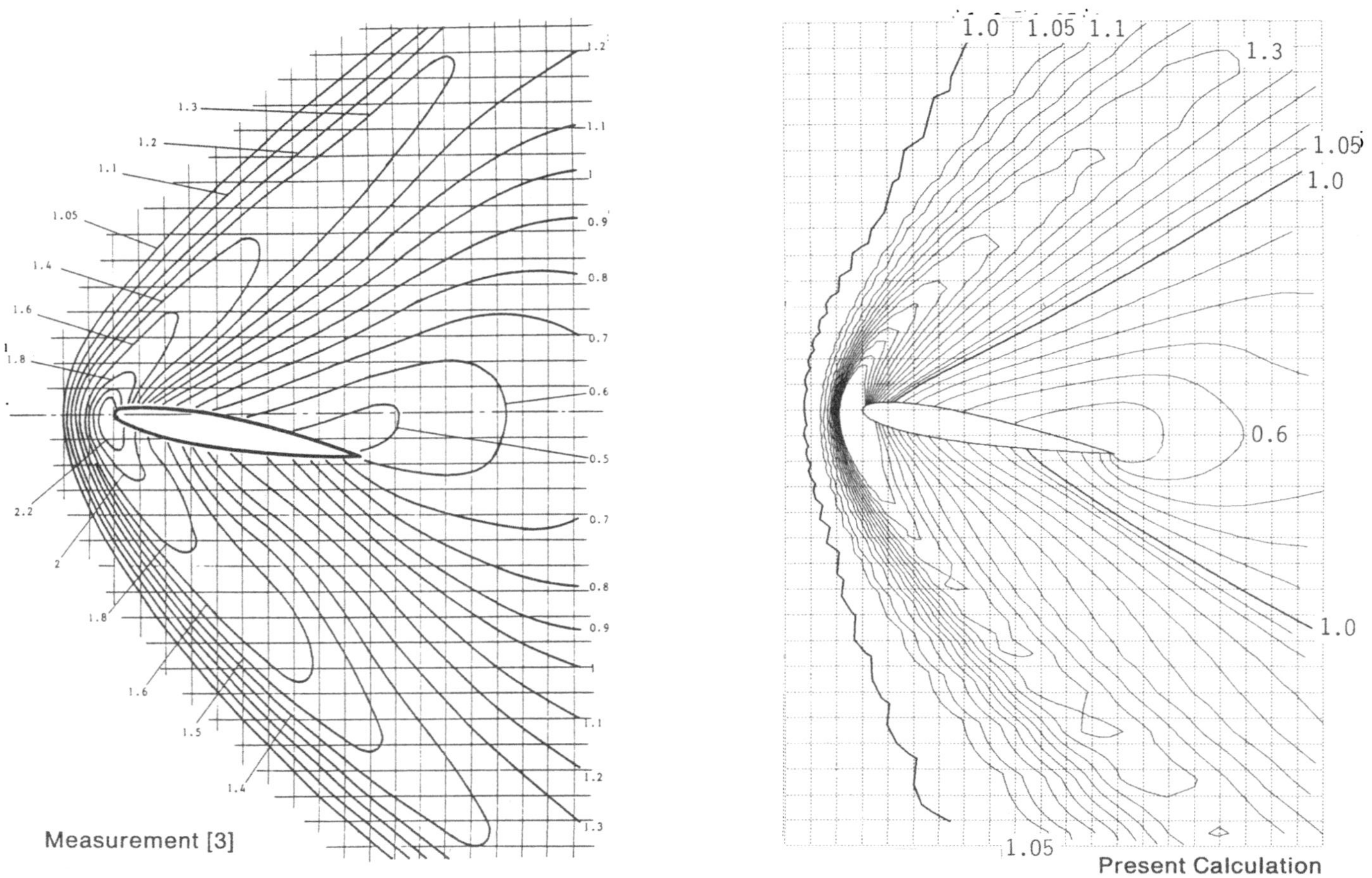

Fig. 14: Density (ρ/ρ_∞) Contours ($\Delta = 0.1$) for Case A3 (Ma = 2, Re = 106, $\alpha = 10°$)

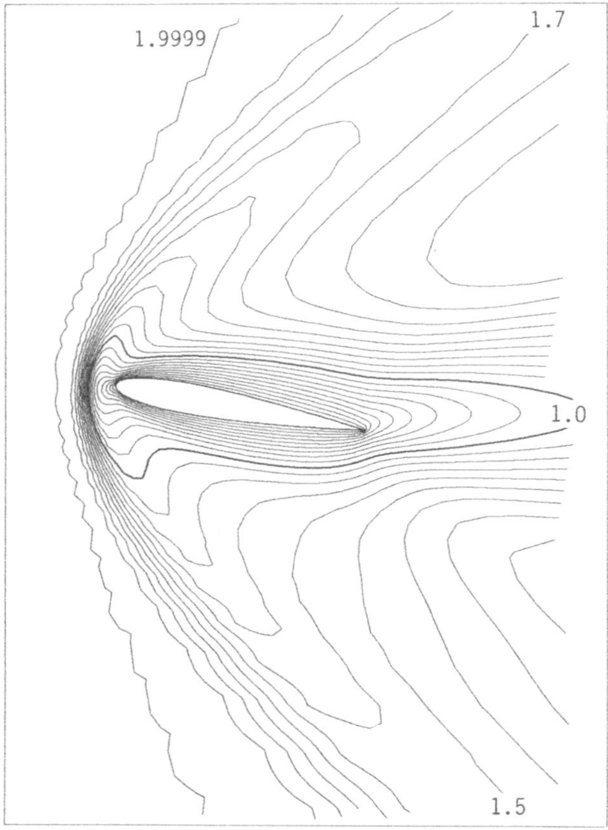

Fig. 15: Mach Number Contours ($\Delta = 0.1$) for Case A3 ($Ma = 2$, $Re = 106$, $\alpha = 10°$)

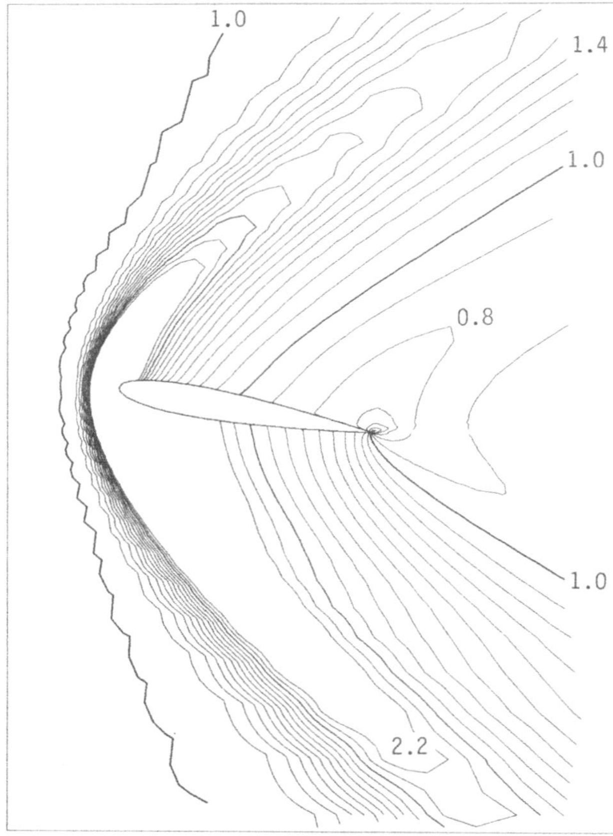

Fig. 16: Pressure (p/p_∞) Contours ($\Delta = 0.1$) for Case A3 ($Ma = 2$, $Re = 106$, $\alpha = 10°$)

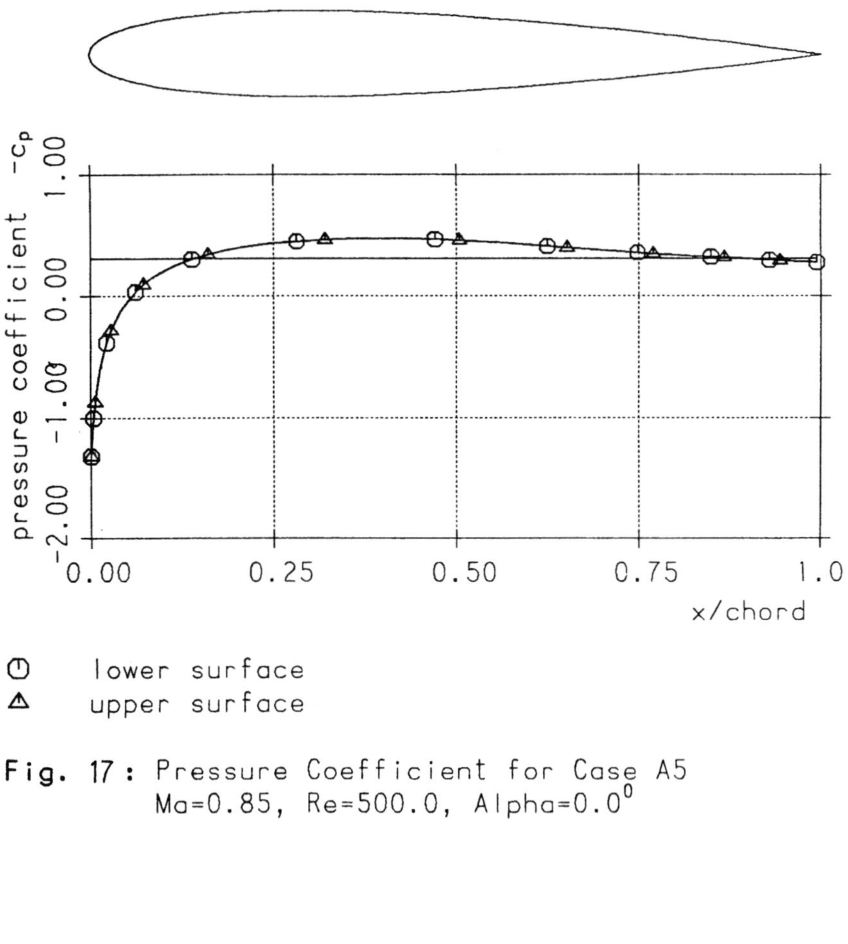

Fig. 17 : Pressure Coefficient for Case A5
Ma=0.85, Re=500.0, Alpha=0.0°

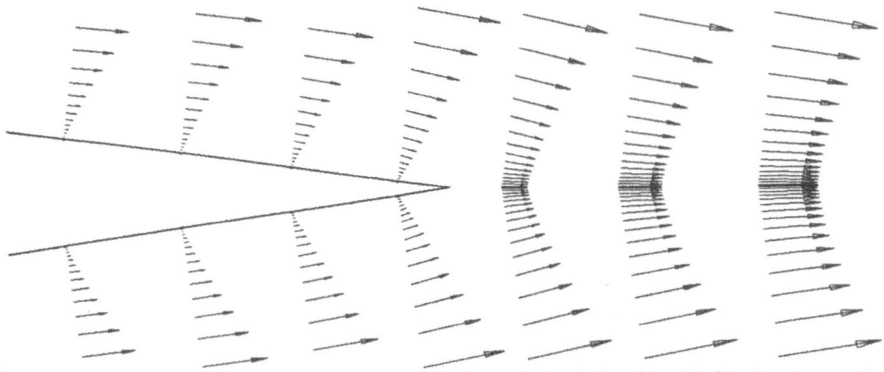

Near Wake Velocity Field for Case A5

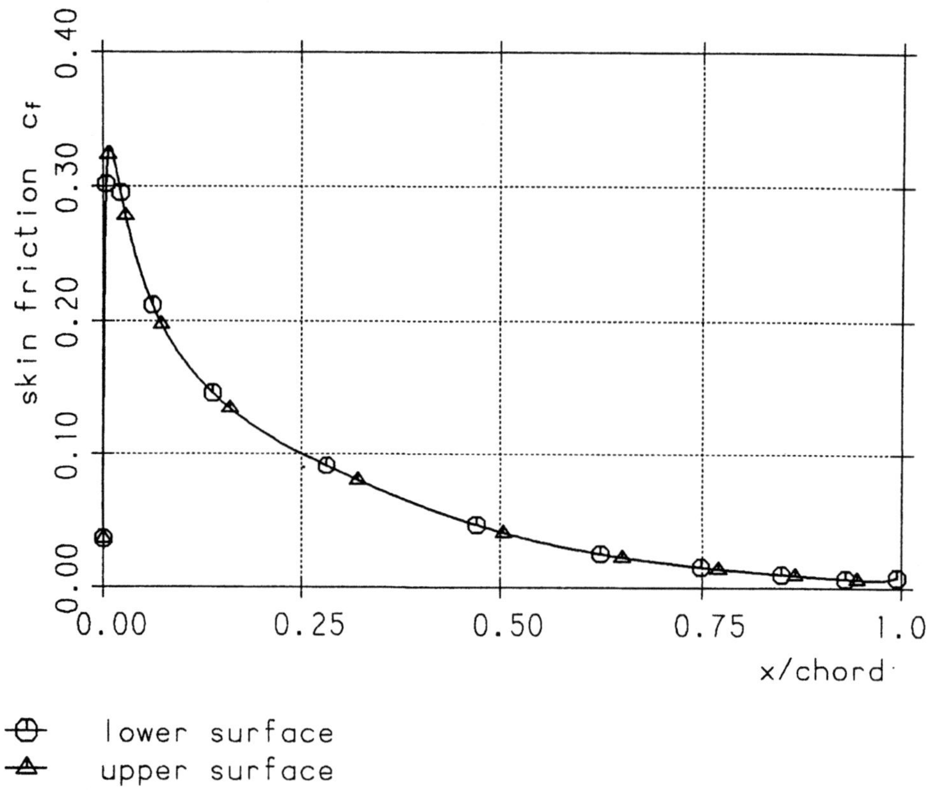

Fig. 18: Skin Friction Distribution for Case A5
Ma=0.85 Re=500.0 Alpha=0.0°

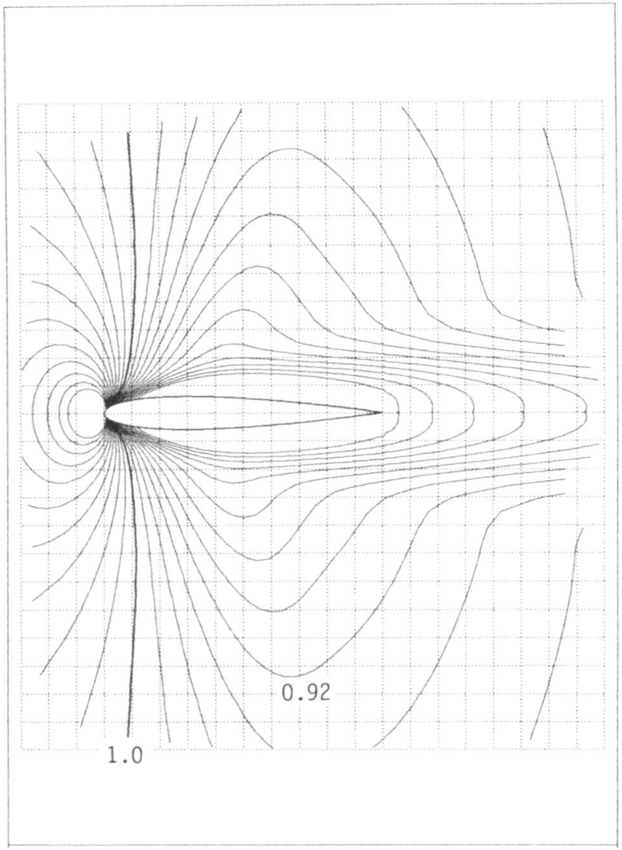

Fig. 19: Density (ρ/ρ_∞) Contours ($\Delta = 0.02$) for Case A5 ($Ma = 0.85$, $Re = 500$, $\alpha = 0°$)

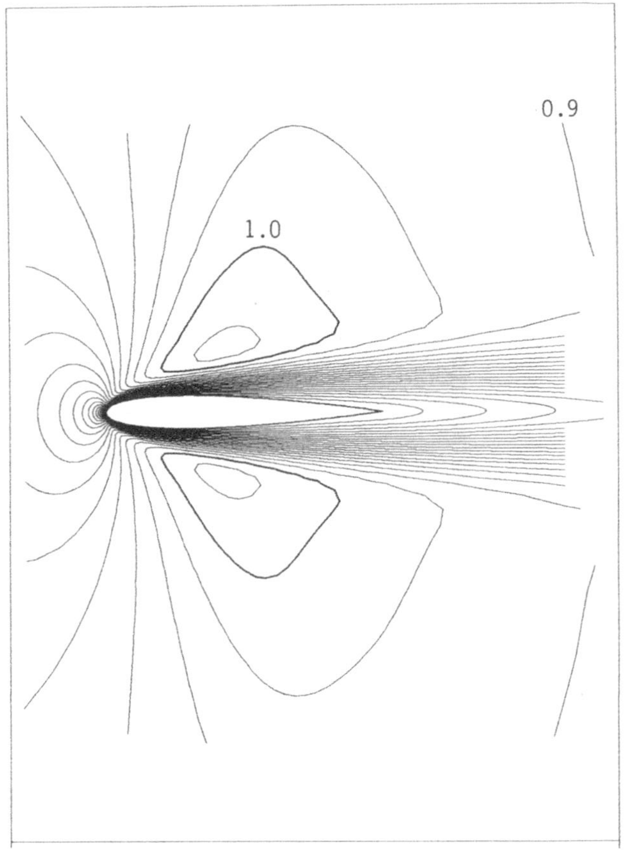

Fig. 20: Mach Number Contours ($\Delta = 0.05$) for Case A5 ($Ma = 0.85$, $Re = 500$, $\alpha = 0°$)

A FINITE ELEMENT METHOD FOR SOLVING

NAVIER-STOKES EQUATIONS

COMPRESSIBLE FLOW AROUND NACA0012

D. KALFON - D. VOLPERT - O.BROCARD
O.N.E.R.A. - C.E.R.T. Groupe d'Analyse Numérique
2, avenue Edouard Belin - B.P. 4025
31055 - TOULOUSE CEDEX (FRANCE)

I. INTRODUCTION

This study concerns problems where the convection perturbation resulting from the compressibility are weak with respect to the dissipativity. The proposed Navier-Stokes equations resolution method aims to directly solve the stationary equations.

II. DESCRIPTION OF THE METHOD

The conservative form of the system is the following :

$$m_i = \varrho u_i \tag{1}$$

$$p^* = \frac{p}{\gamma M_\infty^2} + \frac{2}{3Re} u_{i,i} \tag{2}$$

$$\tau^*_{ij} = p^* \delta_{ij} + \frac{2}{Re} D_{ij} \tag{3}$$

$$\partial_j m_j = 0 \tag{4}$$

$$\partial_j \left(\frac{m_i m_j}{\varrho} - \sigma^*_{ij}\right) = 0 \tag{5}$$

$$\partial_j \left(e m_j - \frac{\gamma}{P_r R_e} \partial_j e\right) = f \tag{6}$$

where f is a divergence form :

$$f = -\partial_j h_j (m, p^*, \varrho)$$

The associed boundary conditions are :

(P_1) $m = 0$

(P_2) $e = e_\infty \left(1 + \frac{\gamma-1}{2} M_\infty^2\right)$

on Γ_p (wall conditions)

(A_1) $m = m_\infty = \begin{pmatrix} \cos\alpha \\ \sin\alpha \end{pmatrix}$

(A_2) $\varrho = \varrho_\infty = 1$

(A_3) $e = e_\infty = \frac{1}{\gamma-1}$

on Γ^-_∞ (upstream conditions)

$$(B_1) \qquad \sigma^*_{ij}\, n_j = - \frac{1}{\gamma M_\infty^2}\, \delta_{ij}\, n_j \qquad \text{on } \Gamma^+_\infty$$

(downstream conditions)

$$(B_2) \qquad \partial_j\, e\, n_j = 0$$

This condition (B_1) on the dowstream "artificial" boundary Γ^+_∞ consists in fixing the normal and tangential component stress to their physical values very far the profile so that we can interpret this condition, for the normal component stress by :

$$p = p_\infty + C \cdot \frac{\partial u_1}{\partial x_1}, \quad \text{with } C = \frac{\gamma M_\infty^2}{3Re}.$$

NUMERICAL ALGORITHM

An iterative method based on the density ϱ has been choosed to be able to use the incompressible code [1].

At step (n), $\varrho^{(n)}$ is known and (m^{n+1}, p^{*n+1}) is computed by solving the system:

$$\partial_j\, m_j^{(n+1)} = 0 \tag{7}$$

$$\partial_j \left(\frac{1}{\varrho(n)}\, m_i^{(n+1)}\, m_j^{(n+1)} - \sigma^*_{ij} \left(\frac{m^{(n+1)}}{\varrho(n)}\right)\right) = 0 \tag{8}$$

with the associated boundary conditions $(P_1),(A_1),(A_2),(B_1)$.

Let

$$\beta = \frac{1}{\varrho\ (n)}, \quad w = m^{(n+1)}, \quad P = p^{*(n+1)}, \quad \Gamma = \partial\,\Omega - \Gamma^+_\infty$$

$$W = \{v \in H^1(\Omega),\quad v = 0|_\Gamma\}$$

$$M = \{q \in L^2\ (\Omega),\quad v = 0|_\Gamma\}.$$

The (W-P) variational formulation is now :

$$\left.\begin{aligned} &\int_\Omega \Big(\frac{2\,\mu}{Re}\, D_{ij}\,(\beta w).D_{ij}(v) - P\,\partial_i\, v_i - \beta\, w_i\, w_j\, \partial_j\, v_i - \\ &\qquad - q\,\partial_j\, w_j\Big)\, dx \\ &\quad + \int_{\Gamma^+_\infty} \beta\, w_i\, w_j\, v_i\, n_j\, ds = \\ &\quad \int_{\Gamma^+_\infty} - \frac{1}{\gamma\, M_\infty^2}\, \delta_{ij}\, n_j\, v_i\, ds \end{aligned}\right\} \tag{9}$$

$$\forall\ v \in W^2, \quad \forall\ q \in M$$

where $W \in H^1(\Omega)^2$ and verify the Dirichlet boundary conditions.

- The (W-P) non-linear problem (9) is solved by a shightly modified Newton Method on the discret problem.

- After computing $(m^{(n+1)}, p^{*(n+1)}$ we can compute the right-hand side of

$$\int_{\Omega}\left(\frac{\gamma}{P_r R_e}\,\partial_j\, e^{(n+1)}\,\partial_j\, v - e^{(n+1)}\, m_j^{(n+1)}\,\partial_j\, v\right) dx = \qquad (10)$$

$$\int_{\Omega} h_j^{n+1}\,\partial_j\, v\, dx - \int_{\Gamma_\infty^+} h_j^{n+1}\, n_j\, v\, ds$$

$\forall\, v \in W^*$.

Computing $p^{(n+1)}$ by :

$$\int_{\Omega} p^{(n+1)}\, q\, dx = \int_{\Omega}\left(p^{*(n+1)} + \lambda\,\partial_j\left(\frac{m_j^{(n+1)}}{\varrho(n)}\right)\right) q\, dx, \qquad (11)$$

$\forall\, q \in M$

and solving (10), we obtain :

$$\varrho^{(n+1)} = \frac{1}{\gamma-1}\,\frac{p^{(n+1)}}{e^{(n+1)}} . \qquad (12)$$

The resolution of (11) is obtained by mass-lumping.

III. FINITE ELEMENT INTERPOLATION

Let

$$W_h = \{w_h \in \mathcal{C}^0(\bar{\Omega}) ,\ w_h|_K \in Q_2(K) ,\ \forall\, K \in \tau_H ,\ w_h|_\Gamma = 0\}$$

$$M_h = \{q_h \in \mathcal{C}^0(\bar{\Omega}) ,\ q_h|_K \in Q_1(K) ,\ \forall\, K \in \tau_h ,\ q_h|_\Gamma = 0\}$$

where K is a quadrilateral element.

The non linear system associed with (9) in $W_h^2 \times M_h$ is solved by a "quasi-Newton method".

$H(W^0)\,(W^{\nu+1} - W^{\nu}) = - F(W^{\nu})$.

In every numerical experiment, this algorithm has converged after few iterations : the first iteration in ϱ is the only one to need an important number of iterations in v.

The assembling, the factorization and the resolution of the system are implemented on the CRAY1-S (CCVR) with a vectorized partition technique.

QUALITY OF THE RESULTS

1) We observe that the solutions in m and e are slightly oscillating.

These numerical oscillations have been cut out by using a Petrov-Galerkin method suggested by T.J.R. HUGUES [2] : <u>only in the convection terms</u>, the test functions are not computed at the Gauss integration points, but at points that are shifted in the direction of the velocity, with an amplitude connected with <u>the velocity module</u> and the <u>dimensions of the element</u>.

2) <u>QUALITY OF THE RESULTS ON THE BOUNDARY</u>

On the boundary Γ_p, no problem arise.

On the numerical boundary Γ_∞ :

- no problem <u>if "upwinding"</u> on the convective terms for equations (5) (m_1, m_2, p^*). The "Neuman condition" B1 seems to be a good numerical boundary condition in Γ_∞^+.

- The calculed pressure p^* is smooth and we obtain :

$$p^*_\infty \simeq \frac{1}{\gamma M^2_\infty} \quad \text{on } \Gamma^+_\infty .$$

CONCLUSION

The numerical approach (which use the Incompressible Navier Stokes solver) gives smooth numerical solution with few iterations in ρ, and few elements. This economic method is yet not able to solve supersonic flow. A new algorithm is studied to solve the full Navier-Stokes equations in the unstationnary case by coupling the four conservatives components, and an accurate treatment of the convective terms.

Cases	Re	M_∞	α	CL	CD	CM	CPU Time (s) CRAY1S (CCVR)
A1	73	0.80	10	0.5584	0.6568	0.1912	$0.7\ 10^3$
A2	500	0.80	10	0.4831	0.2868	01312	$0.84\ 10^3$
A5	500	0.85	0	0.	0.23	0.	$1.0\ 10^3$
A6	2.000	0.85	0	0.	0.136	0.	$1.2\ 10^3$

Computational Mesh M1 (global view)

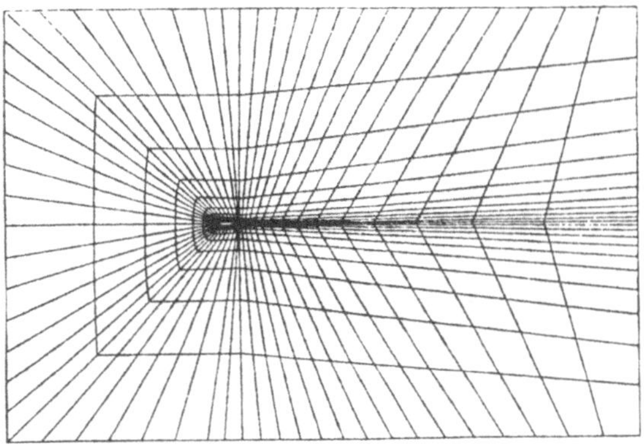

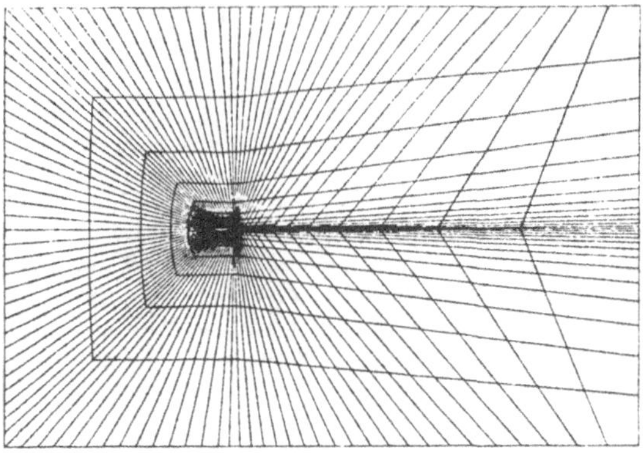

Computational mesh M2

(global view)

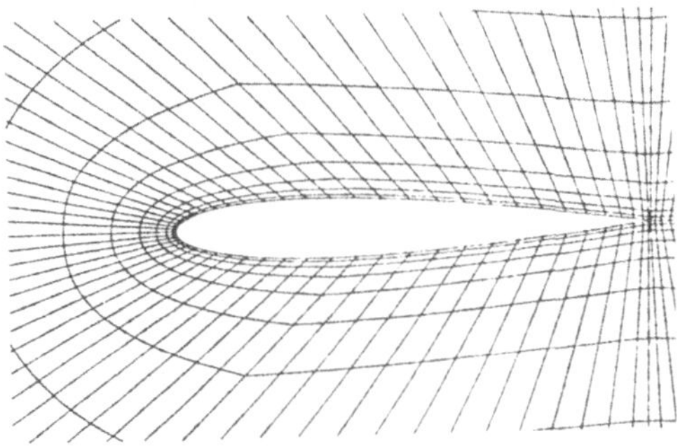

Partial view near the profile (M2)

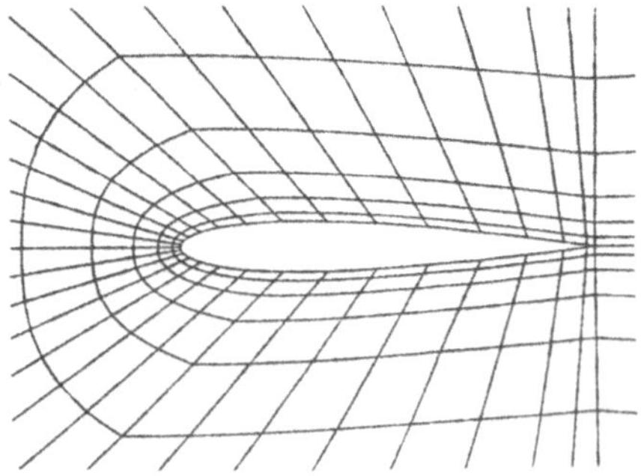

Partial view near the profile (M1)

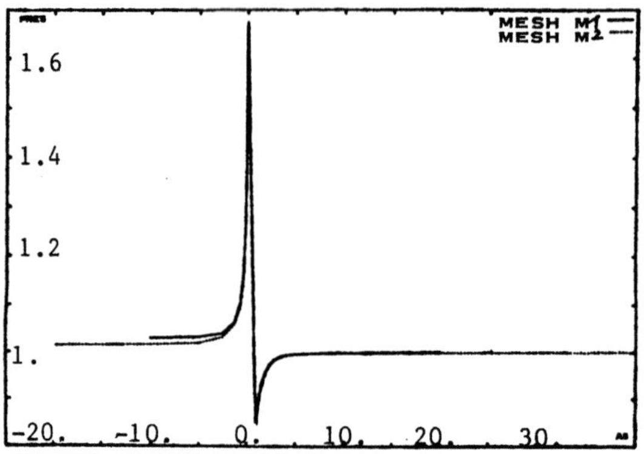

the
pressure
along
x-axis

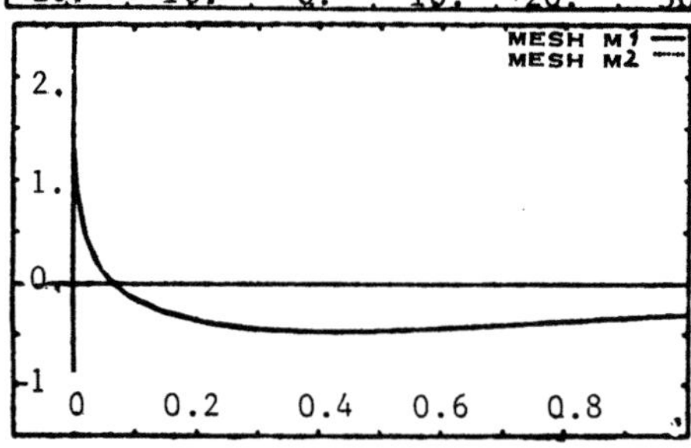

the pressure
coefficient
with M1 and M2

Influence of
mesh and
computational
domain

$\mathrm{Log}(\rho^{n+1} - \rho^{n})$ (A1)

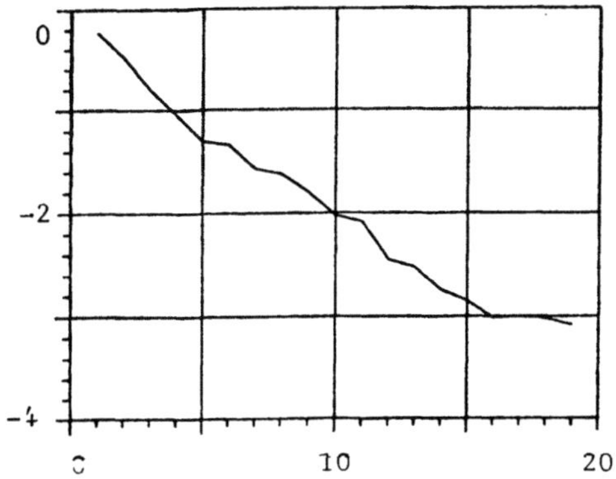

Pressure Coefficient (A1)

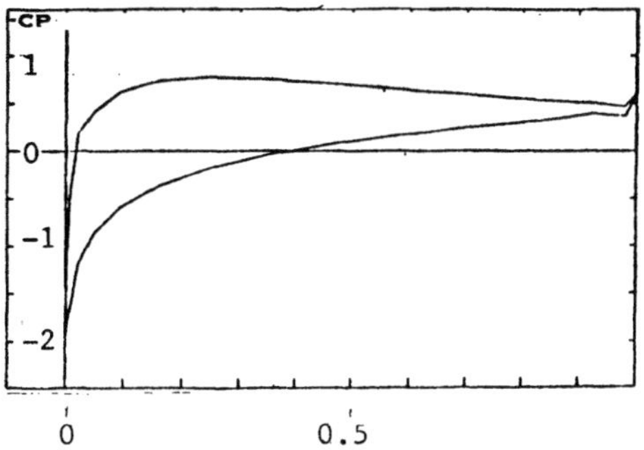

Skin friction coefficient (A1)

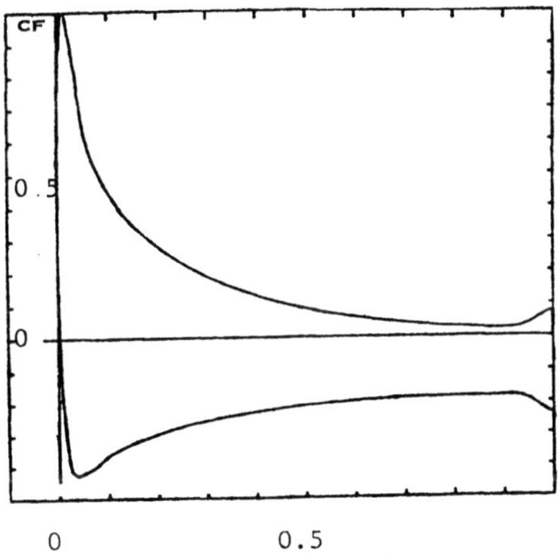

$\mathrm{Log}(\rho^{n+1} - \rho^{n})$ (A2)

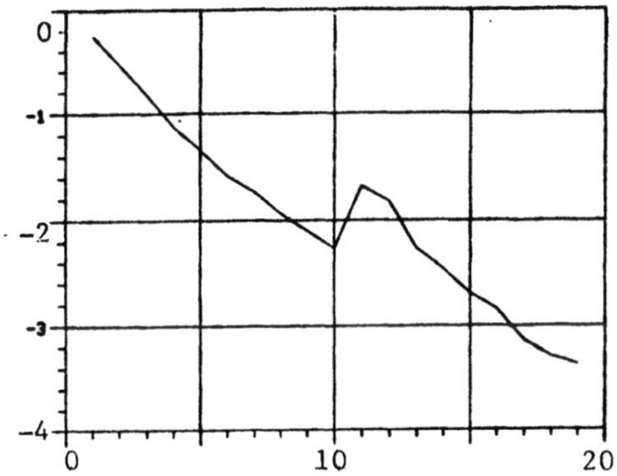

Pressure coefficient (A2)

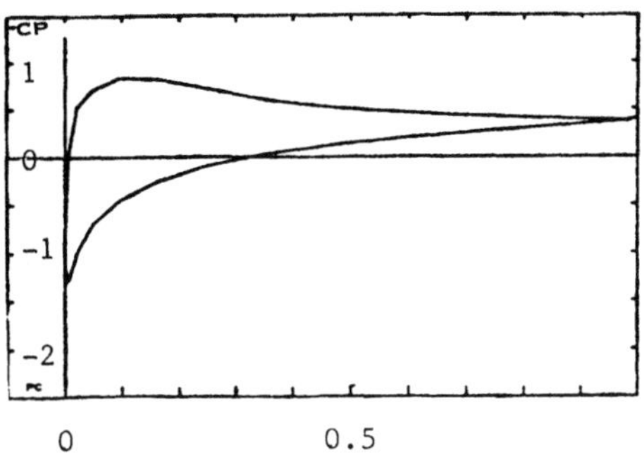

Skin friction coefficient (A2)

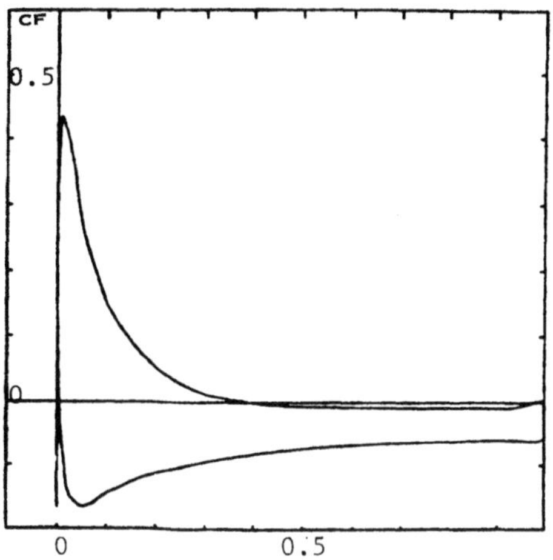

$\mathrm{Log}(\rho^{n+1} - \rho^{n})$ (A5)

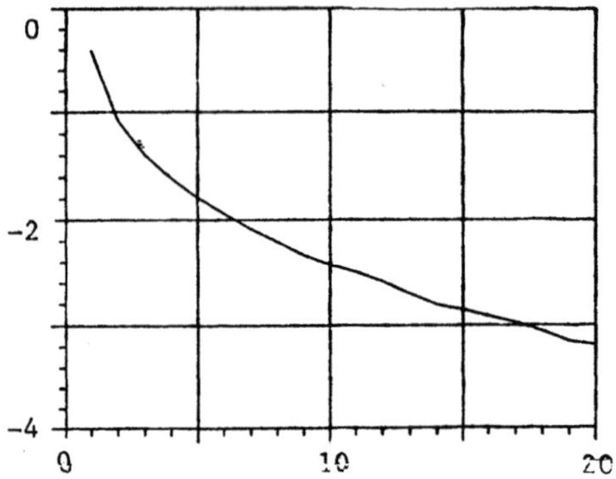

Pressure coefficient (A5)

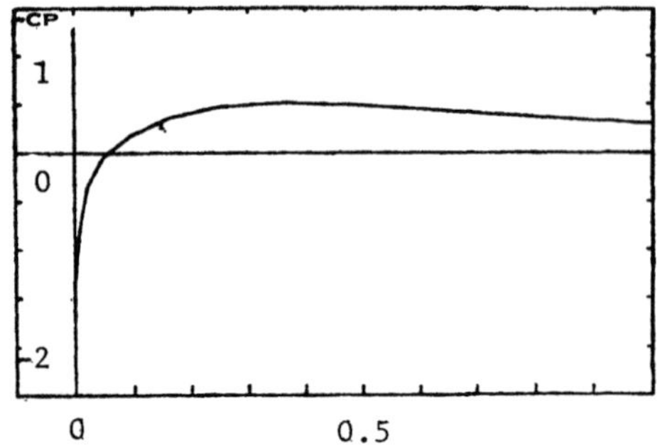

Skin friction coefficient (A5)

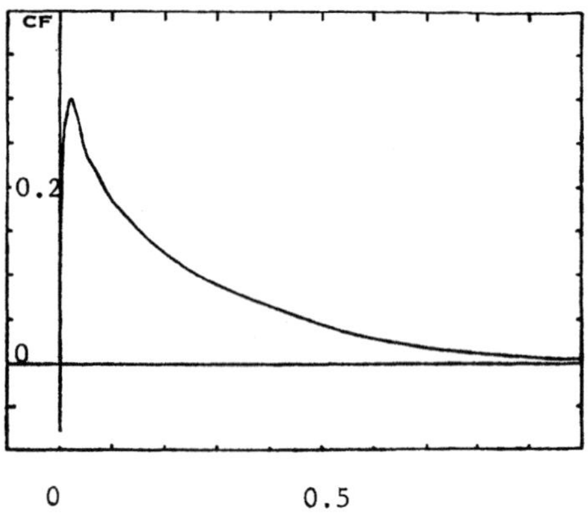

Pressure coefficient (A6)

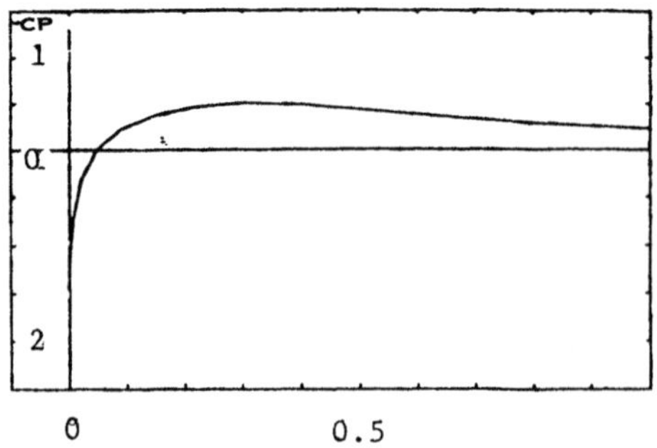

Skin friction coefficient (A6)

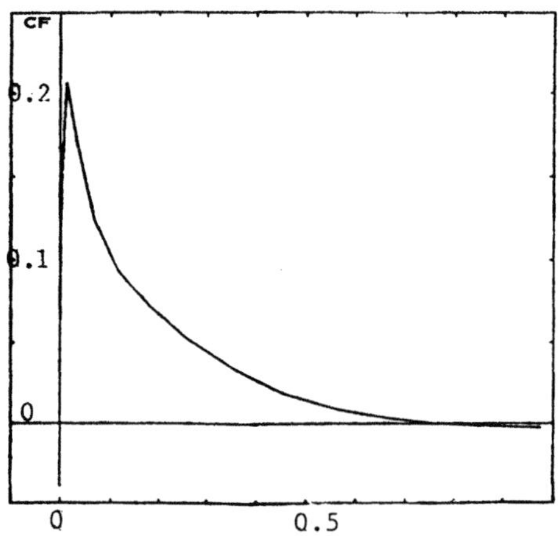

USING AN UNFACTORED IMPLICIT PREDICTOR-CORRECTOR METHOD
- Results with a Research Code

W. Kordulla
DFVLR SM-TS, Bunsenstr.10, D-3400 Göttingen, FRG

The method used is based on MacCormack's recently suggested implicit predictor-corrector scheme [1] in the version described in [2,3] for high-Reynolds number turbulent transonic airfoil flow computations. The method employs an integral ("finite-volume") formulation for general curvilinear coordinates. It is based on MacCormack's well proven second-order accurate explicit two-step version of the Lax-Wendroff scheme, and approximates the full Navier-Stokes equations. Added to the predictor-corrector sequences are implicit integration steps which approximate the thin-layer formulation of the Navier-Stokes equations, and are only first-order accurate with respect to time. This is of no concern, of course, for steady state computations. A constant time step is, however, used in the entire computational domain so that unsteady phenomena are simulated properly, if they occur and the time scale is sufficiently large.

The scheme is characterized by the spatially unfactored solution of the implicit step by means of a relaxation scheme, e.g. a Gauß-Seidel iteration scheme. In order to allow a stable solution the coefficient matrices require diagonal dominance. This is provided by upwind differencing of the convective terms. The viscous terms are discretized with central approximations. To obtain sufficient accuracy second-order accurate approximations are needed in both directions. For the used C-meshes second-order accurate upwinding formula are employed for both the explicit and implicit terms in the C-direction. In the wall normal direction first-order accurate discretization is used for the implicit terms, while the explicit terms are based on second-order accurate, central approximations. Note that no artificial numerical diffusion is needed for the current research code. It is, however, conjectured that some TVD ("total variation diminishing") expressions have to be added in order to inhance stability, see in particular the isobar plots below. It is also noted that with this implicit procedure it seems possible to predict correct stagnation point pressures, lift and drag coefficients [3], if the mesh is properly chosen.

Since the employed code is, clearly, a research code, computation times are virtually meaningless at this time. The current ones can, certainly, be reduced dramatically if the interest is focussed on this topic. One reason is that the vectorization features of the used CRAY-1S are not fully exploited. Note that constant time steps are used in the entire computational domain in order to allow - in principle - for unsteady phenomena.

The present results have been obtained with fairly coarse meshes, see figures 1 to 3. Unfortunately only a mesh generator due to G.S. Deiwert and established more than 10 years ago is currently at the author's disposition. This mesh is wall normal in the dense-grid layer with prescribed thickness around the airfoil and not very smooth in the transition to the remaining mesh in particular near the stagnation point and at the trailing edge. The downstream and outer boundaries extend to roughly 50 chords (location of the control surfaces). In all cases the mesh dimensions are 126 x 32. It is felt that a smoother and finer grid is needed to improve the quality of the solution.

Generally speaking current experience shows that for low Reynolds number laminar flows the convergence is slowed down considerably. The relative error which determines the end of the computation has been reduced to 10^{-4} in order to see how the solution behaves beyond the point ($5 \cdot 10^{-4}$) corresponding to the accuracy due to the mesh (as experienced with high-Reynolds number flows). The relative error is checked with respect to bulk quantities (c_L and c_D) as well as with some L_2 norm for the pressure and shear stress coefficients on the airfoil. The distribution of the L_2 norm of the solution as required is also presented.

The results speak for themselves. The main features are summarized in table 1. For case A3 no results are obtained. It is interesting to note that it was difficult to generate a mesh which allowed for a solution for the case with the lowest Reynolds number. The first step next to the wall is proportional to the inverse of the square root of the Reynolds number. It is also worth pointing out that for the symmetric flow at Re = 500, case A5, the predicted flow is virtually symmetric (considering the possible accuracy given by the resolution of the mesh) but that the distribution of the heat transfer coefficient exhibits some asymmetry near the nose. Also, at Re = 10000, case A7, the flow is oscillating.

Figures 1 to 3 give an impression of the meshes used. In figure 1 an overview over the entire mesh is given, and in figures 2 and 3 a more detailed view of the grids near the airfoil. Figures 4 to 6 present the requested results for case A1 while figure 7 shows the lines of constant pressure and Mach number for the same flow. Case A1 turned out to be the toughest transonic case. Note that lines of constant pressure are shown since these seem to be more sensitive in terms of wiggles than lines of constant density. The latter always yield smoother plots. Figure 8 exhibits the pressure and skin friction coefficients at the wall c_p and c_f for case A2 (for the definition of the requested coefficients see the introduction to the workshop report). The corresponding pressure and Mach number contours are given in figure 9. The heat flux coefficients c_h for cases A2 and A5 are compared in figure 10 (note again the asymmetry near the nose for case A5 which is not understood). Fig. 11 displays the c_p- and c_f-distributions for case A5. There is no visible difference in the values for lower and upper surface. This is also reflected in the isobars and Machnumber contours in figure 12. For case A7 the computation was stopped arbitrarily because convergence could not be reached. Since the corresponding flow is oscillating only isobars and Mach number contours are presented at the end time displayed in table 1 for $\varepsilon = 10^{-4}$. The convergence in terms of the L_2-norms of the components of the solution vector is indicated in figures 14 and 15. Note that the oscillations for case A7 are also found in the lift and moment coefficients, while they are hardly noticed in the distribution of the drag coefficient.

REFERENCES

[1] MacCormack, R.W., "Current Status of Numerical Solutions of the Navier-Stokes Equations". AIAA Paper 85-32, 1985. AIAA Paper 85-1680, 1985.

[2] Kordulla, W., and MacCormack, R.W., "A New Predictor-Corrector Scheme for the Simulation of Three-Dimensional Flows with Separation". AIAA Paper 85-1502-CP, 1985.

[3] Kordulla, W., "Experiences with an Unfactored Implicit Predictor-Corrector Method", Proceedings, 6th GAMM-Conference on Numerical Methods on Numerical Methods in Fluid Mechanics, Notes on Numerical Fluid Mechanics, Vol. 13, pp. 185-192, 1986.

Table 1: Characteristics of the computations of the laminar flows part a NACA0012 airfoil on a C-mesh with 126 x 32 cells. Convergence is determined by the relative error of C_L and C_D and of some L_2 norm like quantity for c_p and c_f on the airfoil.

Case / Topic	A 1	A 2	A 5	A 7
Convergence $5\cdot10^{-4}$ / 10^{-4}	x x	x x	x x	-
$T=t\cdot u_\infty/c$	2.9 6.1	15.6 22.2	32.3 39.1	82.7
ΔT = const	.005	.05	.0517	.0517
# Iterations	580 1224	360 444	624 756	1600
C_L	0.5438 0.5380	0.4281 0.4261	-.0003761 -.0003933	-0.00921
C_D	0.7225 0.7047	0.2852 0.2845	0.2418 0.2420	0.07115
C_M (Nose)	-0.2481 -0.2377	-0.1632 -0.1624	0.0002228 0.0002287	-0.00278
L_2 - norms $\cdot10^2$				
ρ	1.2808 .4582	.0863 .0647	.0384 .0277	1.0000
ρu	2.7281 .9548	.1908 .1402	.0710 .05018	3.5159
ρw	1.1824 .3812	.0573 .0448	.0208 .0133	.7000
e	1.3517 .4694	.0928 .0699	.0407 .0296	.9978
Separation Location x/c	- -	.362 .362	- -	≅ 0.973

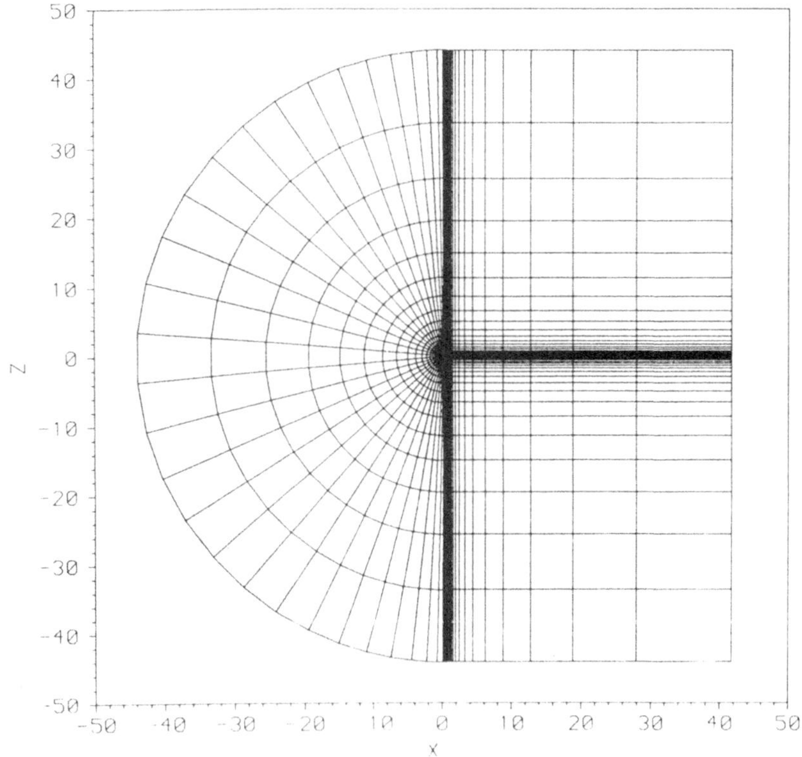

Figure 1: An overall view of a typical C-grid. All meshes have the same size 126 x 32. The nodes are the cell centers.

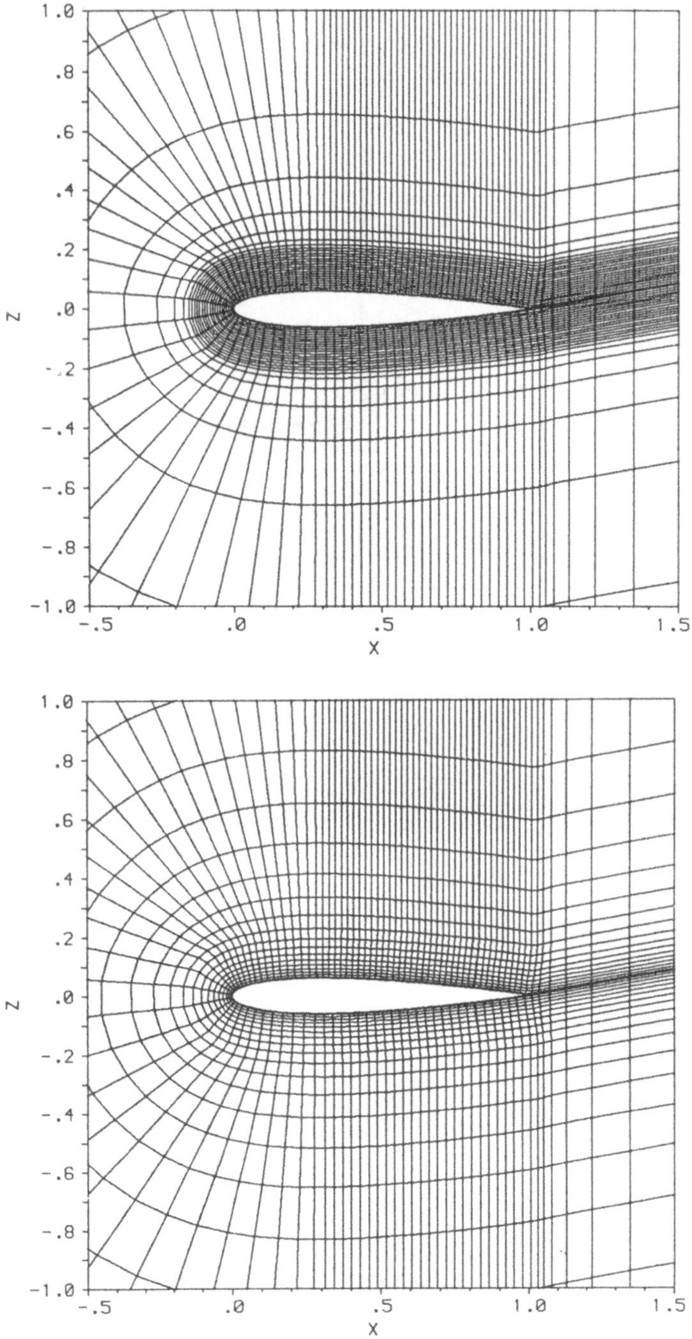

Figure 2: C-grids chosen for cases A1 (top) and A2 (bottom) near the airfoil. The nodes are the cell centers.

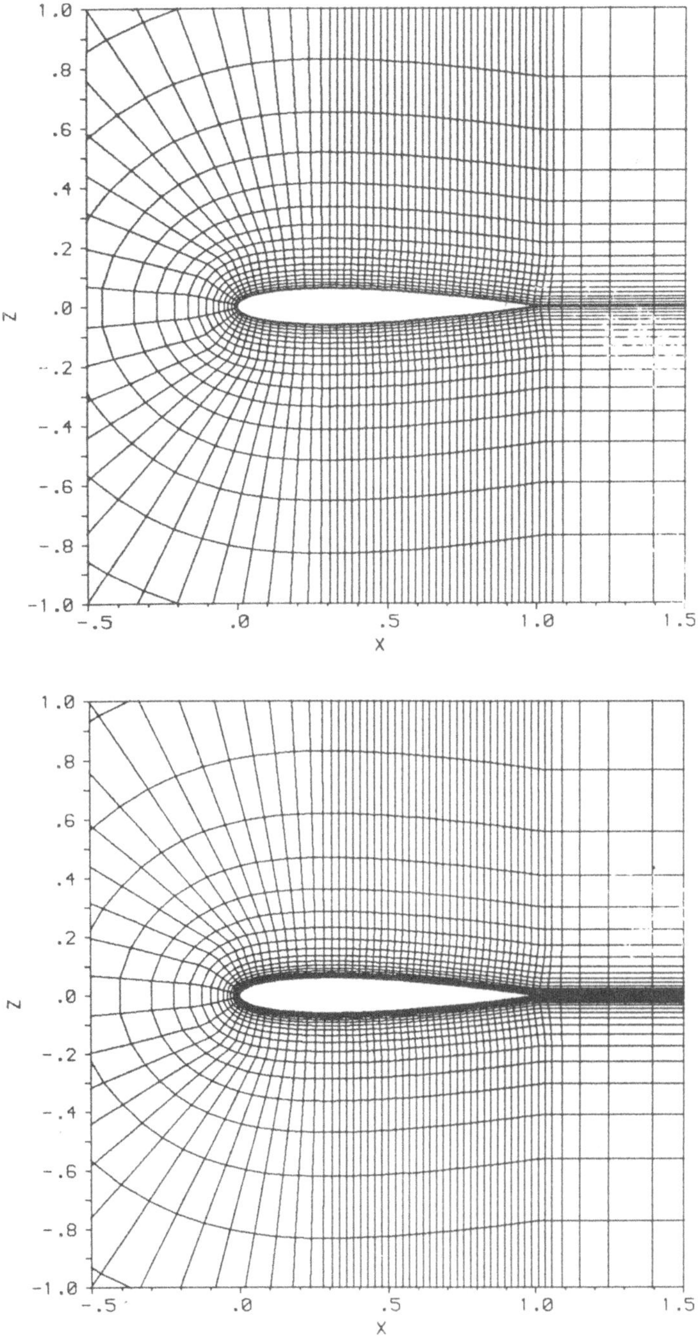

Figure 3: C-grids chosen for cases A5 (top) and A7 (bottom) near the airfoil. The nodes are the cell centers.

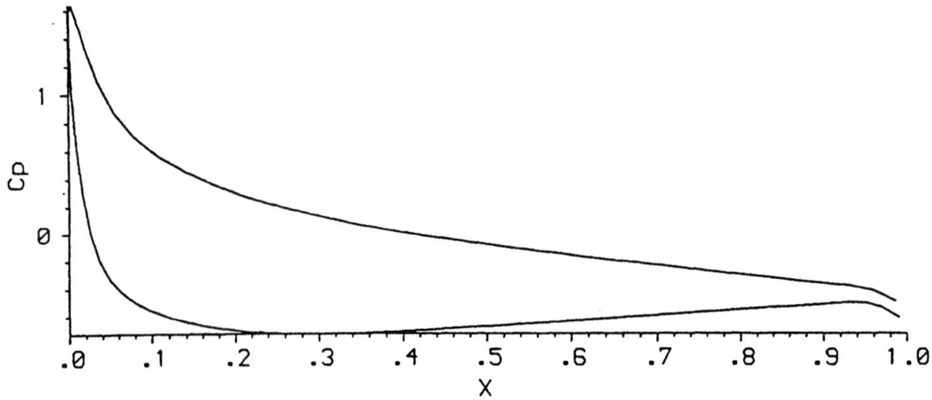

Figure 4: C_p-distribution for case A1.
$M_\infty = 0.80$, $\alpha = 10°$, $Re = 73$.

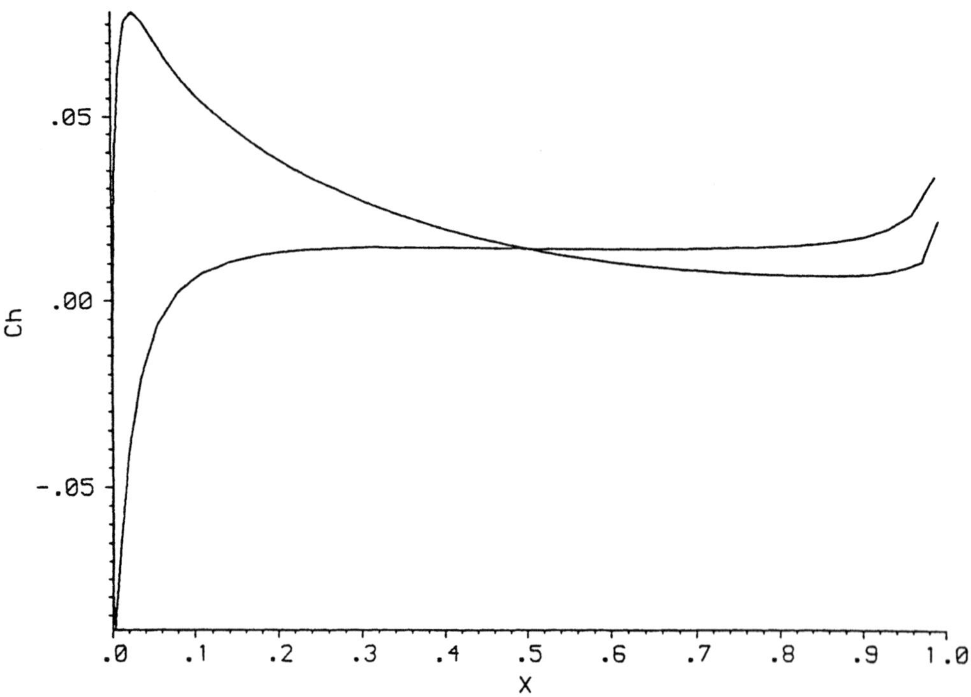

Figure 5: Distribution of the heat flux coefficient for case A1.
$M_\infty = 0.80$, $\alpha = 10°$, $Re = 73$.

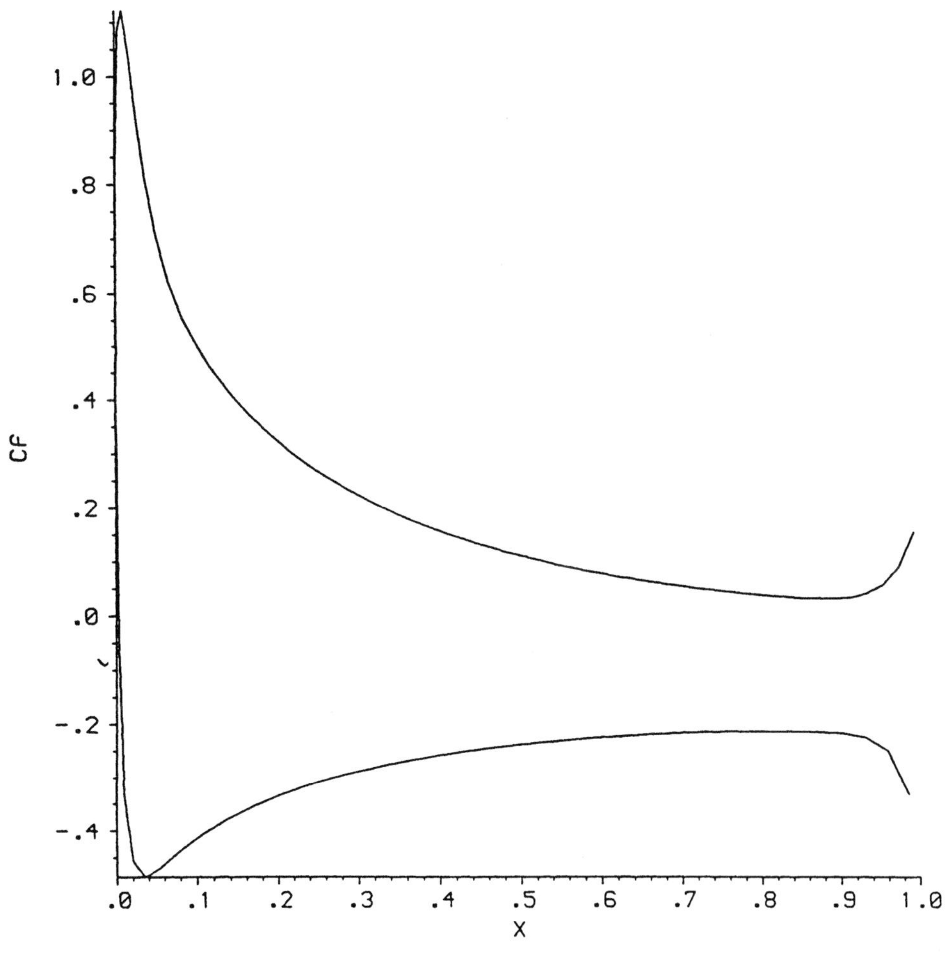

Figure 6: C_f-distribution for case A1.
$M_\infty = 0.80$, $\alpha = 10°$, $Re = 73$.

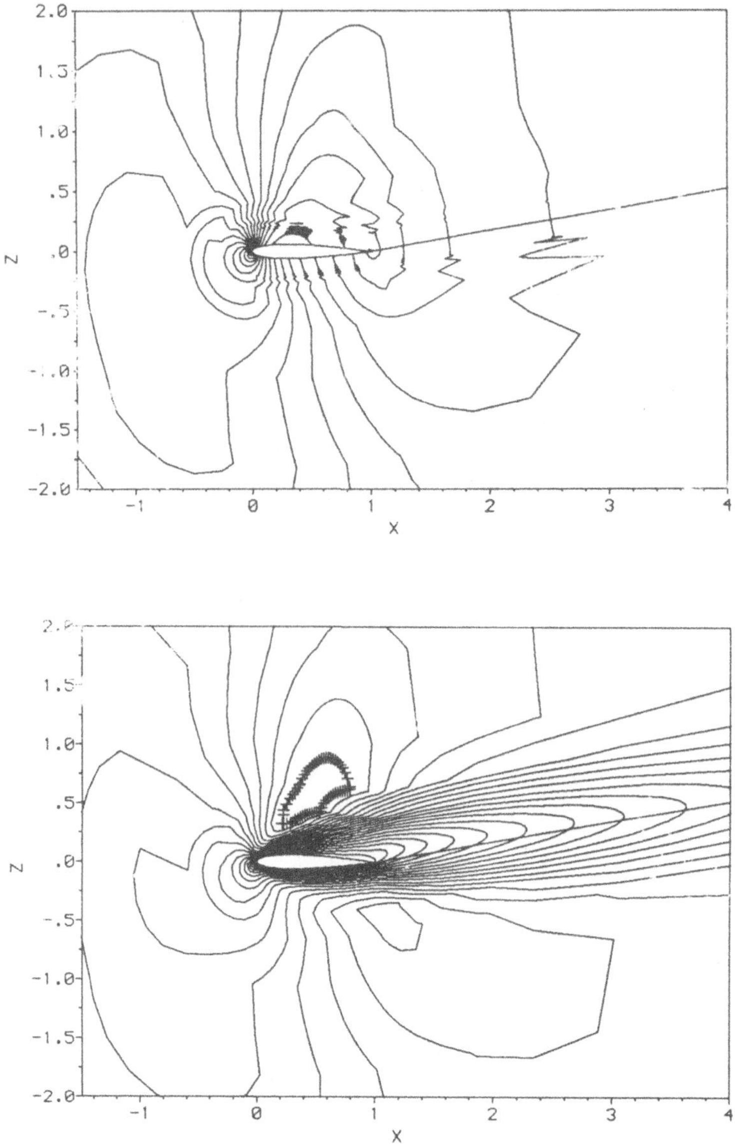

Figure 7: Pressure (top) and Mach number (bottom) contours for case A1.
$M_\infty = 0.80$, $\alpha = 10°$, Re = 73, +++ : sonic line.

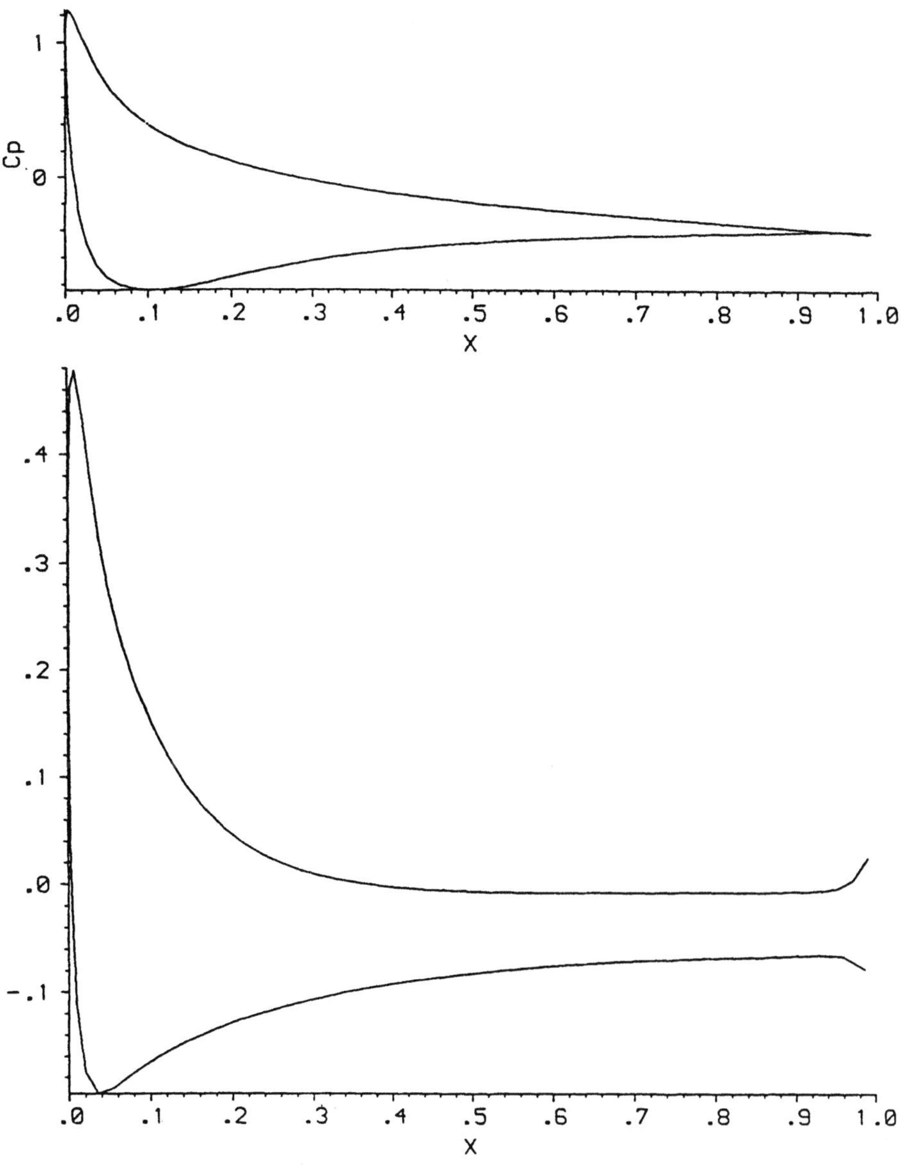

Figure 8: C_p- and c_f-distributions for case A2.
$M_\infty = 0.80$, $\alpha = 10°$, Re = 500.

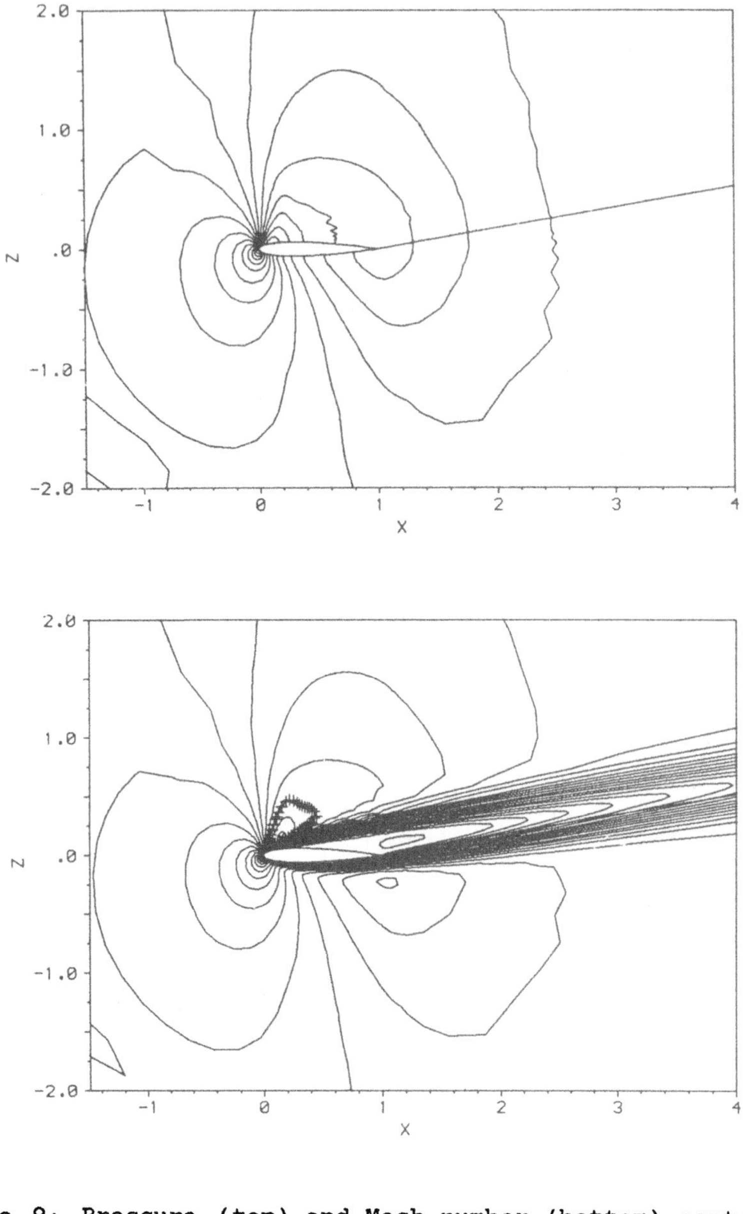

Figure 9: Pressure (top) and Mach number (bottom) contours for case A2.
$M_\infty = 0.80$, $\alpha = 10°$, Re = 500, +++ : sonic line.

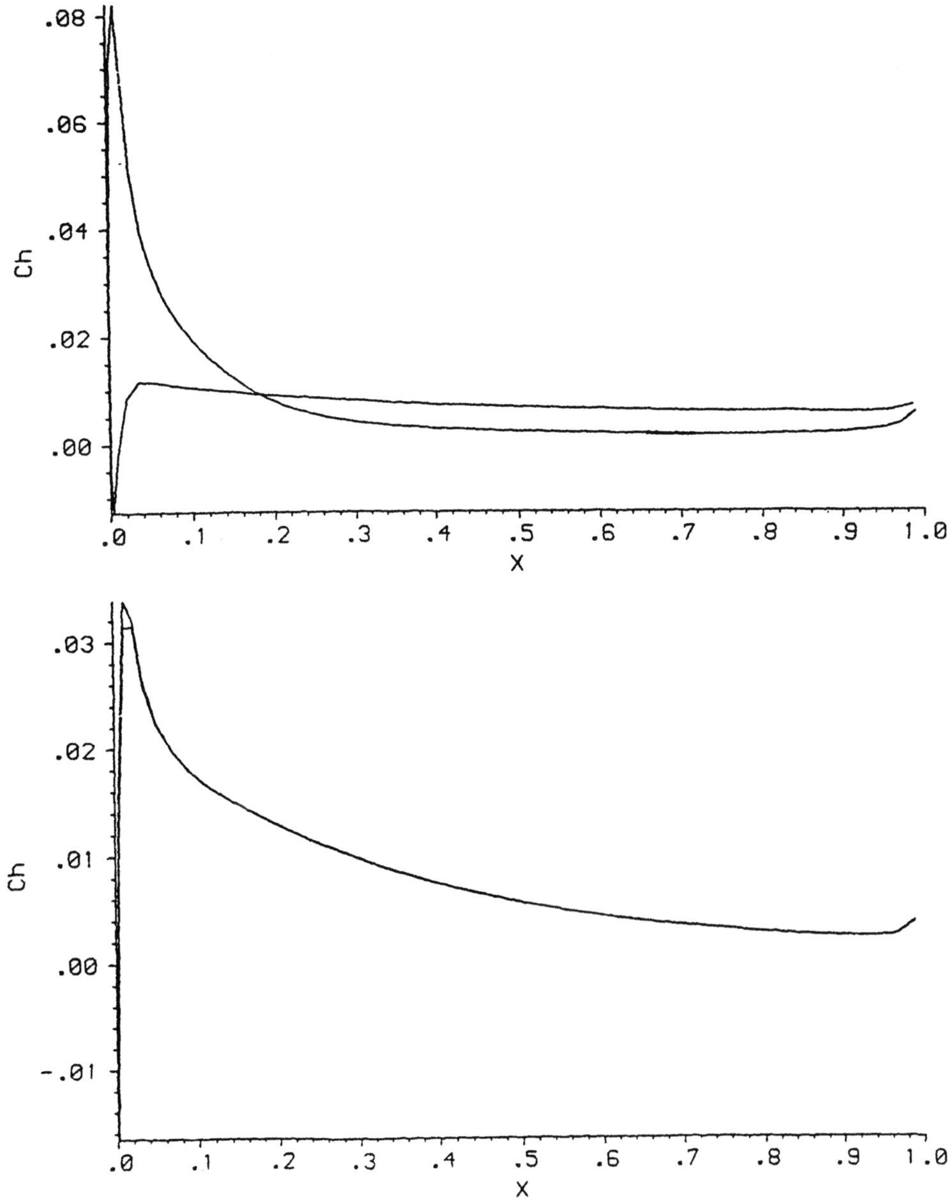

Figure 10: Distributions of the heat flux coefficient for cases A2 (top) and A5 (bottom).
M_∞ = 0.80 and 0.85, α = 10° and 0°, Re = 500.

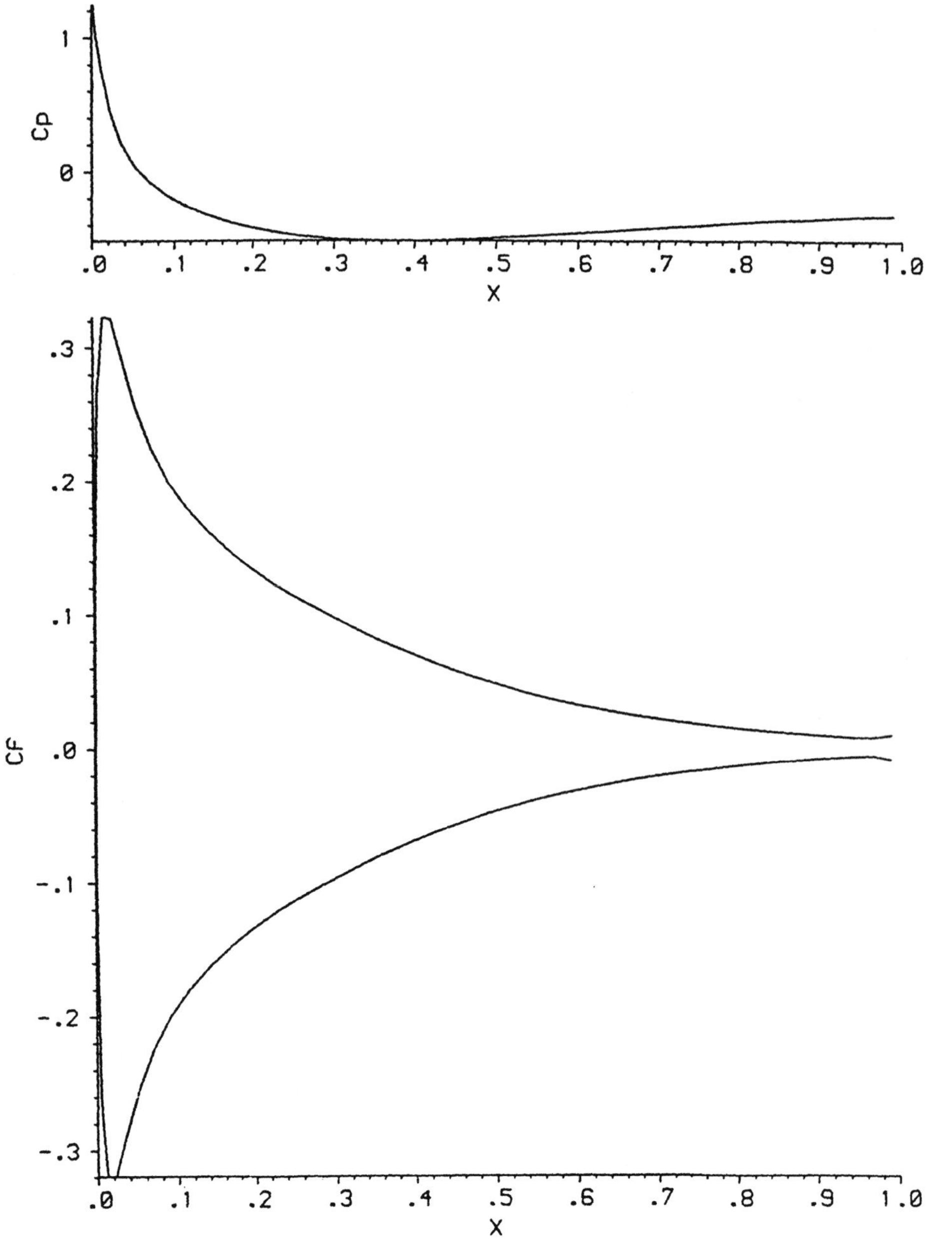

Figure 11: C_p- and c_f-distributions for case A5.
$M_\infty = 0.85$, $\alpha = 0°$, Re = 500.

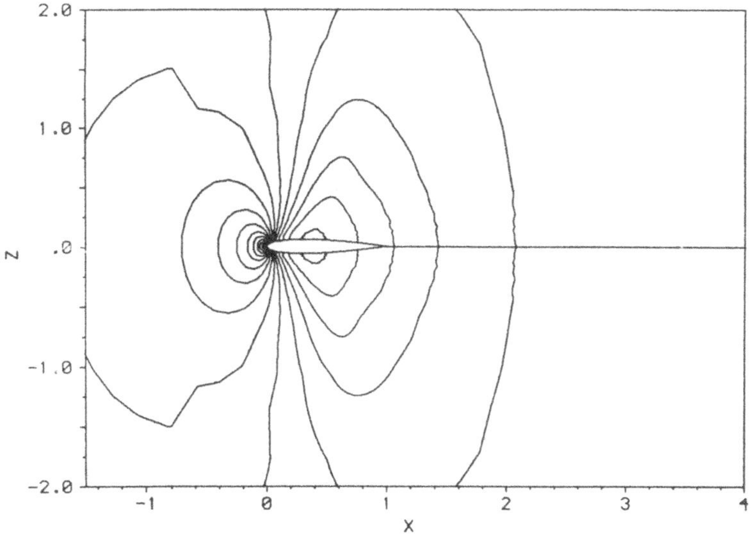

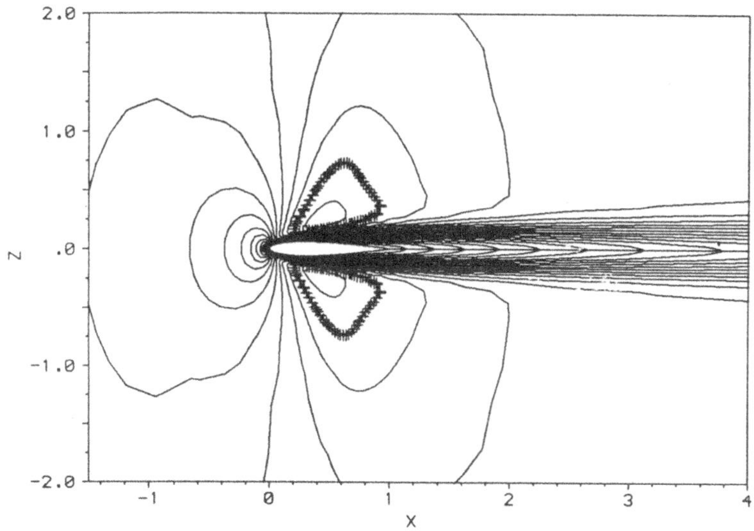

Figure 12: Pressure (top) and Mach number (bottom) contours for case A5.
M_∞ = 0.85, α = 0°, Re = 500, +++ : sonic line.

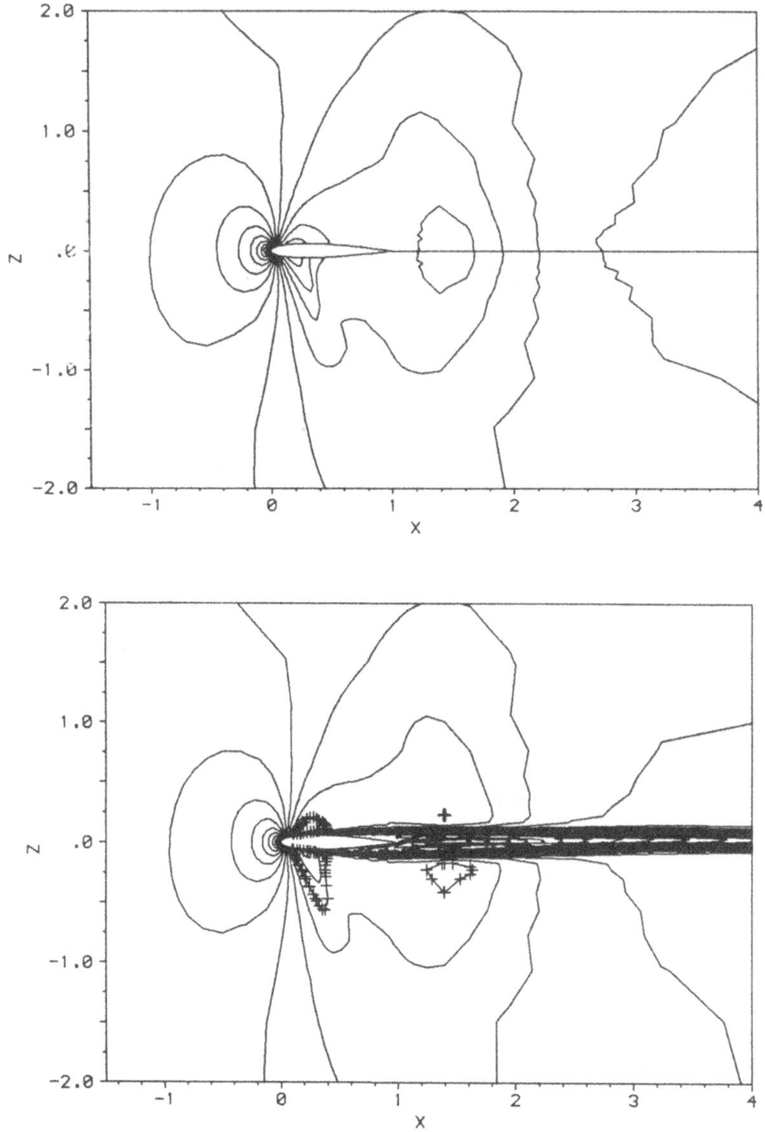

Figure 13: Pressure (top) and Mach number (bottom) contours for case A7 at a dimensionless time T = 82.7.
$M_\infty = 0.85$, $\alpha = 0°$, Re = 10000, +++ : sonic line.

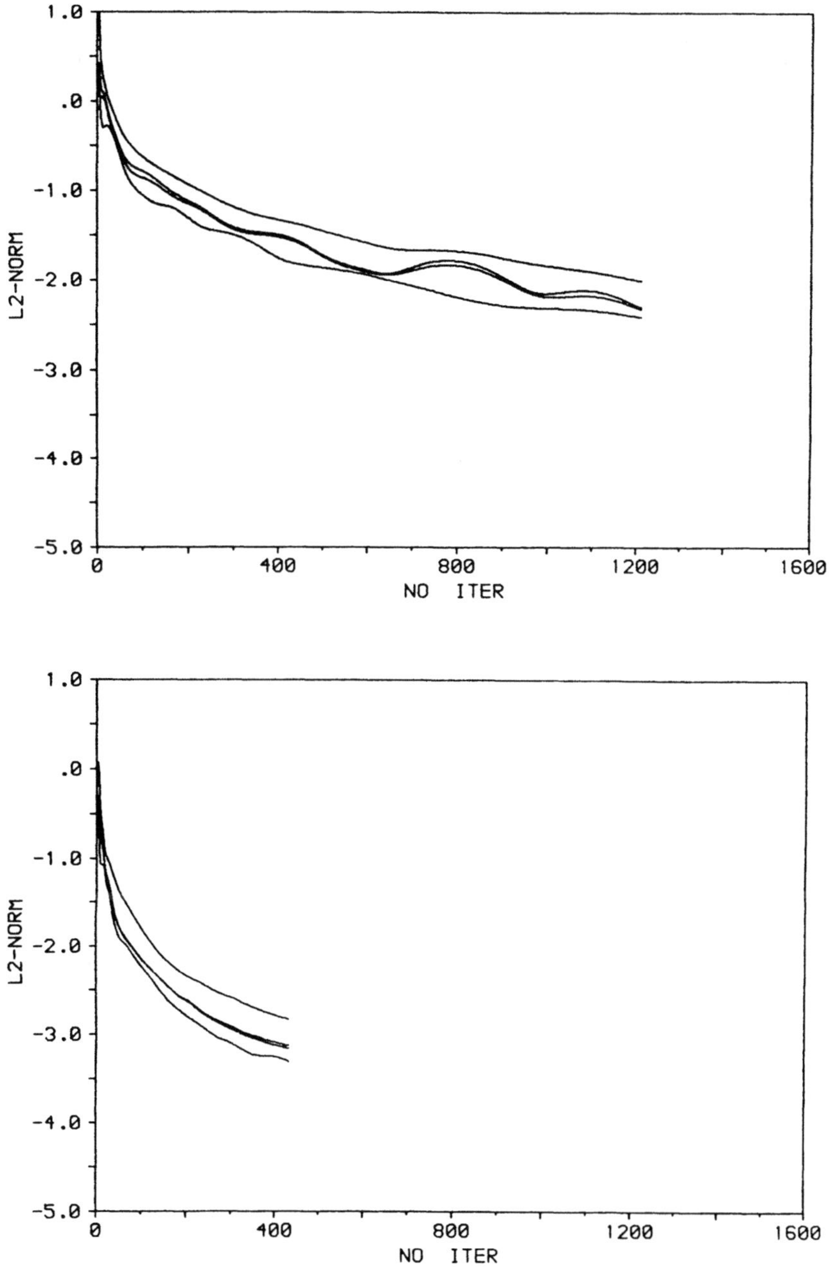

Figure 14: Convergence history in terms of the L_2 norm of the solution vector for cases A1 (top) and A2 (bottom).

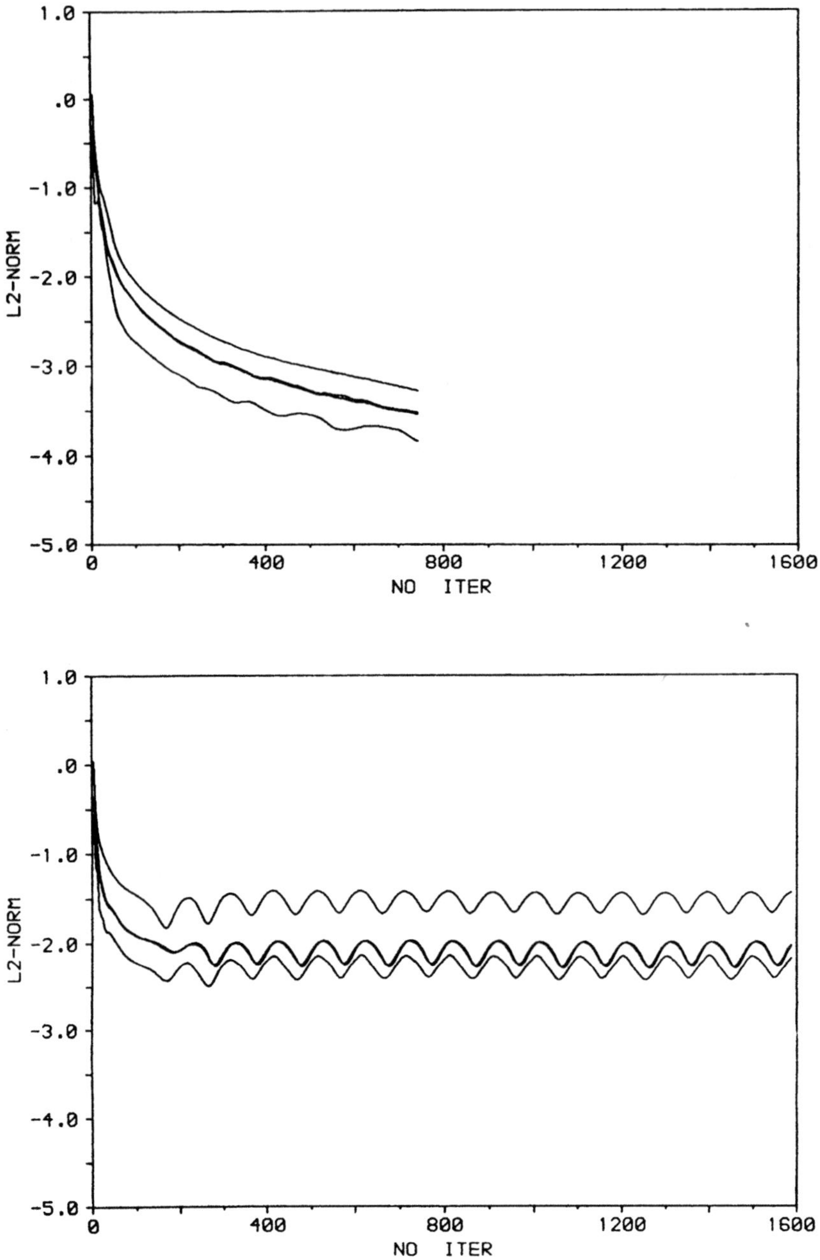

Figure 15: Convergence history in terms of the L_2 norm of the solution vector for cases A5 (top) and A7 (bottom).

IMPLICIT CENTRAL DIFFERENCE SIMULATION OF COMPRESSIBLE NAVIER-STOKES FLOW OVER A NACA0012 AIRFOIL

B. Müller, T. Berglind and A. Rizzi
FFA, The Aeronautical Research Institute of Sweden,
Box 11021, S-161 11 BROMMA, Sweden

SUMMARY

Transonic laminar flow at low Reynolds numbers over a NACA0012 airfoil is numerically simulated by solving the two-dimensional compressible Navier-Stokes equations by means of the implicit central difference scheme of Beam and Warming.

GOVERNING EQUATIONS

The independent variables of the two-dimensional compressible Navier-Stokes equations in Cartesian coordinates are transformed such that the region of interest is mapped on a rectangle (Fig. 1) and the conservation-law form of the equations is maintained [1].

The investigated fluid is a perfect gas with constant specific heats their ratio being $\gamma=1.4$. The viscosity coefficients are related by Stokes's hypothesis, and the Prandtl number is constant with a value of $Pr=0.72$. According to the specifications of the GAMM-Workshop on Numerical Simulation of Compressible Navier-Stokes Flows, the viscosity coefficients and the thermal conductivity coefficient are assumed to be constant.

ALGORITHM

The transformed Navier-Stokes equations are solved by means of the implicit factorized Beam and Warming scheme [2]. The time derivative is approximated by the first-order Euler implicit formula, except for test case A7 where the second-order three-point-backward formula is used, because unsteady flow has been suspected. The spatial derivatives are second-order central differenced. The metric terms are also second-order central-differenced, except for the airfoil contour, farfield and outflow boundaries where second-order three point formulae are used. The linearization of the viscous flux vectors follows the approach of Steger [3], except for the mixed derivatives which are treated explicitly to allow for the factorization. For efficiency reasons, a one-leg method is used instead of the corresponding linear multistep method [2], [4]. Fourth- order numerical damping terms are added explicitly and second-order smoothing terms implicitly for convergence and stability reasons [5]. The fourth-order terms are modified near boundaries [6] to ensure their dissipative property also there [7]. Constant time stepping is employed with the time step linearly increased to its final value during the first 100 time levels [4].

Oscillations near shocks can be reduced by replacing the fourth-order numerical damping terms by a combination of second-order differences sensed by the discretized second derivative of the pressure and fourth-order differences switched off near shocks [8], [6]. This modified smoothing operator is employed for test case A3 only together with local time stepping to enhance convergence [9].

MESH

A C-type mesh with a wake-cut extending from the trailing edge (=TE) of the NACA0012 airfoil along the x-axis is generated by transfinite interpolation [10], [11] (Figs. 2 and 3). The outer boundary consists of an ellipse to the left of TE and straight lines to the right chosen 5 chord lengths off the airfoil. For test cases A5 and A6, the straight lines in the farfield are prescribed at an angle of 5 deg. with respect to the x-axis to ensure inflow along that boundary for zero angle of attack. The grid points are clustered near the leading edge (=LE) and to a less extent also near TE with less resolution in the mid-section. On the wake-cut and on the straight lines in the farfield, the points are stretched towards the outflow boundary with a slight clustering near the outflow boundary itself. On the ellipse, the grid points are distributed at equal arclength increments. The mesh is orthogonal at and clustered near airfoil and wake-cut. For $Re=10^4$, i.e. test case A7, the distribution in the wall-normal direction is refined resulting in a distance of ~ 0.00025 of the first point in the boundary layer from LE. The number of points on the airfoil is 129 for the 193×65 mesh.

INITIAL CONDITIONS

The flowfield is initialized by the freestream values, except for the airfoil, on which the no-slip condition and the freestream total temperature are imposed from the very beginning and the pressure is set equal to its freestream value.

For the 193×65 mesh, the flowfield is initialized by averaging the result obtained with 97×33 grid points.

BOUNDARY CONDITIONS

On the airfoil, the required physical boundary conditions, i.e. no-slip and freestream total temperature, are imposed. The wall pressure is calculated from the wall-normal momentum equation to second-order accuracy, except for the viscous terms which are evaluated by central differencing at the first grid line in the boundary layer [12].

For subsonic inflow or outflow at the farfield boundary, the appropriate locally one-dimensional Riemann invariants, tangential velocity component and entropy are either extrapolated by employing the second-order three-point-backward formula or fixed to their freestream values [9]. For subsonic outflow at the outflow boundary, the density and mass flux are extrapolated linearly and the pressure is fixed to its freestream value.

For supersonic inflow, the flow variables are kept constant to their freestream values. They are linearly extrapolated at a supersonic outflow.

On the wake-cut, the conservative variables are obtained by averaging the values at the first adjacent grid points above and below.

PROGRAMMING

The basic algorithm requires storage for 15 two-dimensional arrays of IMAX×JMAX words, i.e. for 4 conservative variables, 4 delta-variables and 7 metric terms including 2 time-dependent ones needed for moving grids. In addition 3×IMAX 4×4 matrices and IMAX 4-component vectors have to be stored temporarily for the solution of the resulting block-tridiagonal linear systems.

The code was run on the scalar computer VAX 11 of FFA taking 0.007 seconds of CPU-time per grid point and per time step. The program is vectorized for the vector computer CRAY-1S [13]. From previous experience with the code for blunt body calculations, the CPU-time on the CRAY-1S will be lower by a factor of about 60.

RESULTS

For test cases A1 to A6, steady results were obtained with the L_2-norm of the discrete time deriatives of the non-dimensional conservative variables lower than 2.5×10^{-3}. Test case A7 gave unsteady results with an asymmetric wake structure similar to a Karman vortex street.

Mach number contours and pressure, skin friction and heat flux coefficients are presented for the mandatory test cases A1, A2, A3 and A5 in Figs. 4 to 7, 8 to 11, 12 to 15, and 16 to 19, respectively. For A2, the flow separates at x/c= 0.371 from the upper surface and reattaches at x/c= 0.973. Integral values and CPU-times are given in Table 1 for all results (at $t \cdot |\vec{u}_\infty|/c$= 19.515 for A7).

More detailed results including the optional test cases may be found in [14].

REFERENCES

[1] Peyret, R., and Viviand, H.: "Computation of Viscous Compressible Flows Based on the Navier-Stokes Equations", AGARD-AG-212 (1975).

[2] Beam, R.M., and Warming, R.F.: "An Implicit Factored Scheme for the Compressible Navier-Stokes Equations II. The Numerical ODE Connection", AIAA Paper 79-1446 (1979).

[3] Steger, J.L.: "Implicit Finite-Difference Simulation of Flow About Arbitrary Two-Dimensional Geometrics", AIAA J. 16 (1978), pp. 679-686.

[4] Müller, B.: "Navier-Stokes Solution for Hypersonic Flow over an Indented Nosetip", AIAA Paper 85-1504, Proceedings of AIAA 7th CFD Conference, Cincinnati, Ohio (1985) pp. 141-149.

[5] Pulliam, T.H., and Steger, J.L.: "Implicit Finite-Difference Simulations of Three-Dimensional Compressible Flow", AIAA J. 18 (1980), pp. 159-167.

[6] Pulliam, T.H.; "Artificial Dissipation Models for the Euler Equations", AIAA Paper 85-0438 (1985).

[7] Eriksson, L.-E., and Rizzi, A.: "Computer-Aided Analysis of the Convergence to Steady State of Discrete Approximations to the Euler Equations", J. Computational Physics 57 (1985), pp. 90-128.

[8] Jameson, A., Schmidt, W., and Turkel, E.: "Numerical Solutions of the Euler Equations by Finite Volume Methods Using Runge-Kutta Fime-Stepping Schemes", AIAA Paper 81-1259, AIAA 14th Fluid and Plasma Dynamics Conference, Palo Alto (1981).

[9] Pulliam, T.H., and Steger, J.L.: "Recent Improvements in Efficiency, Accuracy, and Convergence for Implicit Approximate Factorization Algorithms", AIAA Paper 85-0360 (1985).

[10] Eriksson, L.-E.: "Generation of Boundary-Conforming Grids Around Wing-Body Configurations Using Transfinite Interpolation", AIAA J. 20 (1982), pp. 1313-1320.

[11] Berglind, T.: "Grid Generation around Car Configurations Above a Flat Ground Plane Using Transfinite Interpolation", FFA Report 139, 1985.

[12] Müller, B.: "Berechnung abgelöster laminarer Uberschallströmungen um nichtangestellte stumpfe Rotationskörper", DFVLR-FB 85-30 (1985).

[13] Müller, B.: "Vectorization of the Implicit Beam and Warming Scheme", in: Gentzsch, W., "Vectorization of Computer Programs with Applications to Computational Fluid Dynamics", Vieweg: Braunschweig, 1984, pp. 172-194.

[14] Müller, B.: "Navier-Stokes Solution for Laminar Transonic Flow over a NACA0012 Airfoil", FFA Report 140, 1986. (To appear).

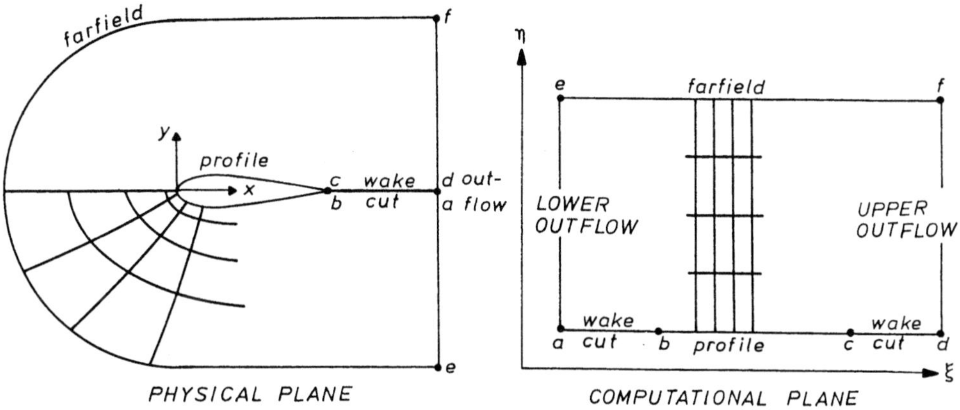

Fig. 1 Transformation of physical to computational plane for a C-type mesh.

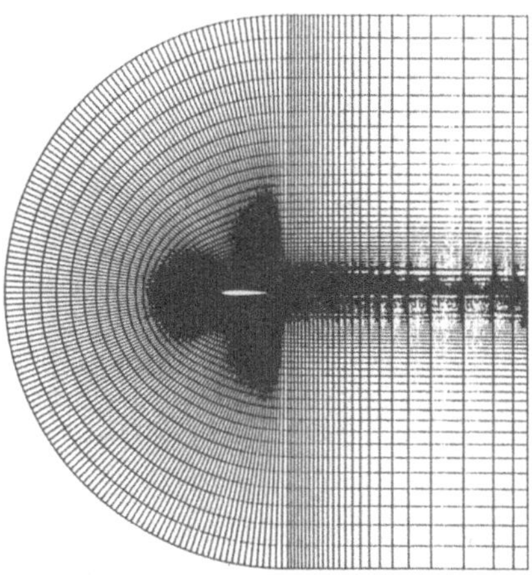

Fig. 2

193×65 mesh for test cases A1 to A4, NACA0012, chord = 0.735 cm.

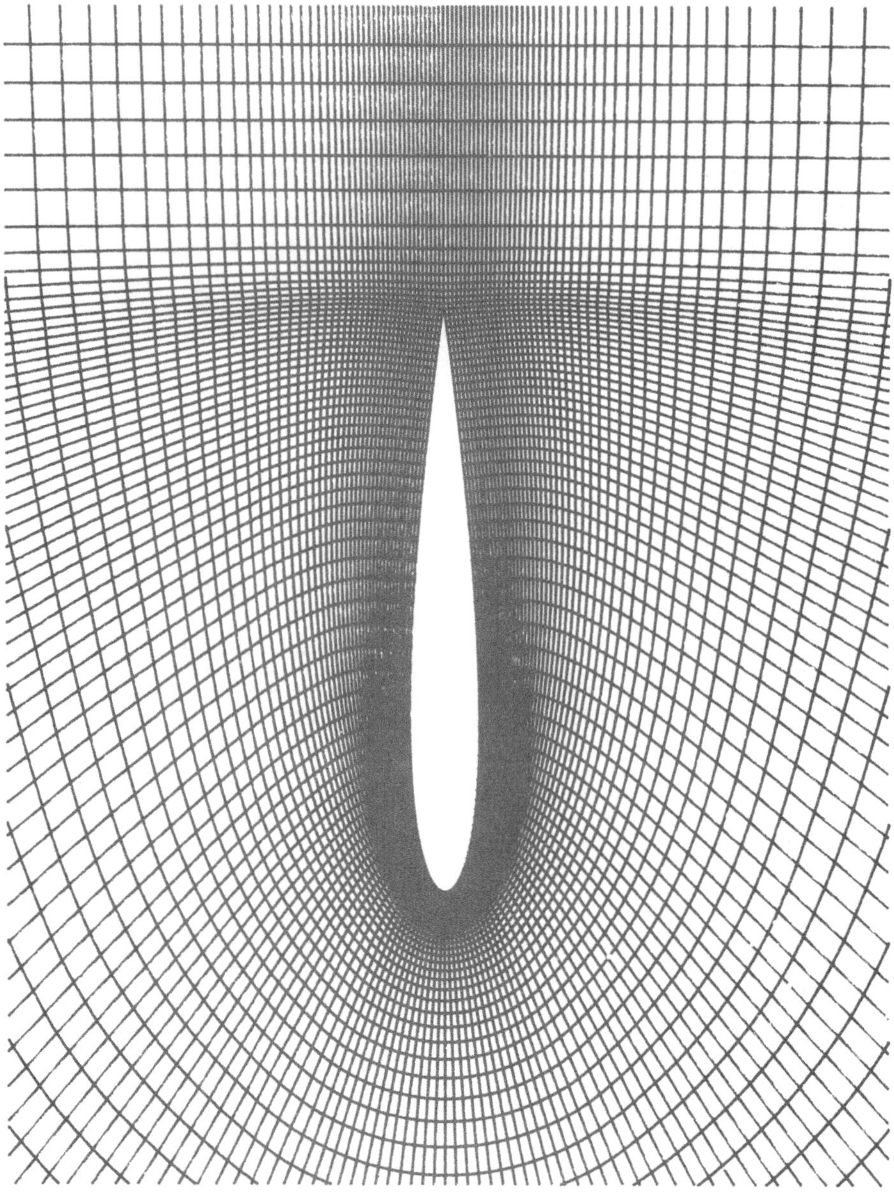

Fig. 3 Enlargement of Fig. 2, chord = 10 cm.

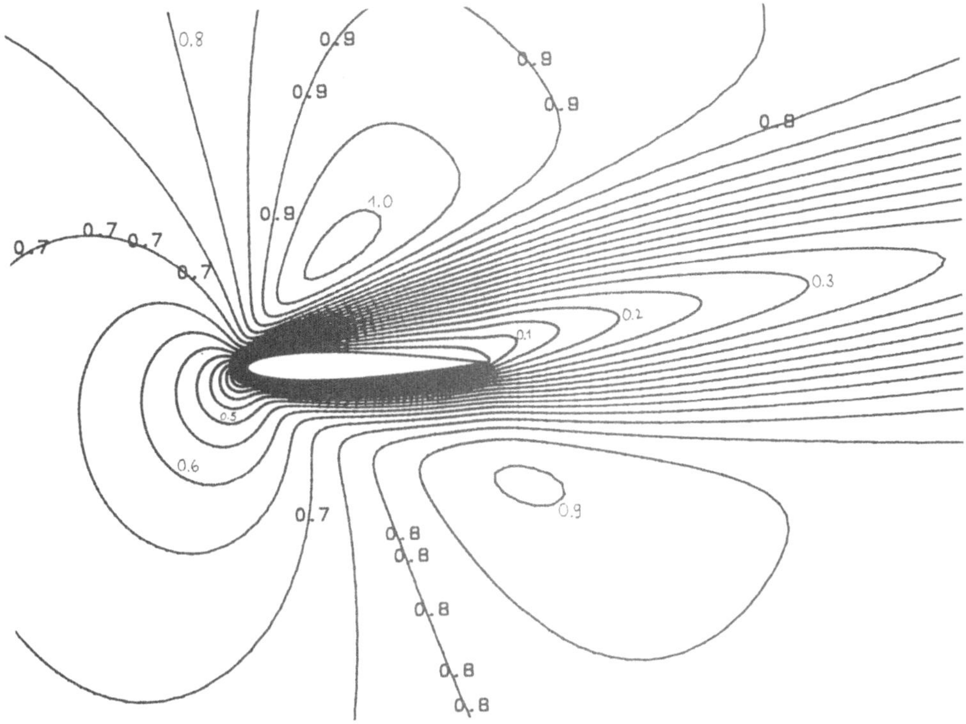

Fig. 4 Mach number contours for test case A1
$\Delta M_\infty = 0.05$, chord = 4 cm.

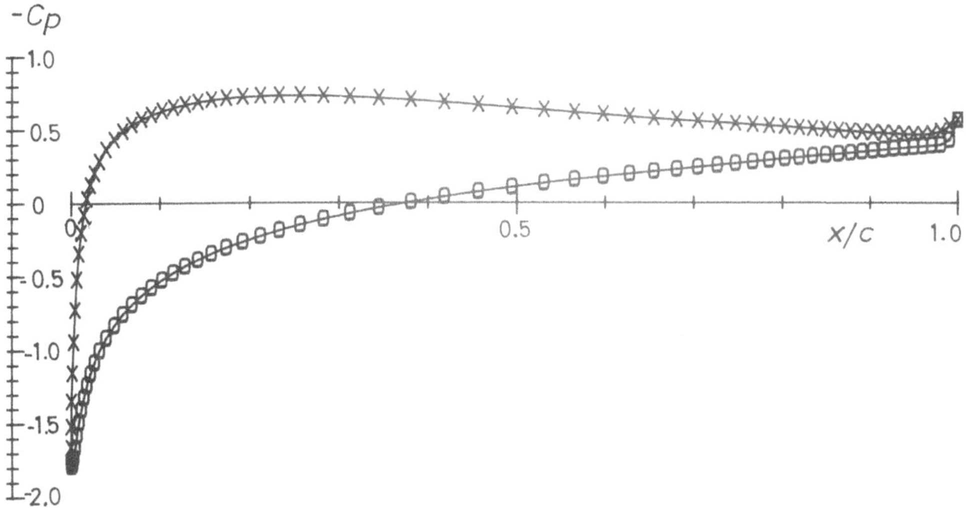

Fig. 5 Pressure coefficient for test case A1
NACA0012, $M_\infty = 0.8$, $\alpha = 10^\circ$, $Re_\infty = 73$, 193×65 mesh
X= upper surface, 0= lower surface

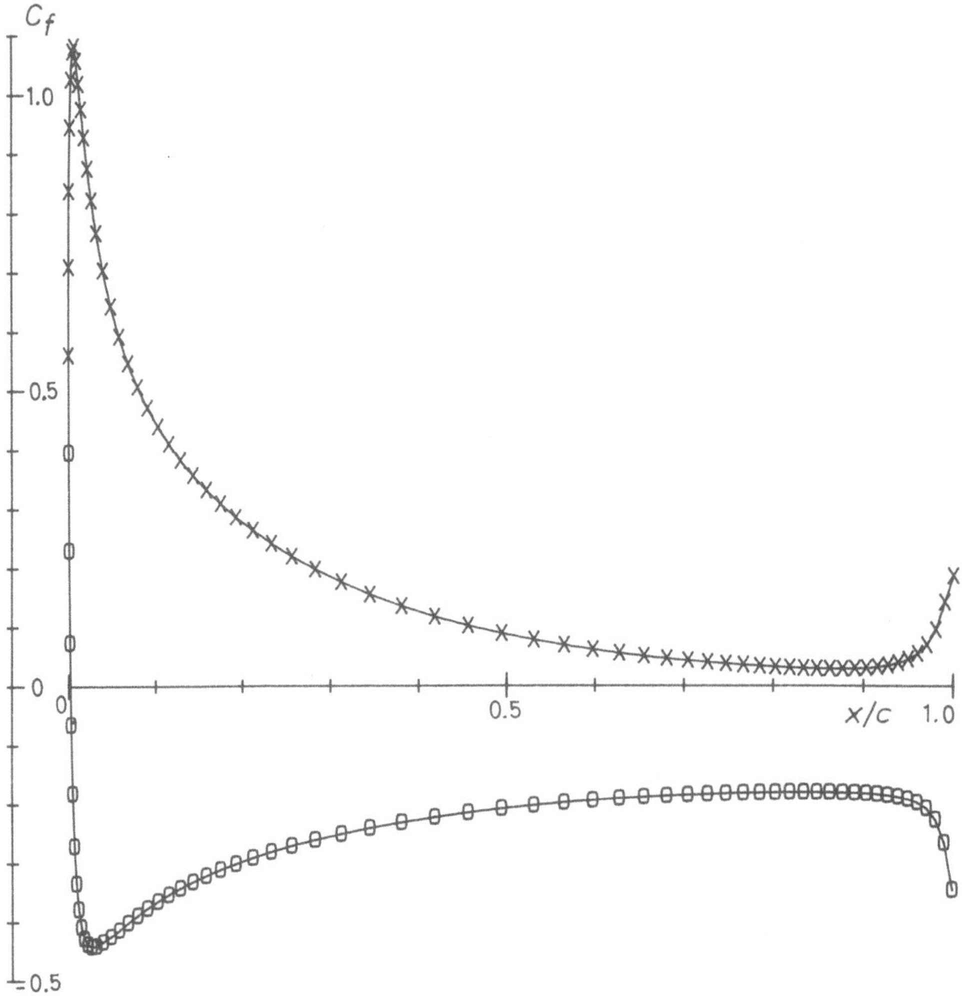

Fig. 6 Skin friction coefficient for case A1,
NACA 0012, M_∞=0.8, α=10°, Re_∞=73, 193×65 mesh
X= upper surface, 0= lower surface

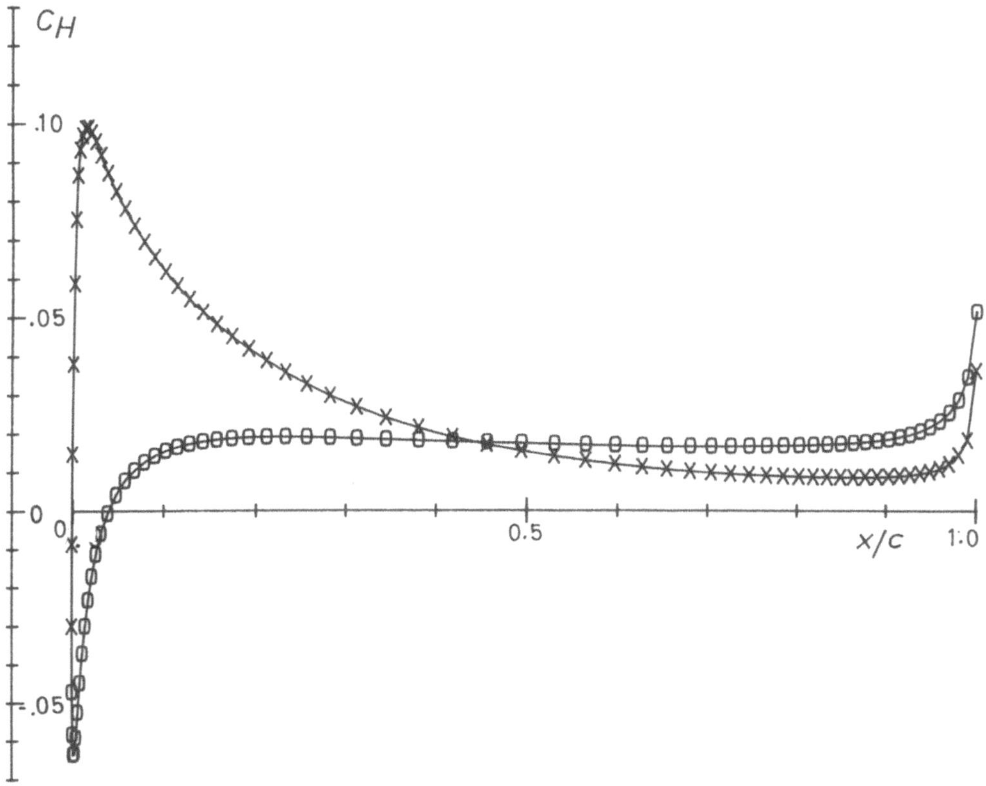

Fig. 7 Heat flux coefficient for test case A1
NACA0012, M_∞=0.8, α=10°, Re_∞=73, 193×65 mesh
X= upper surface, 0= lower surface

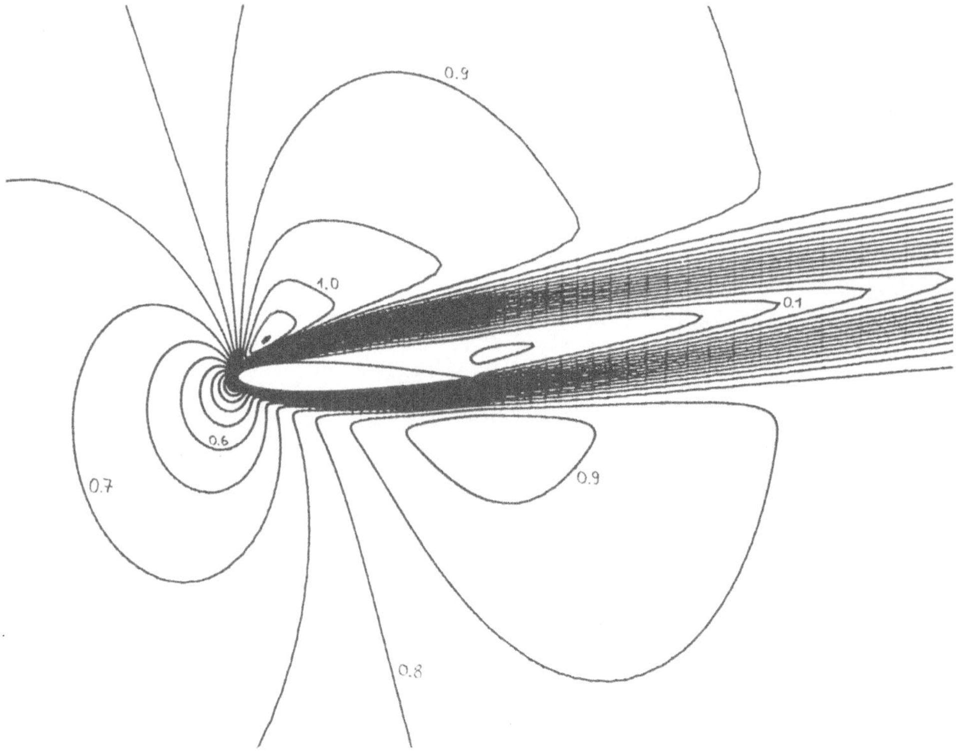

Fig. 8 Mach number contours for test case A2
$\Delta M = 0.05$, chord = 4 cm

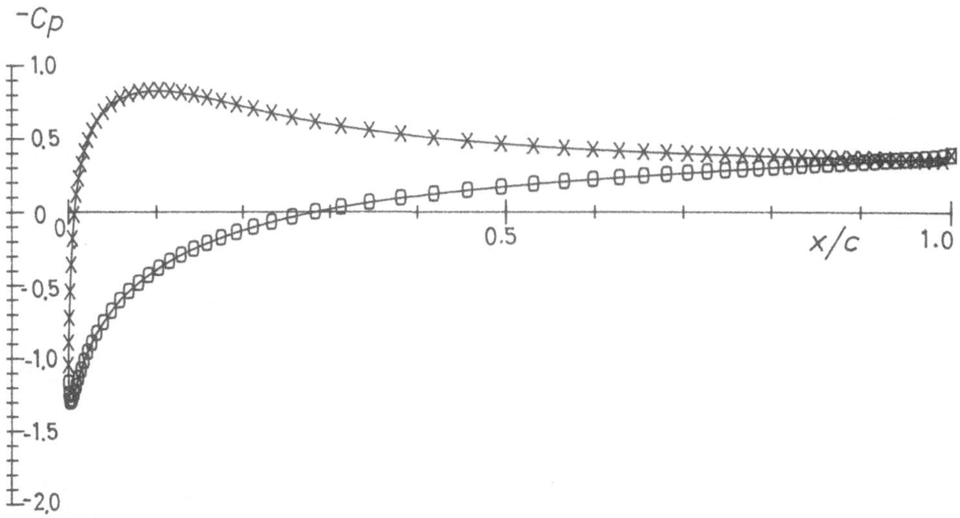

Fig. 9 Pressure coefficent for test case A2
NACA0012, $M_\infty=0.8$, $\alpha=10^\circ$, $Re_\infty=500$, 193×65 mesh
X= upper surface, 0= lower surface

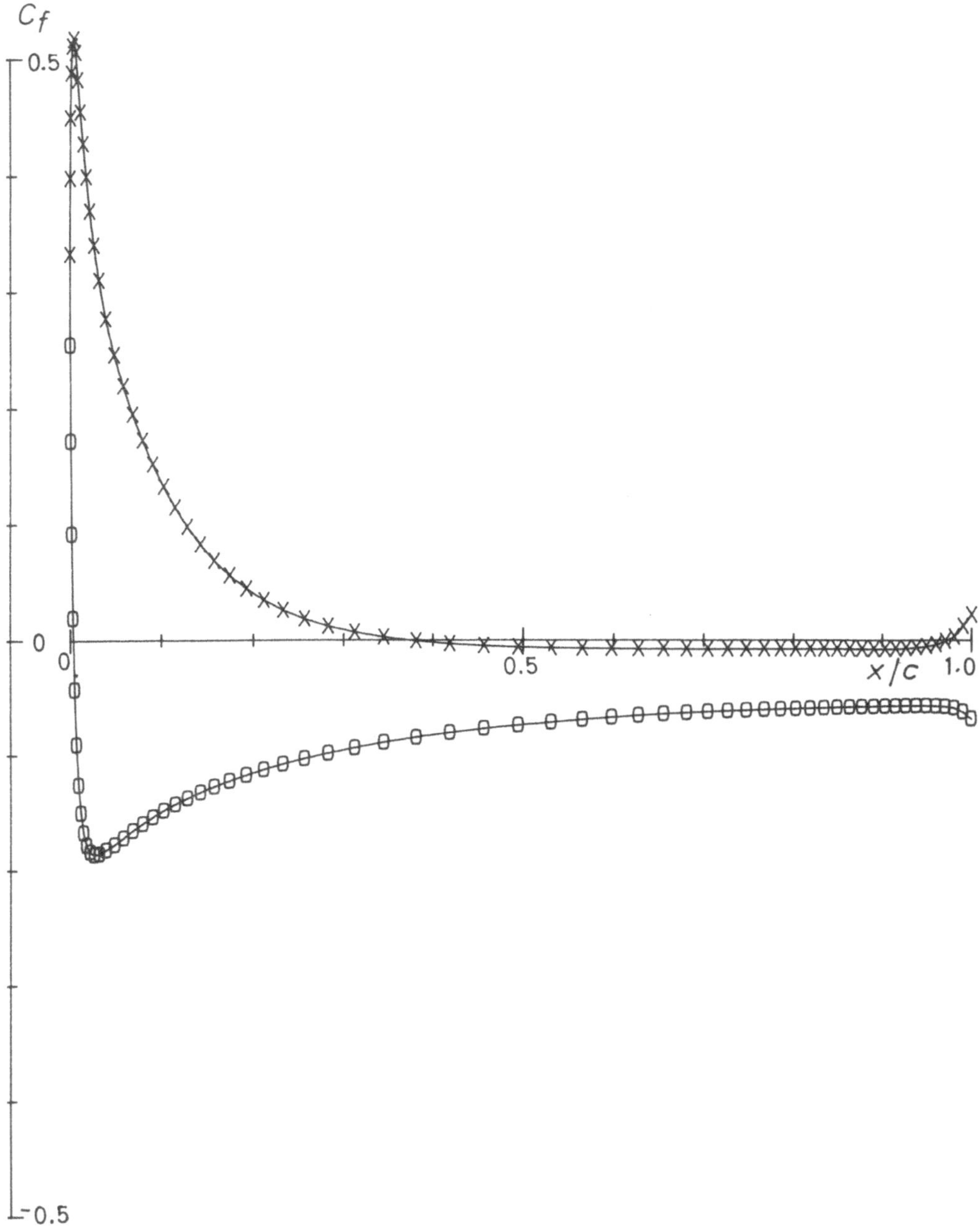

Fig. 10 Skin friction coefficient for case A2
NACA0012, M_∞=0.8, α=10°, Re_∞=500, 193×65 mesh
X= upper surface, 0= lower surface

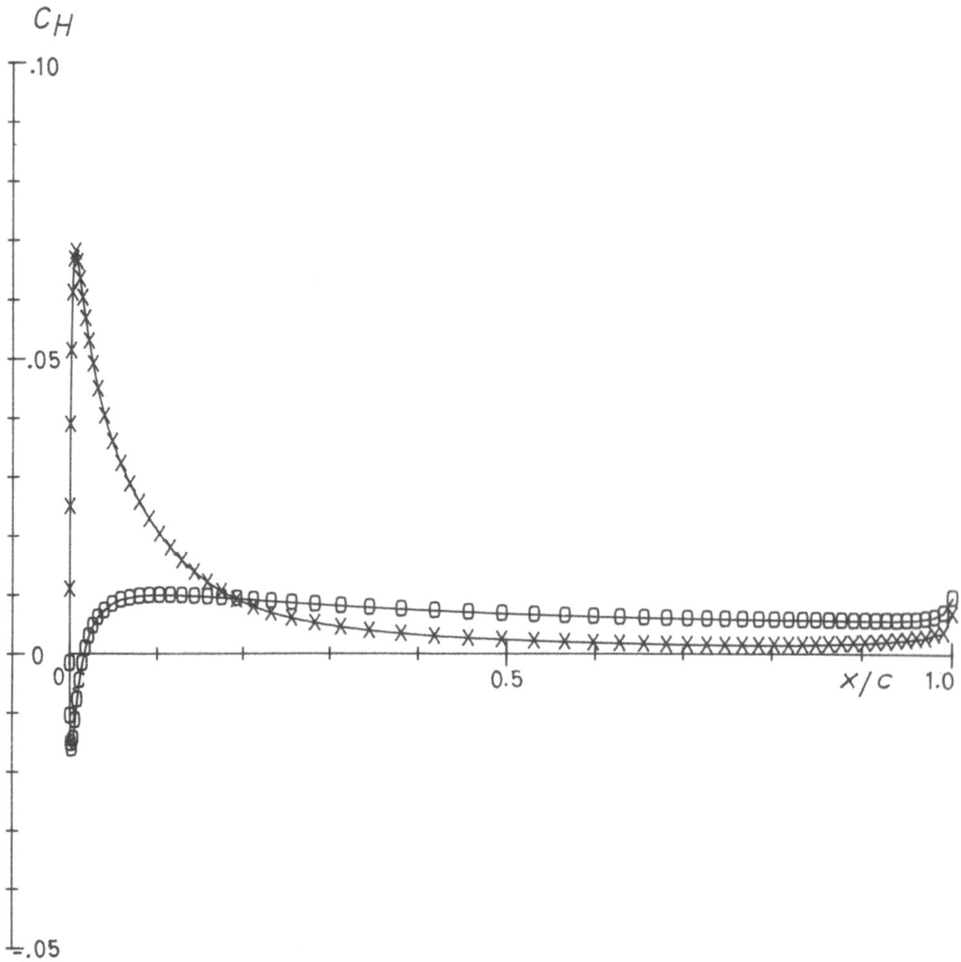

Fig. 11 Heat flux coefficient for test case A2
NACA0012. $M_\infty=0.8$, $\alpha= 10^o$, $Re_\infty=500$, 193×65 mesh
X = upper surface, 0 = lower surface

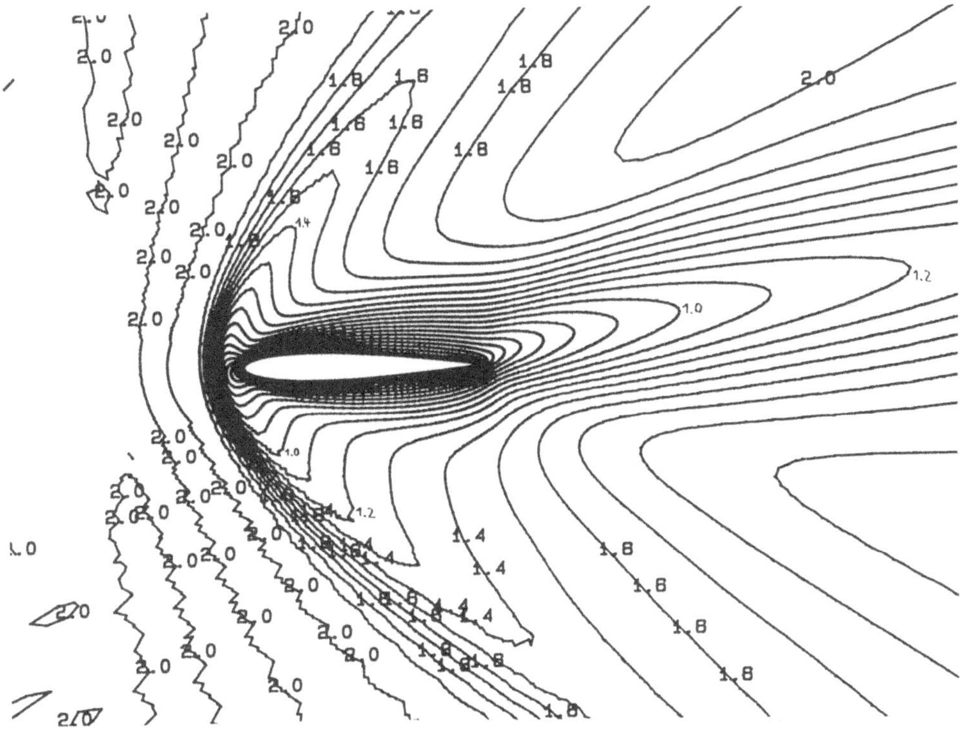

Fig. 12 Mach number contours for test case A3
$\Delta M = 0.1$, chord = 4 cm.

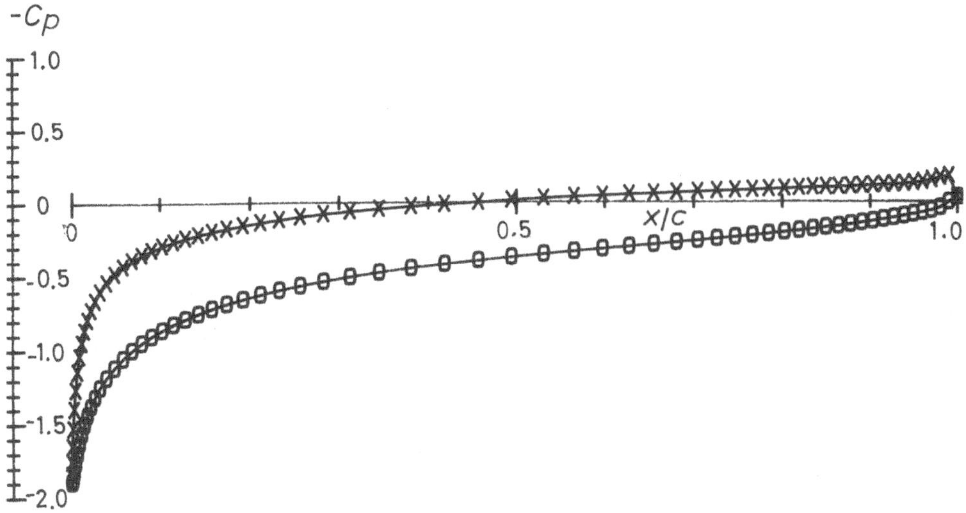

Fig. 13 Pressure coefficient for test case A3
NACA0012, $M_\infty=2$, $\alpha = 10^\circ$, $Re_\infty=106$, 193×65 mesh
X = upper surface, 0 = lower surface

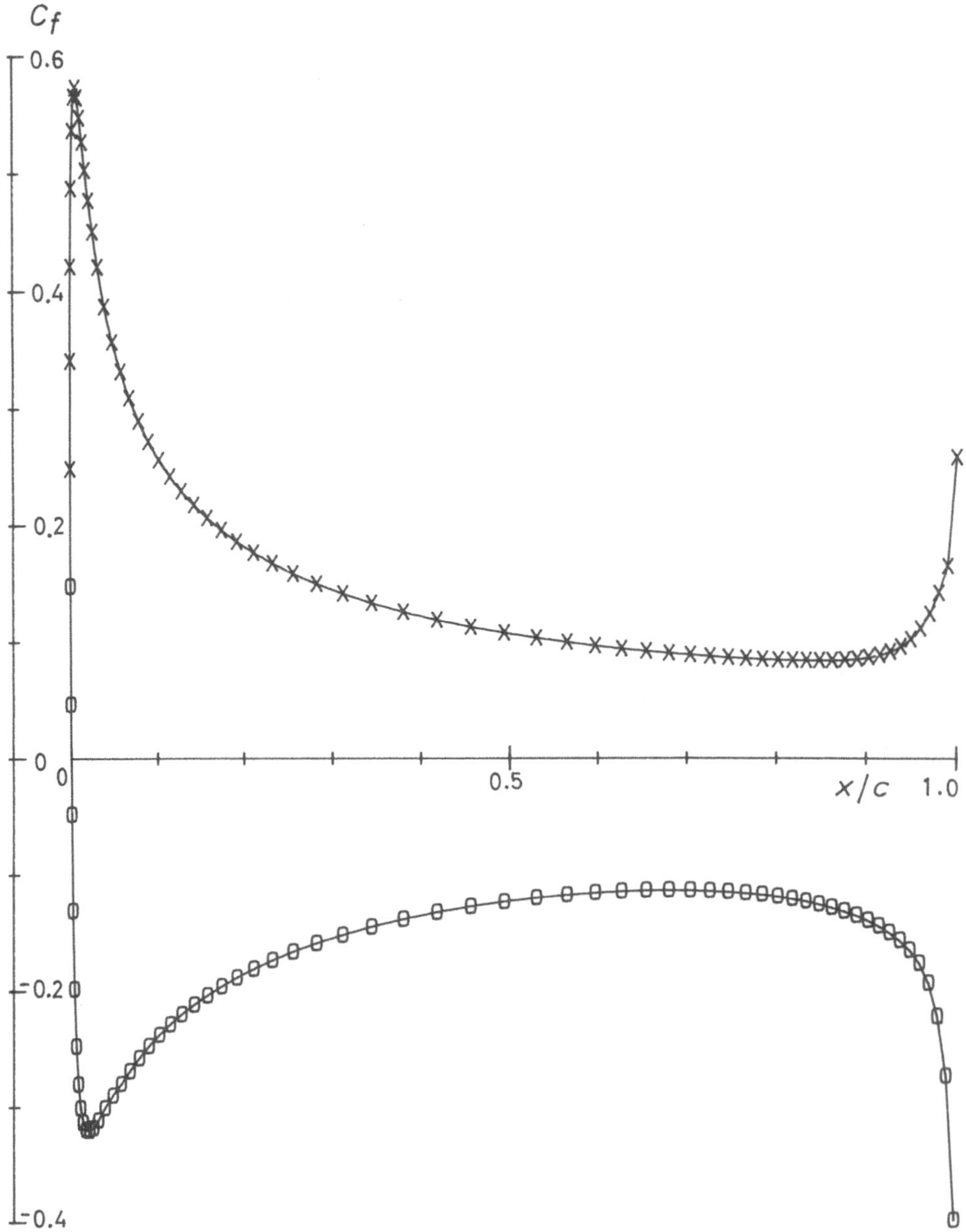

Fig. 14 Skin friction coefficient for case A3
NACA0012, $M_\infty=2$, $\alpha = 10^o$, $Re_\infty=106$, 193×65 mesh
X = upper surface, 0 = lower surface

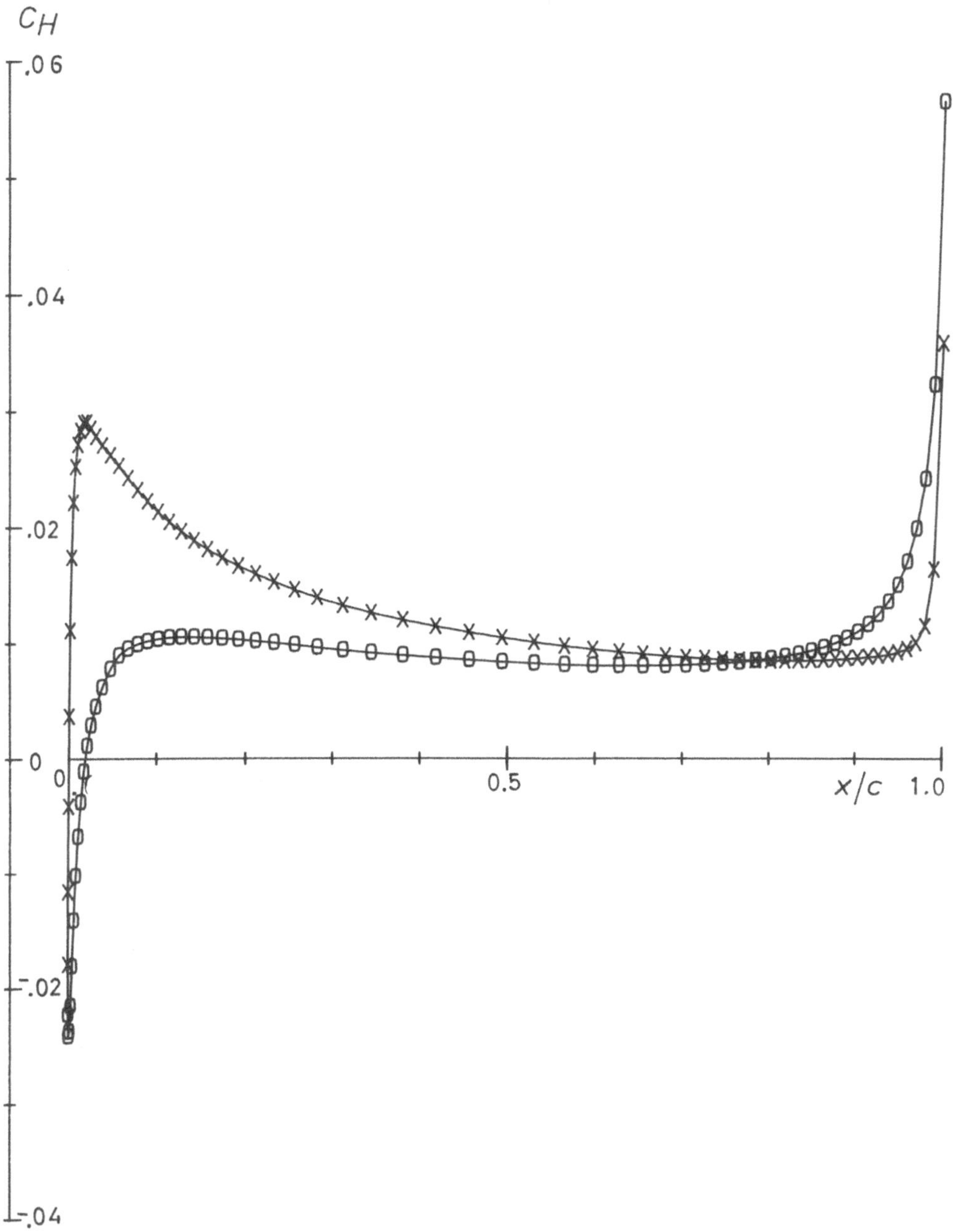

Fig. 15 Heat flux coefficient for test case A3
NACA0012, $M_\infty=2$, $\alpha = 10^o$, $Re_\infty=106$, 193×65 mesh
X = upper surface, 0 = lower surface

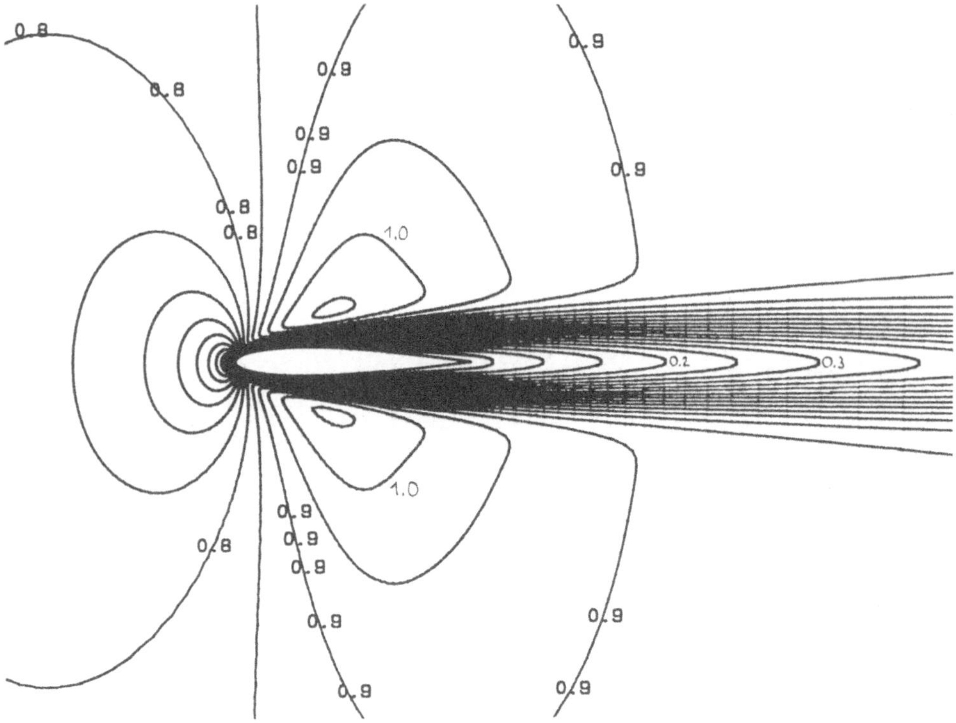

Fig. 16 Mach number contours for test case A5
$\Delta M = 0.05$, chord = 4 cm.

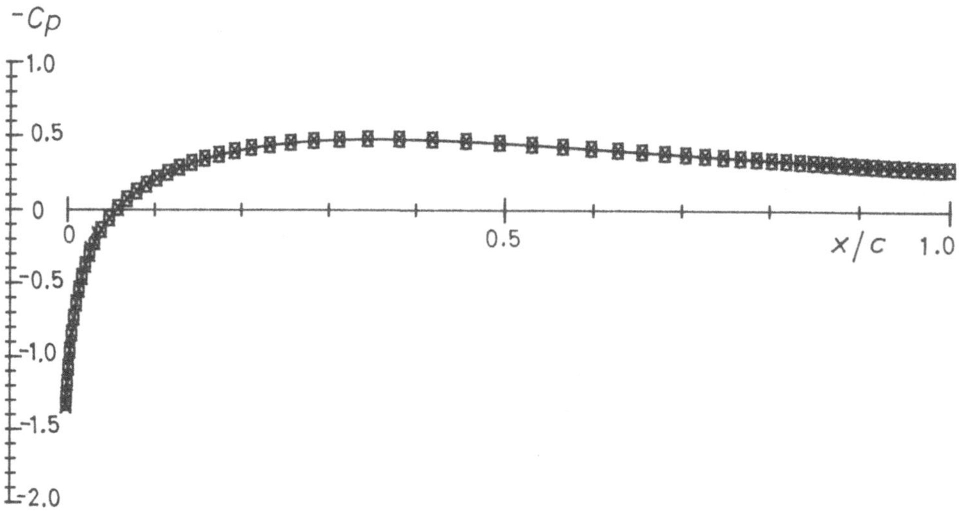

Fig. 17 Pressure coefficient for test case A5
NACA0012, $M_\infty=0.85$, $\alpha=0^\circ$, $Re_\infty=500$, 193×65 mesh
X = upper surface, 0 = lower surface

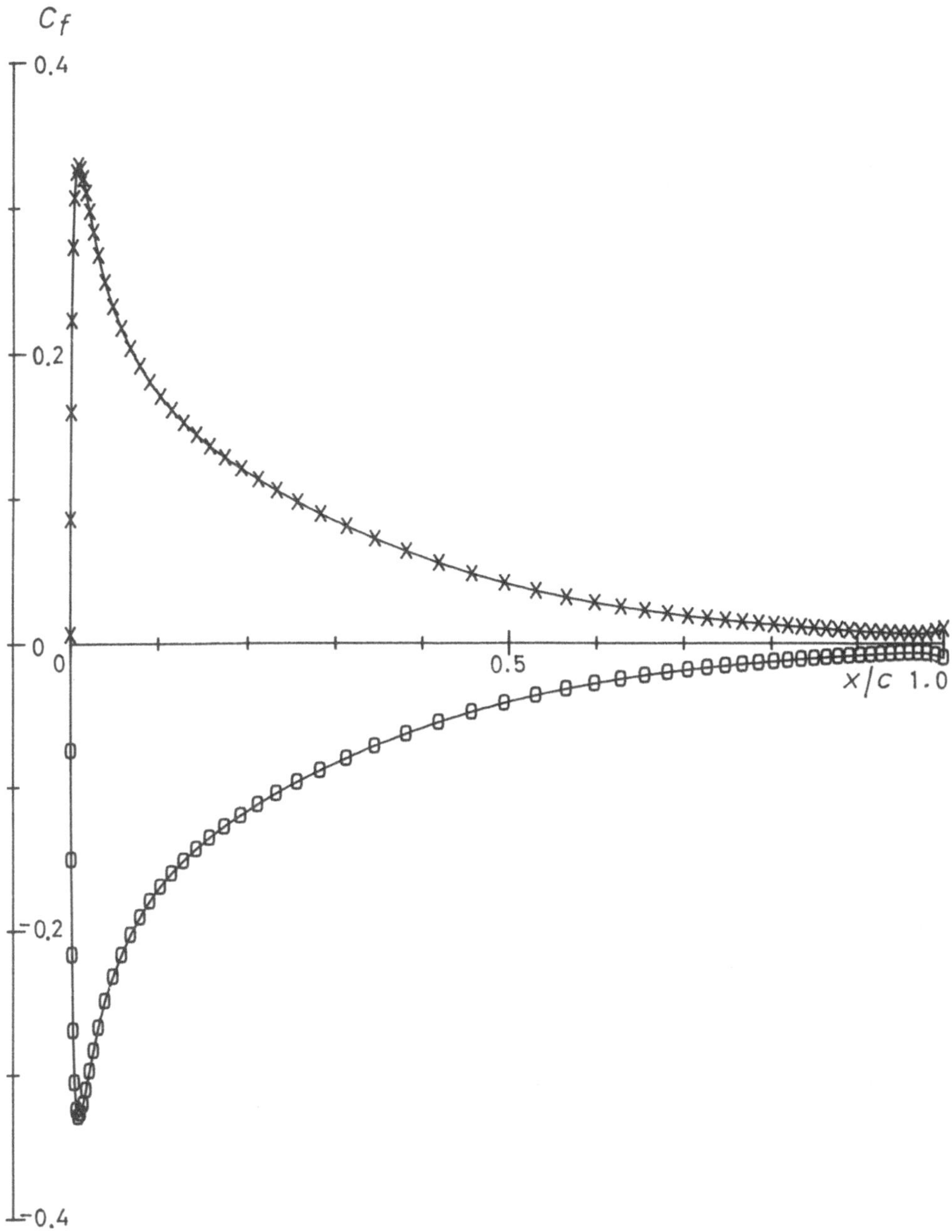

Fig. 18 Skin friction coefficient for case A5
NACA0012, $M_\infty=0.85$, $\alpha=0^\circ$, $Re_\infty=500$, 193×65 mesh
X = upper surface, 0 = lower surface

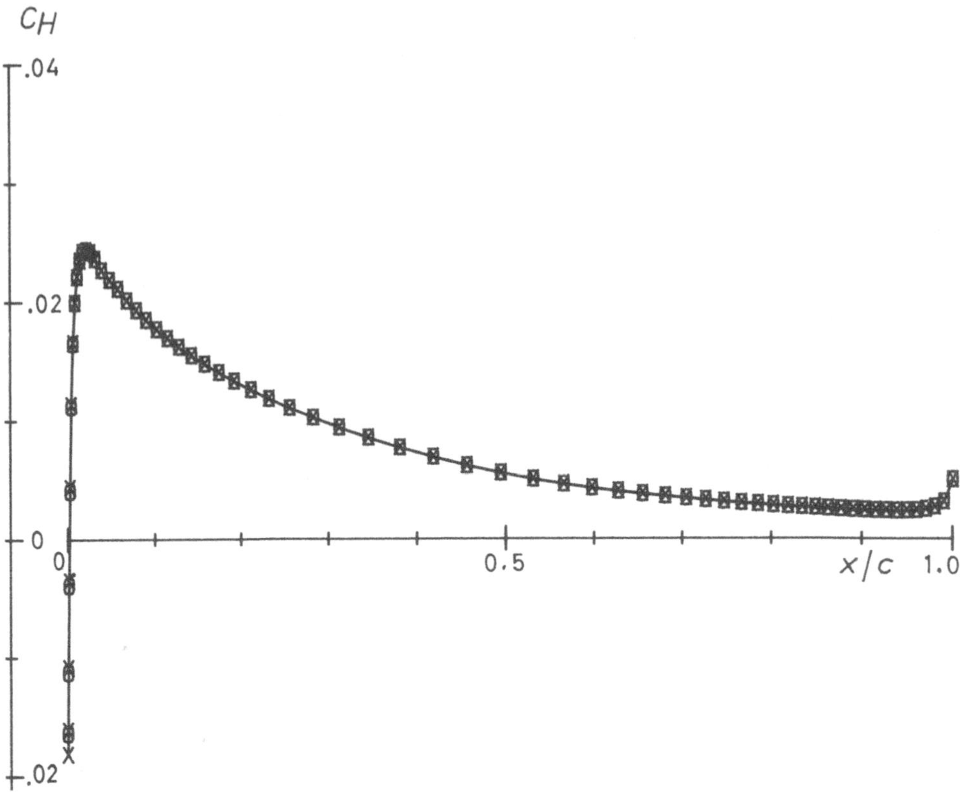

Fig. 19 Heat flux coefficient for test case A5
NACA0012, M_∞=0.85, α=0°, Re_∞=500, 193×65 mesh
X = upper surface, 0 = lower surface

Table 1 Integral values and CPU-minutes on VAX-11 for test cases A1 to A7.

TEST CASE	MESH	C_L	C_D	C_m	CPU MIN.
A1	193×65	0.5242	0.6276	0.2019	1394
A2	193×65	0.4199	0.2632	0.1145	989
A3	193×65	0.3261	0.4771	0.1652	547
A4	193×65	0.3388	0.2515	0.1549	703
A5	193×65	0.00004	0.2199	-0.00001	958
A6	193×65	0.000006	0.1160	-0.000004	698
A7	193×65	0.00033	0.0936	0.00061	1160

NUMERICAL SOLUTION OF TWO-DIMENSIONAL COMPRESSIBLE NAVIER-STOKES EQUATIONS USING RATIONAL RUNGE-KUTTA METHOD

Nobuyuki Satofuka, Koji Morinishi, and Yusuke Nishida

Department of Mechanical Engineering
Kyoto Institute of Technology
Matsugasaki, Sakyo-ku
Kyoto 606, Japan

1. INTRODUCTION

In this paper, the method of lines approach is proposed for solving viscous compressible flows. In the method of lines, semi-discretization of independent variables reduces the governing partial differential equations to a set of ordinary differential equations (ODEs) in time, which are integrated by using an appropriate time integration scheme. This separation of the space and time discretization assures a steady state solution independent of time step. As a time stepping procedure, we propose to use rational Runge-Kutta (RRK) method. The RRK method proposed by Wambecq [1] is fully explicit, requires no matrix inversion, and is stable at much larger time step than the usual explicit methods. The RRK method has been applied to solve both the Euler [2,3] and the Navier-Stokes equations [4,5]. Local time stepping and implicit residual averaging [6] techniques have been employed to accelerate convergence of solution to steady state.

Computed results for both mandatory and optional test cases of problem A for the GAMM Workshop on Numerical Simulation of Compressible Navier-Stokes Flows. The computations were performed on a vector computer FACOM VP-200 at the Computer Center of Kyoto University.

2. GOVERNING EQUATIONS

The compressible Navier-Stokes equations subject to general transformation can be written in dimensionless, conservation-law form as:

$$\frac{\partial \vec{q}}{\partial t} + \frac{\partial \vec{E}}{\partial \xi} + \frac{\partial \vec{F}}{\partial \eta} = \frac{1}{Re}\left(\frac{\partial \vec{R}}{\partial \xi} + \frac{\partial \vec{S}}{\partial \eta} \right) , \tag{1}$$

where $\vec{q}$, $\vec{E}$, $\vec{F}$, $\vec{R}$, and $\vec{S}$ are

$$\vec{q} = \frac{1}{J}\begin{pmatrix} \rho \\ \rho u \\ \rho v \\ e \end{pmatrix}, \quad \vec{E} = \frac{1}{J}\begin{pmatrix} \rho u \\ \rho u U + \xi_x p \\ \rho v U + \xi_y p \\ (e+p)U \end{pmatrix}, \quad \vec{F} = \frac{1}{J}\begin{pmatrix} \rho v \\ \rho u V + \eta_x p \\ \rho v V + \eta_y p \\ (e+p)V \end{pmatrix},$$

$$\vec{R} = \frac{1}{J}\begin{pmatrix} 0 \\ \xi_x \sigma_x + \xi_y \tau_{xy} \\ \xi_x \tau_{xy} + \xi_y \sigma_y \\ \xi_x R_4 + \xi_y S_4 \end{pmatrix}, \quad \vec{S} = \frac{1}{J}\begin{pmatrix} 0 \\ \eta_x \sigma_x + \eta_y \tau_{xy} \\ \eta_x \tau_{xy} + \eta_y \sigma_y \\ \eta_x R_4 + \eta_y S_4 \end{pmatrix}, \tag{2}$$

with

$$\sigma_x = \lambda\left(\frac{\partial u}{\partial x} + \frac{\partial v}{\partial y} \right) + 2\mu\frac{\partial u}{\partial x}, \quad \sigma_y = \lambda\left(\frac{\partial u}{\partial x} + \frac{\partial v}{\partial y} \right) + 2\mu\frac{\partial v}{\partial y},$$

$$\tau_{xy} = \mu\left(\frac{\partial u}{\partial y} + \frac{\partial v}{\partial x} \right), \tag{3}$$

$$R_4 = u\sigma_x + v\tau_{xy} + \frac{\gamma}{\gamma-1}\frac{\kappa}{Pr}\frac{\partial T}{\partial x}, \quad S_4 = u\tau_{xy} + v\sigma_y + \frac{\gamma}{\gamma-1}\frac{\kappa}{Pr}\frac{\partial T}{\partial y}.$$

In the conservative variables of Eq. (2), the pressure p is nondimensionalized by p_0, the density ρ by ρ_0, and the velocity components u and v in the x and y directions by $a_0/\sqrt{\gamma}$. Re represents the reference Reynolds number defined as $Re = a_0\rho_0 L/(\sqrt{\gamma}\mu_0)$, Pr the Prandtl number, κ the coefficient of thermal conductivity, μ the viscosity coefficient, and γ the ratio of specific feats. The coefficient of thermal conductivity and the viscosity coefficient are nondimensionalized with respect to their reference values. The pressure, density and velocity components are related to the total energy per unit volume e by the following equation for an ideal gas

$$e = p/(\gamma - 1) + \rho(u^2 + v^2)/2 . \tag{4}$$

The so-called contravariant velocity components U and V in the ξ and η directions, respectively, are given by

$$U = \xi_x u + \xi_y v, \quad V = \eta_x u + \eta_y v . \tag{5}$$

In the general coordinate transformation, ξ varies around the body surface, and η varies away from it. The transformation Jacobian in Eq. (2) is defined by

$$J = \xi_x\eta_y - \xi_y\eta_x . \tag{6}$$

The geometrical factors (metrics) in Eqs. (2)-(6) resulting from the coordinate transformation are defined as follows in terms of the derivatives of the Cartesian coordinates of the grid points;

$$\xi_x = Jy_\eta, \quad \xi_y = -Jx_\eta, \quad \eta_x = -Jy_\xi, \quad \eta_y = Jx_\xi . \tag{7}$$

In general, the metrics of Eq. (7) are not known analytically and must be determined numerically at the beginning of the calculation and stored.

3. DESCRIPTION OF THE METHOD

Spatial Discretization

The Navier-Stokes equations (1) are first discretized by the conventional central finite difference approximation to the spatial derivatives. This spatial discretization leads to a system of ODEs,

$$\begin{aligned}\frac{d}{dt}\vec{q}_{i,j} = &- (\vec{E}_{i+1,j} - \vec{E}_{i-1,j})/(2\Delta\xi) - (\vec{F}_{i,j+1} - \vec{F}_{i,j-1})/(2\Delta\eta) \\ &+ [(\vec{R}_{i+1/2,j} - \vec{R}_{i-1/2,j})/(\Delta\xi) \\ &+ (\vec{S}_{i,j+1/2} - \vec{S}_{i,j-1/2})/(\Delta\eta)]/Re ,\end{aligned} \tag{8}$$

where subscripts i, j are the grid index such that $\vec{q}_{i,j} = \vec{q}(i\Delta\xi, j\Delta\eta)$. The viscous flux terms $\vec{R}$ and $\vec{S}$ are evaluated at the midpoint of the grids. In the present computation, no artificial dissipation term is added since we intend to solve the Navier-Stokes flows dominated by physical viscosity.

Time Integration

The spatial derivatives are approximated by the conventional central finite differences. Then the Navier-Stokes equations (1) yield the system of ordinary differential equations as,

$$\frac{d\vec{q}}{dt} = \vec{G}(\vec{q}) \ . \tag{9}$$

This system of equations is integrated by the 2-step rational Runge-Kutta (RRK) scheme [1] with appropriate initial and boundary conditions as,

$$\begin{aligned} &\vec{g}_1 = \Delta t\ \vec{G}(\vec{q}^{\,n}) \ , \quad \vec{g}_2 = \Delta t\ \vec{G}(\vec{q}^{\,n} + c_2\vec{g}_1) \ , \\ &\vec{q}^{\,n+1} = \vec{q}^{\,n} + [\ 2\vec{g}_1(\vec{g}_1,\vec{g}_3) - \vec{g}_3(\vec{g}_1,\vec{g}_1)\ ] \ / \ (\vec{g}_3,\vec{g}_3) \ , \\ &\vec{g}_3 = b_1\vec{g}_1 + b_2\vec{g}_2 \ , \quad b_1 + b_2 = 1 \ . \end{aligned} \tag{10}$$

where $(\vec{d},\vec{e})$ denotes the inner product of $\vec{d}$ and $\vec{e}$. In Eq. (10), $\vec{q}^{\,n}$ and $\vec{q}^{\,n+1}$ indicate the values of $\vec{q}$ at the n and n+1 time step, respectively. The RRK scheme is generally of order 1 but is of order 2 if in addition $b_2c_2 = -1/2$. The scheme is computationally explicit and yet unconditionally stable for some class of parabolic PDEs. Hairer [7] has proved that for a system of equations, the scheme, Eq. (10), is A_0-stable if $b_2c_2 \leq -1/2$ and $A(\alpha)$-stable if $b_2c_2 \leq -1/[2\cos\alpha(2-\cos\alpha)]$. According to his stability definition, a method is said to be A_0-stable if $\{x \in R | x \leq 0\}$ is the stability region and $A(\alpha)$-stable if $\{z| |\pi-\arg z| \leq \alpha\}$ is the stability region. The scheme is also I-stable if $b_2c_2 \leq -1/2$.

Acceleration of Convergence

Although the method presented in the previous section is in itself more efficient than the classical explicit method by factor of 2. The rate of convergence to steady state solution is not so fast as the implicit scheme. Therefore, we incorporated a convergence acceleration technique into the basic scheme. As the acceleration technique, we chose the local time stepping and the residual averaging proposed by Jameson and Baker [6].

In the residual averaging approach, we replace the residuals at each point by a weighted average of residuals at neighbouring points. Consider the RRK scheme described by Eq. (10), in the form,

$$\vec{q}^{\,n+1} = \vec{q}^{\,n} + r \ . \tag{11}$$

In the two-dimensional case, we might replace the residual r by the implicit average

$$(1 - \varepsilon\delta_\xi^2)(1 - \varepsilon\delta_\eta^2)\ \bar{r} = r \ , \tag{12}$$

where δ_ξ^2 and δ_η^2 are second difference operators in the ξ and η directions, and ε is the smoothing parameter. Thus, in case of implicit averaging, it is necessary to solve a sequence of scalar tridiagonal equations and requires relatively small amount of computational effort per time step.

A spatially variable time step $\Delta t_{i,j}$ is used to accelerate convergence to steady state. Time steps are chosen at each grid point so that the Courant number is constant everywhere, typically 4.0, as

$$\Delta t_{i,j} = C_N \min(\Delta t_\xi, \Delta t_\eta)_{i,j} \ , \tag{13}$$

with

$$(\Delta t_\xi)_{i,j} = [1/(U + a \cdot R_\xi)]_{i,j} ,$$

$$(\Delta t_\eta)_{i,j} = [1/(V + a \cdot R_\eta)]_{i,j} ,$$

in which C_N is CFL number, a is the sound speed, and R_ξ and R_η are defined as

$$R_\xi = (\xi_x^2 + \xi_y^2)^{1/2} , \quad R_\eta = (\eta_x^2 + \eta_y^2)^{1/2} .$$

Computational Grid

The grid used for the computation is a body-fitted C-type grid generated by solving a set of Poisson equations. Forcing terms in the Poisson equation are chosen such that the desired grid spacing and grid angle may be maintained at the inner and outer boundaries as proposed by Sorenson and Steger [8]. The computational domain is typically discretized in 257X65 grid points as shown in Fig. 1, however computation has also been performed on a finer (257X257) grid for a special case. The far field boundaries are located approximately 8 chords away from the airfoil.

Boundary Conditions

If the free stream Mach number is supersonic, all the unknowns are fixed at the inflow boundary and extrapolated at the outflow boundary. If the flow is subsonic, the treatment of the far field boundary condition is of inviscid type based on the introduction of Riemann invariants for a one-dimensional flow normal to the boundary. For viscous flow, the exit Riemann invariant is not constant across the wake, therefore, the characteristic treatment is abandoned. In the part of outflow boundary, the conservation variables are extrapolated. On the airfoil surface, we impose a no-slip condition and we set the temperature value. Wall pressure is derived from the momentum equation at the wall. Density is calculated from the equation of state using prescribed wall temperature. The periodic wake boundary is treated by taking average of the two boundary-adjacent points.

4. COMMENTS ON CALCULATIONS

Computations related to the four mandatory and three optional test cases of the external flow problem A have been performed using our new Navier-Stokes code. The calculations have been carried out on a vector computer FACOM VP-200 at the Computer Center of Kyoto University.

Mandatory Cases (A1, A2, A3, and A5)

As shown in Fig. 1, C-type grid with 257X65 points are used for the mandatory cases. The grid refinement near the wall is chosen according to the Reynolds number value. Therefore we use the same grid for case A2 and A5. The far field boundaries for case A3, which is supersonic at infinity, locate at approximately two chord lengths. Grobal view of the grid is shown in Fig. 2. The wall distributions of pressure coefficient Cp and skin friction coefficient C_f for four mandatory test cases are shown in Figs. 3-6. Case A1 shows no separated flow region since C_f is everywhere positive, while case A2 indicates flow separation at upper surface. The precise x position of separation point is x = 0.345.

To investigate the computational efficiency of our code, convergence histories for cases A1 and A3 are shown in Figs. 7 and 8, which show the log of L2 norm of residuals of three conservation variables. It can be recognized that our code requires less than 2000 time steps to converge based on a criterion of three order of magnitude drop of residuals. With the computational grid of 257X65 points, CPU time for 2000 time steps is about 186s on a FACOM VP-200.

Computed Mach number contours are shown in Figs. 9-12. Thick low Mach number regions near the airfoil surfaces indicate the characteristic feature of low Reynolds number flows. A thick bow shock wave appearing in Fig. 11 is a results of large diffusion due to the low Reynolds number of this case. Integral values of lift coefficient C_L, drag coefficient C_D, and momentum coefficient C_M of the four mandatory cases are tabulated in Table 1.

Optional Cases (A4, A6, and A7)

The wall distributions of pressure and skin friction coefficients for cases A4 and A6 are shown in Figs. 13 and 14. To resolve very thin shock structure, a fine grid with 257X257 points is used for case A4. The same 257X65 grid as for cases A2 and A5 is used for case A6 except further refinement near the wall according to higher Reynolds number of 2000.

Mach number contours for these cases are shown in Figs. 15 and 16. Since Reynolds number for these cases are higher than the mandatory cases, viscous layers near the wall becomes thinner. Owing to very fine grid the sharp bow shock wave is clearly resolved without any added artificial viscosity.

For the test case A7, the computed solution dose not reach a steady state and shows an oscillatory flow field. Figure 17 shows computed isodensity contours and pressure coefficients for some arbitrary period during the iterations. To suppress the spurious oscillation caused by a lack of resolution for small eddies shed from the trailing egde, a fourth order artificial dissipation is added for this case.

5. CONCLUSIONS

All seven test cases of the external flow problem A are solved by using a new explicit Navier-Stokes code based on a combination of central finite differencing and rational Runge-Kutta time stepping. We conclude from the numerical experiments that our code is accurate and fast enough to compute compressible Navier-Stokes flows on vector computers.

REFERENCES

[1] Wambecq, A., Computing, Vol. 20, 1978, pp. 333-342.

[2] Satofuka, N., Morinishi, K., and Tokunaga, H., Notes on Numerical Fluid Mechanics, Vol. 13, Vieweg, 1986, pp. 319-326.

[3] Angrand, F., Billey, V., Dervieux, A., Desideri, J.A., Periaux, J., and Stoufflet, B., Lecture Notes on Physics, Vol. 218, Springer-Verlag, 1985, pp. 77-81.

[4] Satofuka, N. and Tsujita, M., Computational and Asymptotic Methods for Boundary and Interior Layers, Boole Press, 1982, pp. 364-369.

[5] Satofuka, N., Notes on Numerical Fluid Mechanics, Vol. 7, Vieweg, 1984, pp. 291-298.

[6] Jameson, A. and Baker, T.J., AIAA Paper 83-1929, 1983.

[7] Hairer, E., Numerische Mathematik, Vol. 35, 1980, pp. 57-68.

[8] Sorenson, R.L. and Steger J.L., Journal of Computational Physics, Vol. 33, 1979, pp. 405-410.

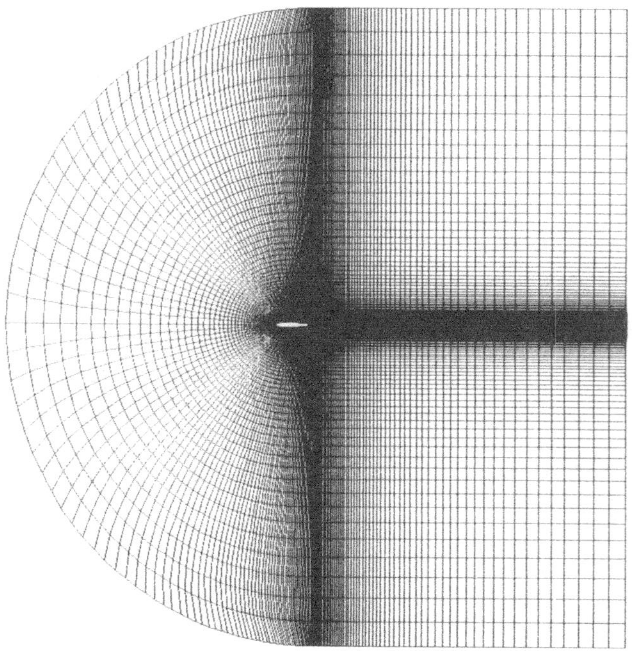

Figure 1 Global view of the computational grid for case A1.

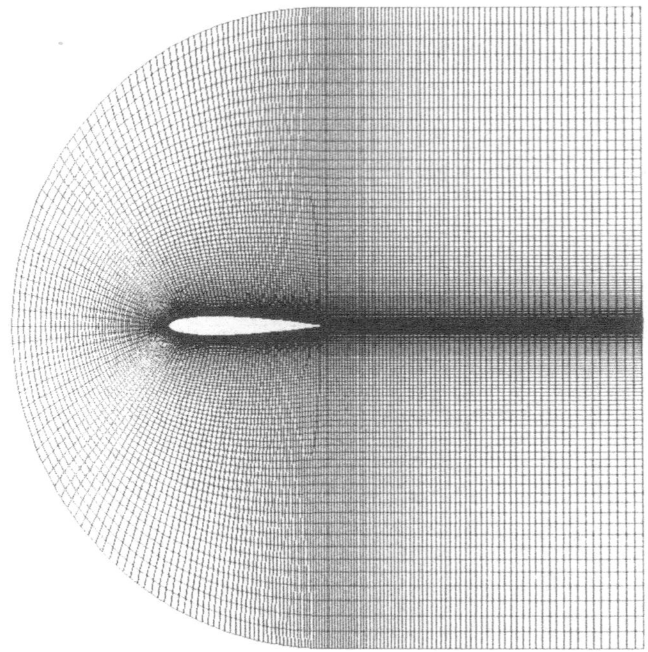

Figure 2 Global view of the computational grid for case A3.

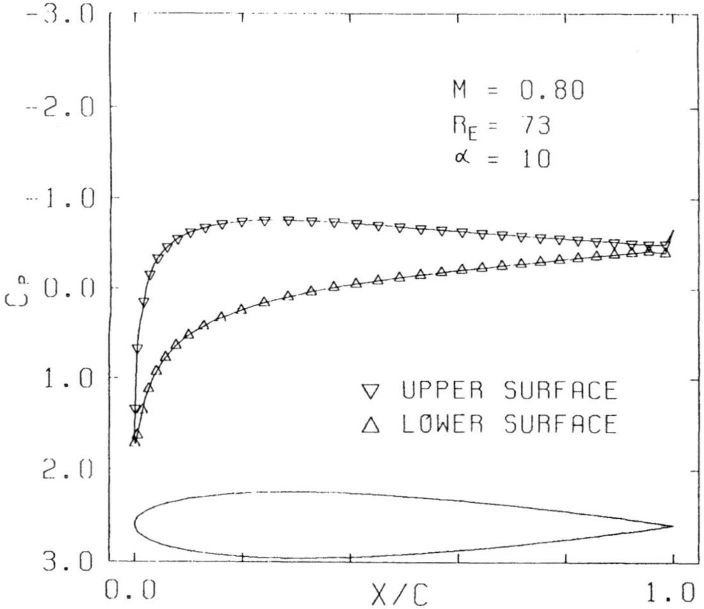

Figure 3.a Pressure coefficient for case A1.

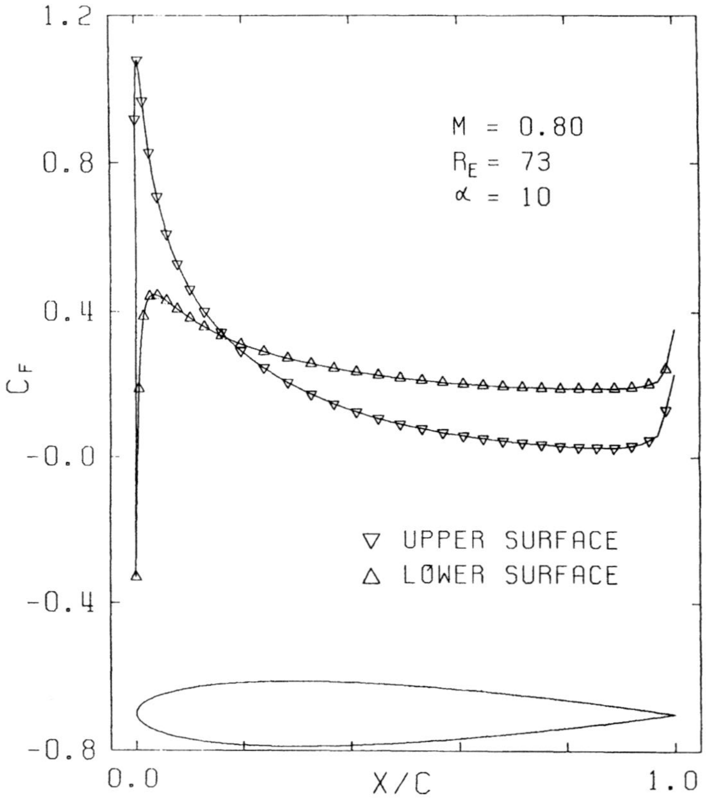

Figure 3.b Skin friction coefficient for case A1.

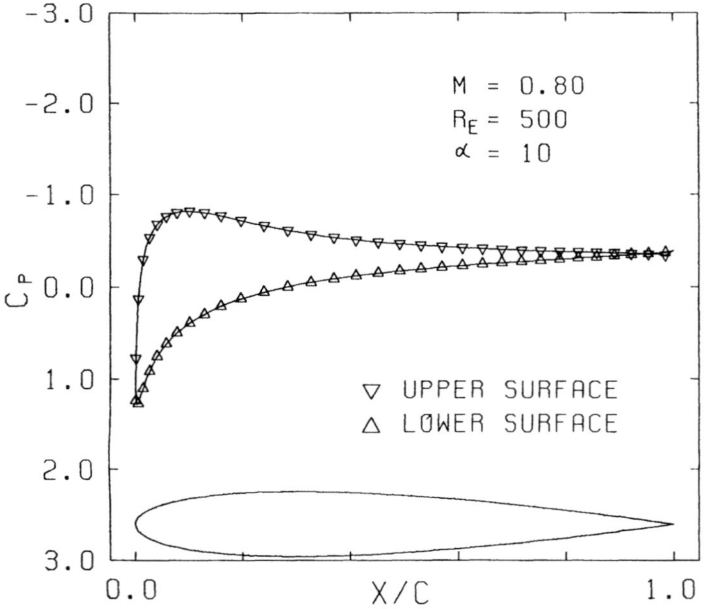

Figure 4.a Pressure coefficient for case A2.

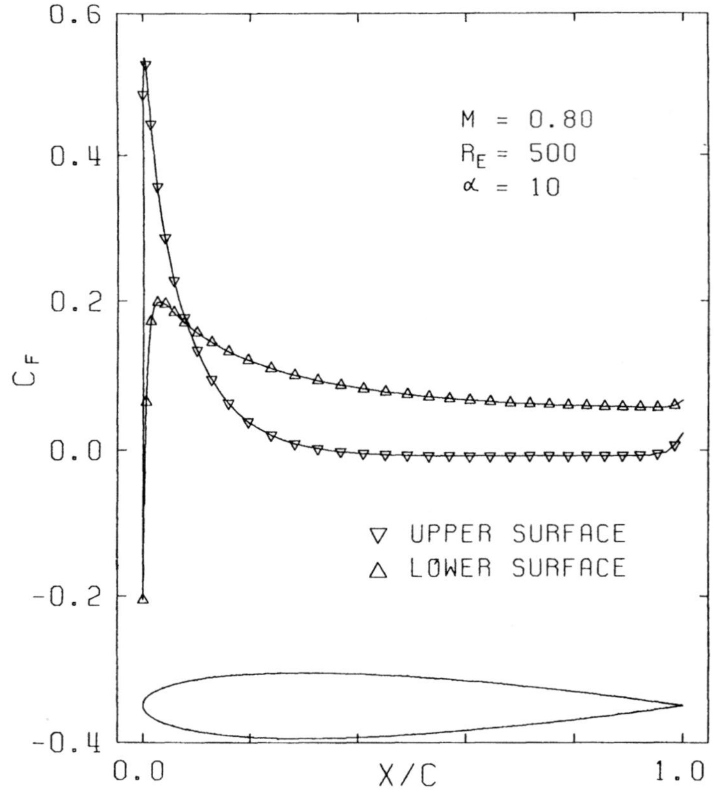

Figure 4.b Skin friction coefficient for case A2.

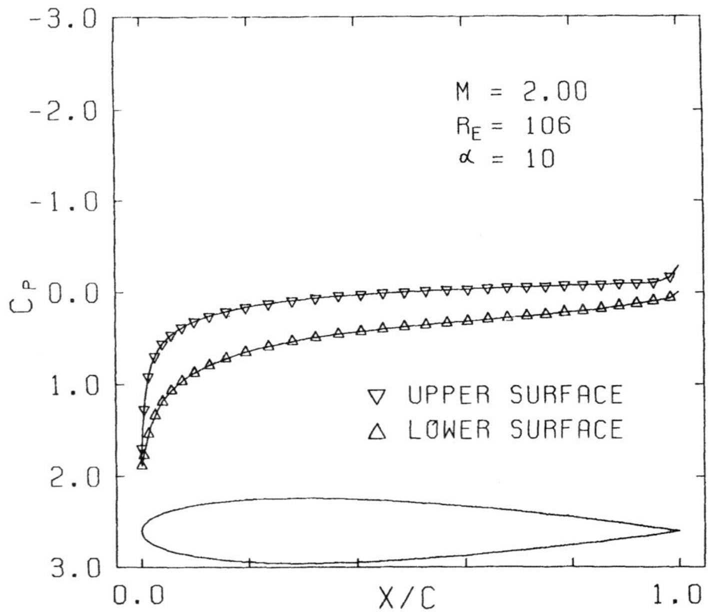

Figure 5.a Pressure coefficient for case A3.

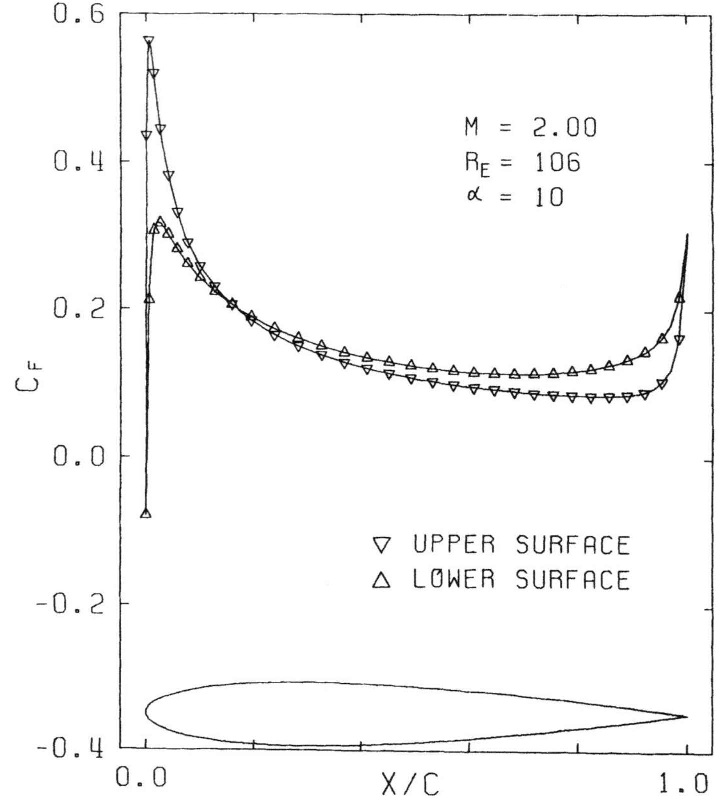

Figure 5.b Skin friction coefficient for case A3.

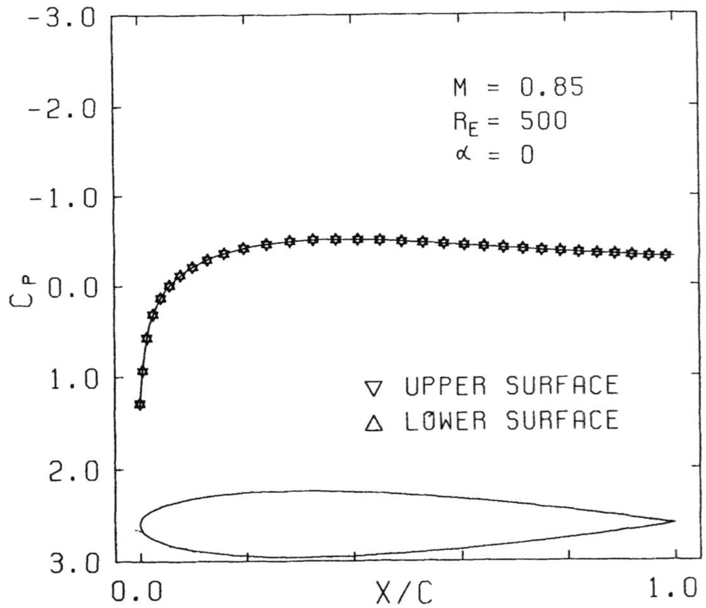

Figure 6.a Pressure coefficient for case A5.

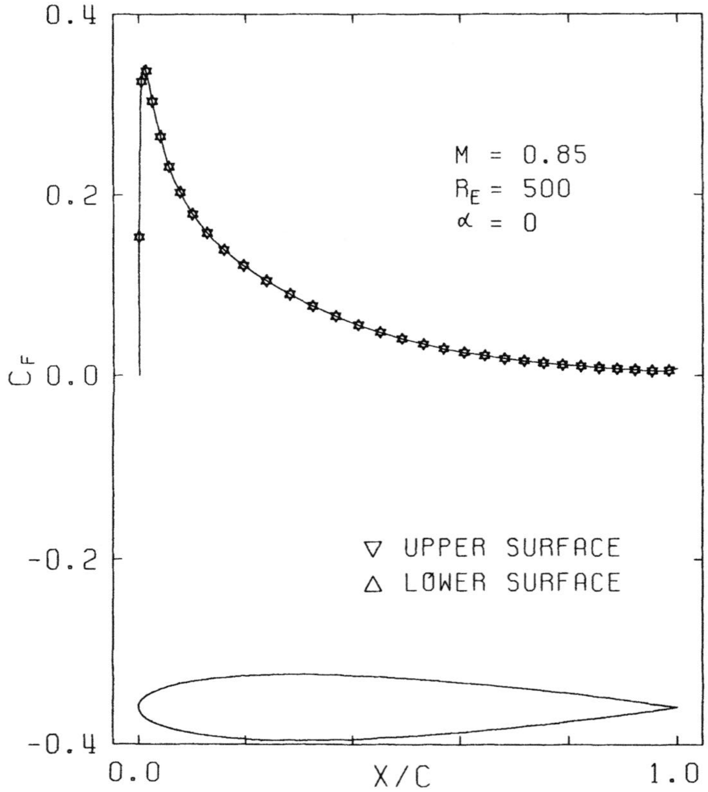

Figure 6.b Skin friction coefficient for case A5.

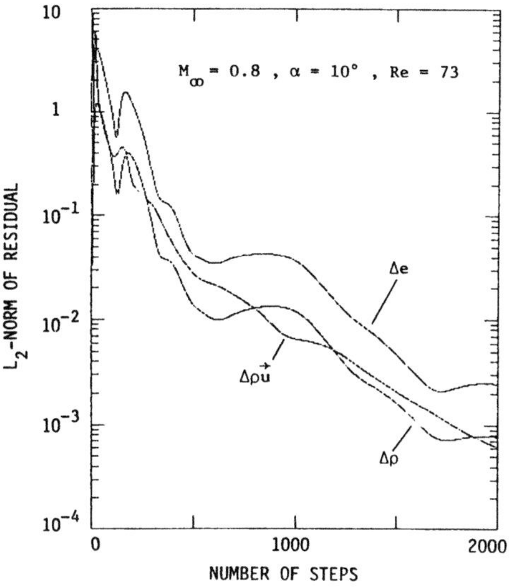

Figure 7 Convergence history for case A1.

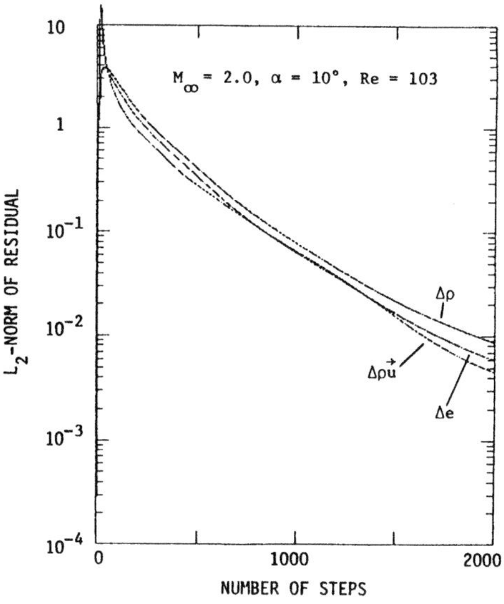

Figure 8 Convergence history for case A3.

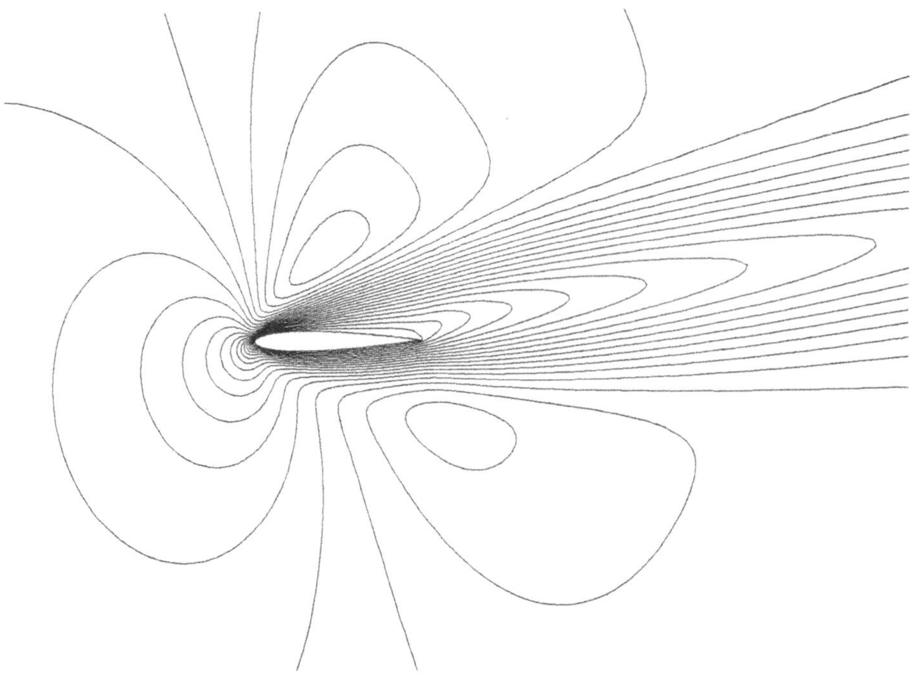

Figure 9 Iso-Mach lines for case A1.

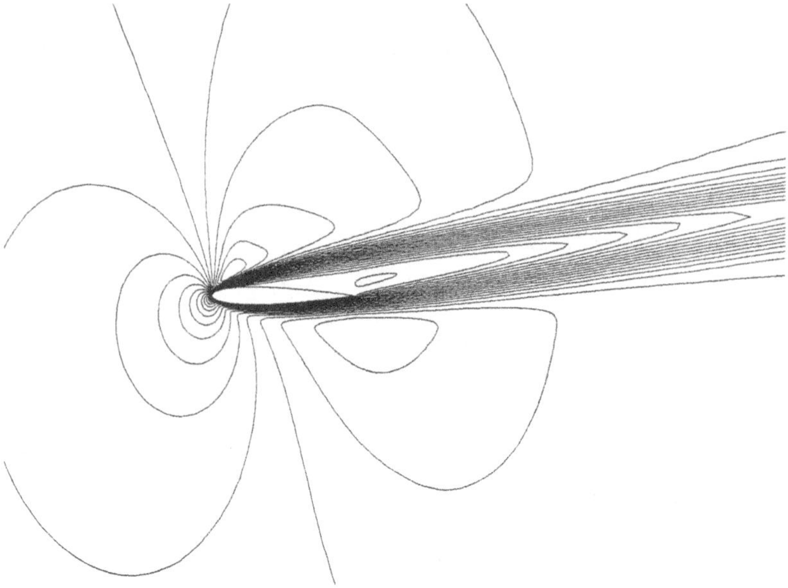

Figure 10 Iso-Mach lines for case A2.

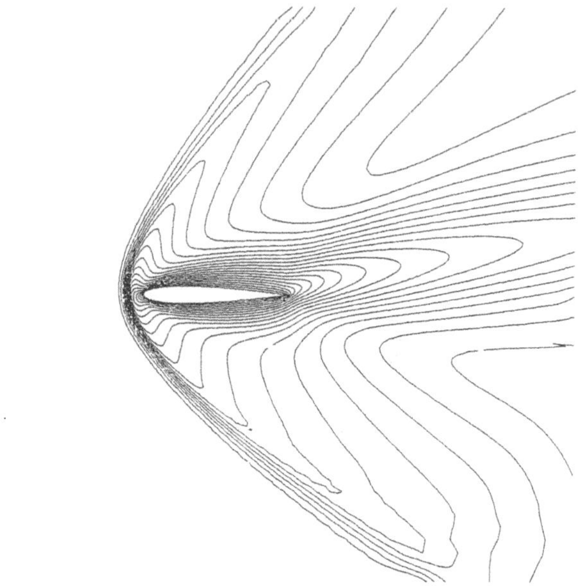

Figure 11 Iso-Mach lines for case A3.

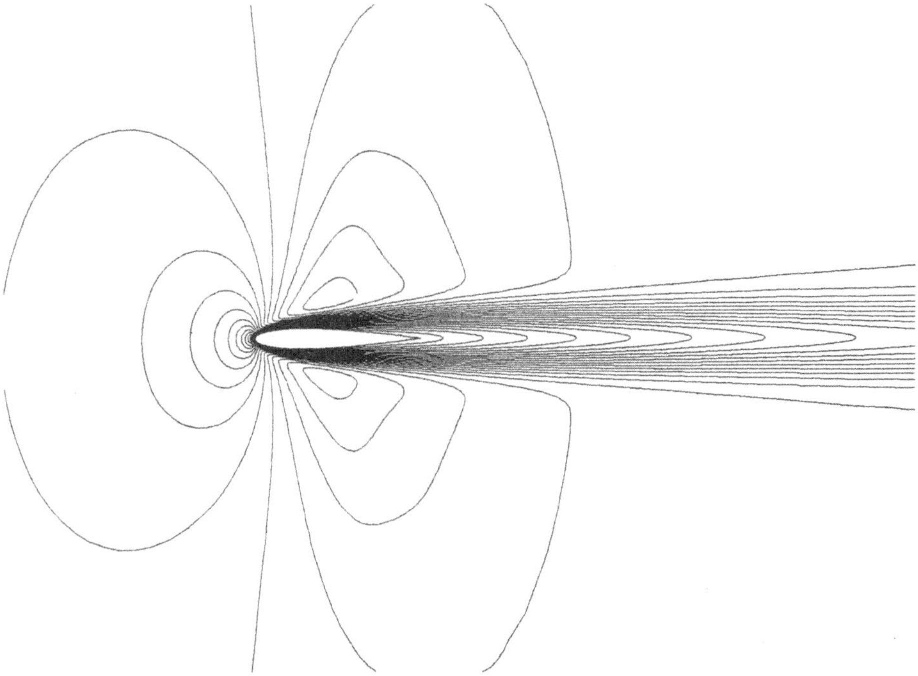

Figure 12   Iso-Mach lines for case A5.

Table 1   Integral values of the four mandatory case.

Case	A1	A2	A3	A5
C_L	0.5268	0.4145	0.3173	0.0
C_D	0.6447	0.2644	0.4910	0.2261
C_M	0.2050	0.1124	0.1640	0.0

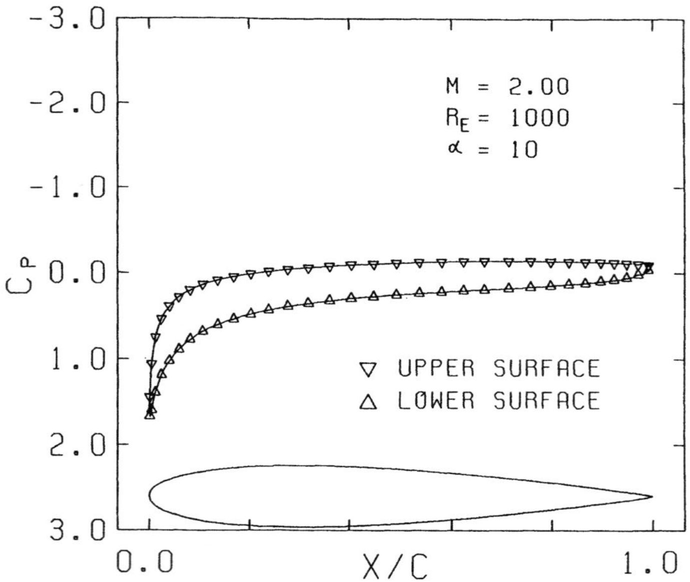

Figure 13.a Pressure coefficient for case A4.

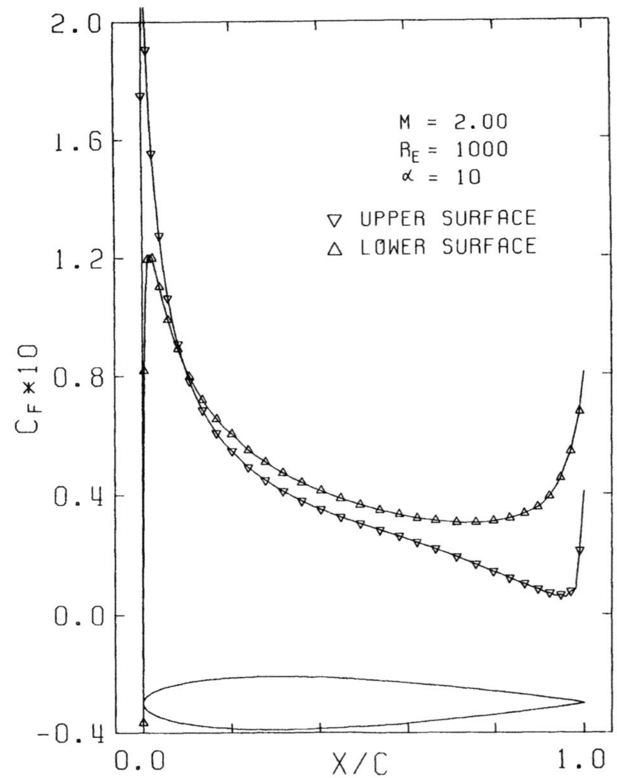

Figure 13.b Skin friction coefficient for case A4.

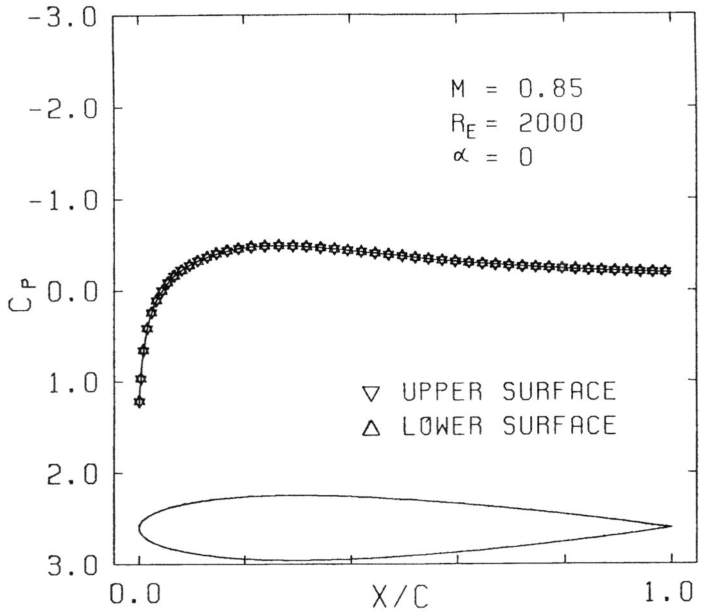

Figure 14.a Pressure coefficient for case A6.

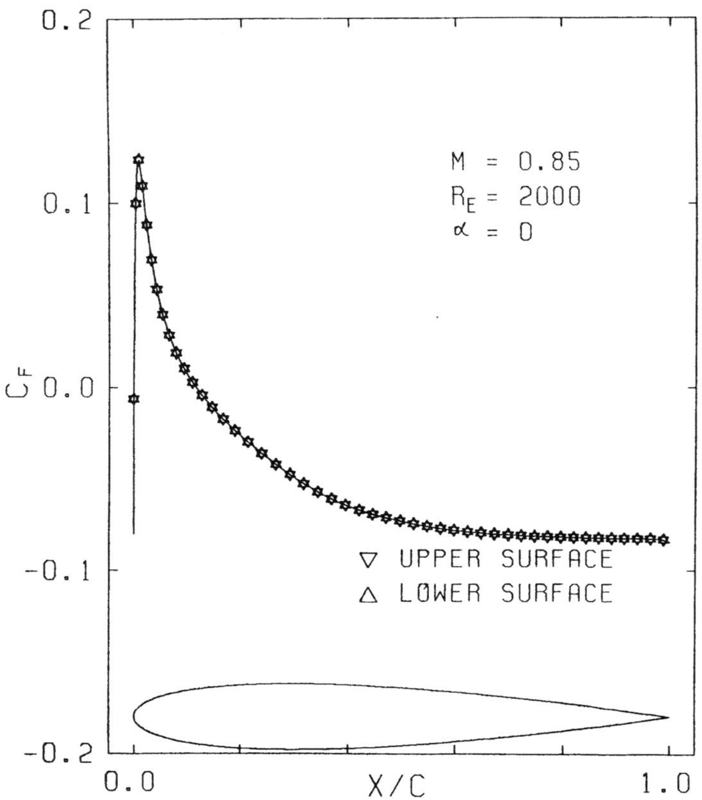

Figure 14.b Skin friction coefficient for case A6.

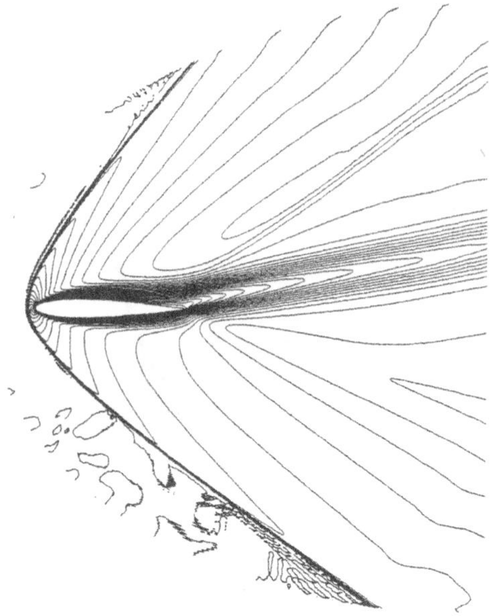

Figure 15   Iso-Mach lines for case A4.

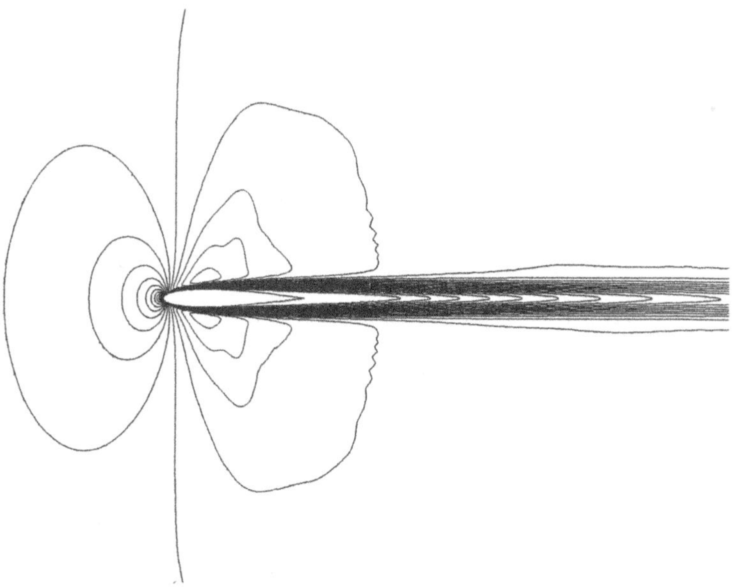

Figure 16   Iso-Mach lines for case A6.

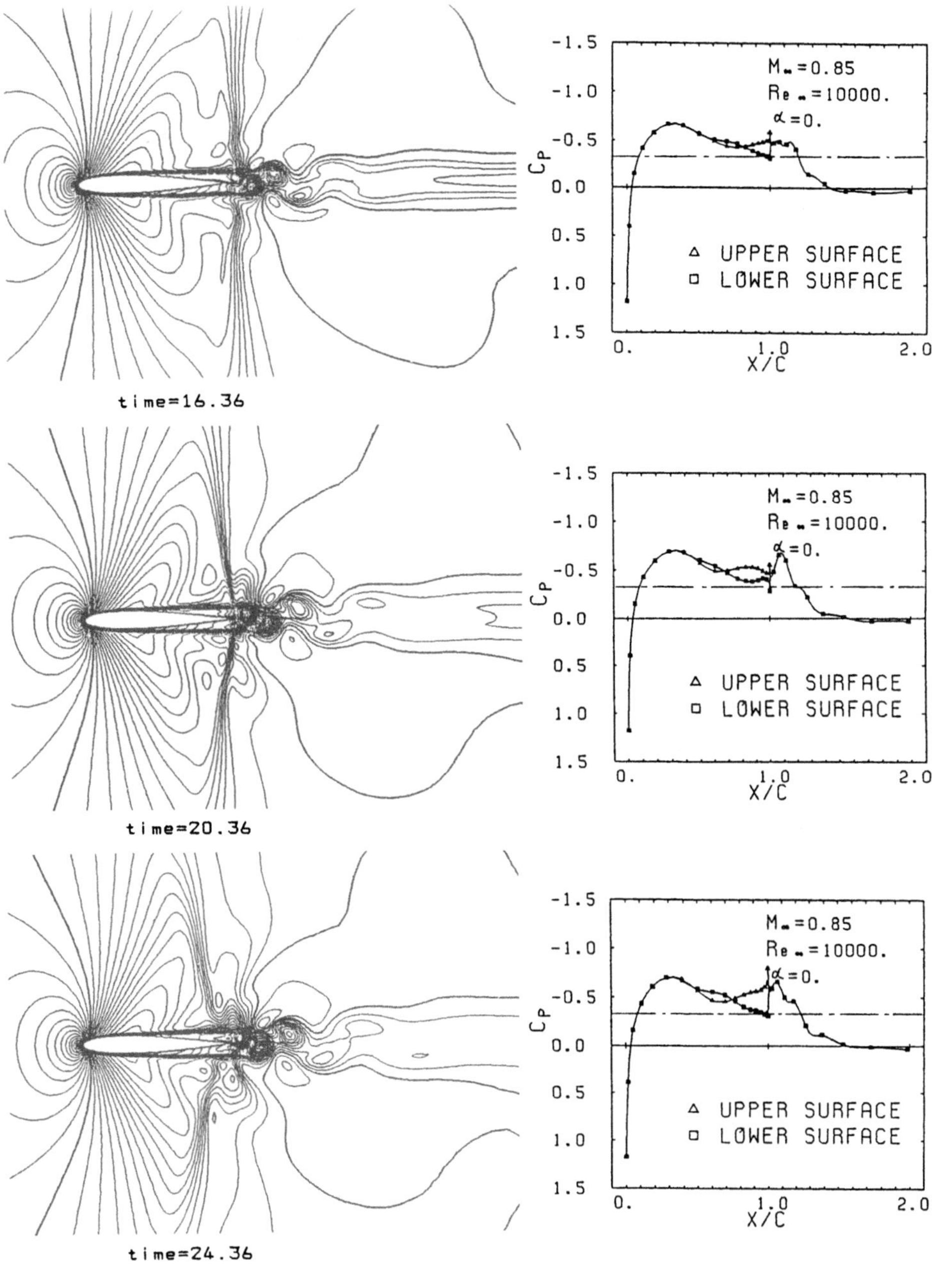

Figure 17 Iso-density lines and pressure coefficient for case A7.

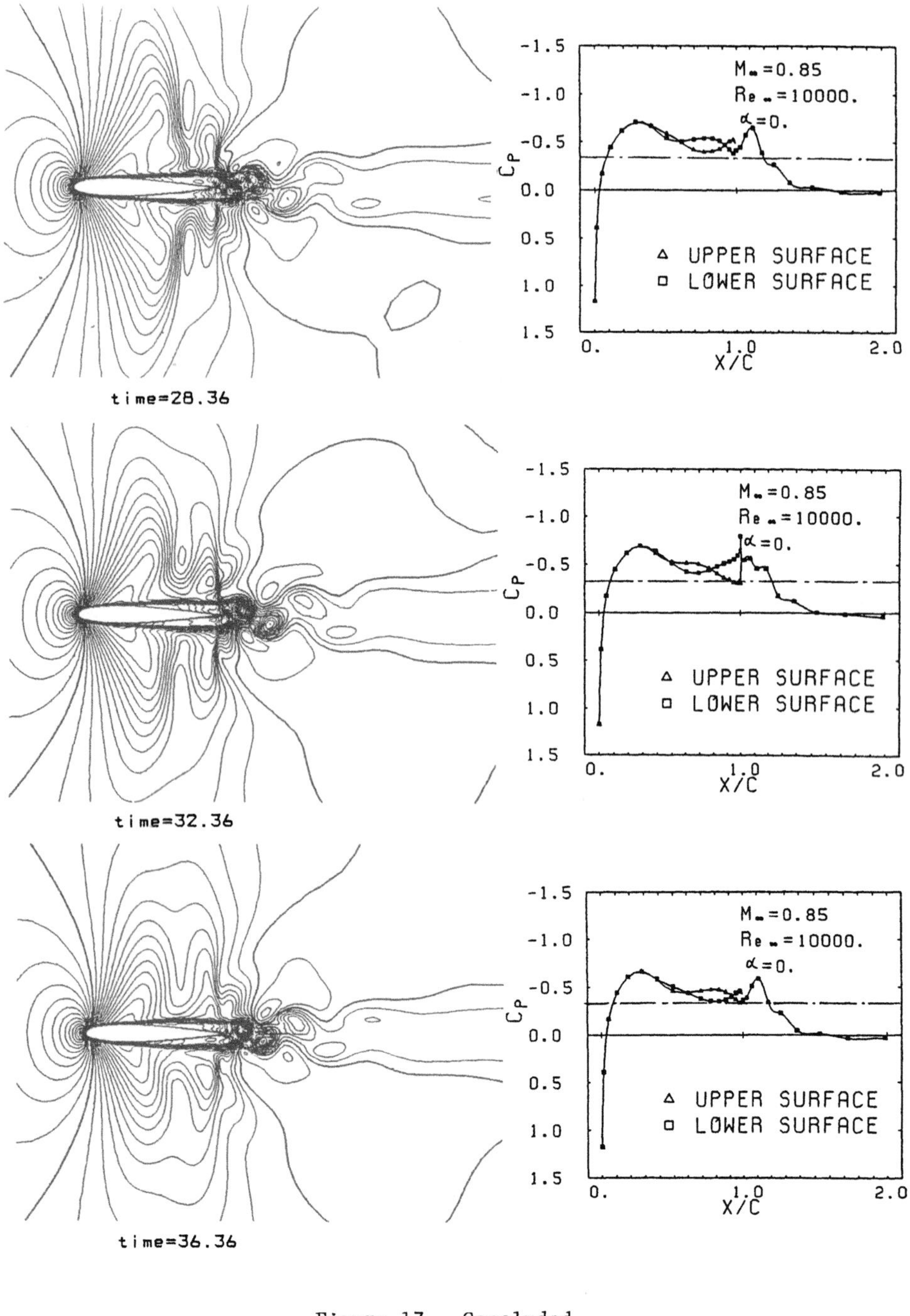

Figure 17 Concluded.

COMPRESSIBLE VISCOUS FLOW AROUND A NACA-0012 AIRFOIL

Yves Secretan*, Gouri Dhatt** and Dinh Nguyen***,

Université Laval

Québec, P.Q., Canada G1K 7P4

INTRODUCTION

We are interested to develop numerical solutions of problem A which concern the external 2D compressible viscous flow around a NACA-0012 airfoil, by the Galerkin Finite Element Method. A multi-mesh strategy coupled with a load increment is used.

NOMENCLATURE

Cl	global lift coefficient
C_d	global drag coefficient
Cm	global momentum coefficient
c_p	specific heat at constant pressure
p	pressure
t	time
$\underline{v}$	velocity vector (u,v)
$\overline{Pr}$	Prandtl number
R	gas constant
Re	Reynolds number
T	temperature
α	angle of attack
γ	ratio of specific heat
λ	thermal conductivity
μ	coefficient of viscosity
ρ	density
τ	stress tensor
"0"	(subscript) free stream values

MATHEMATICAL MODEL

The following assumptions are made :

- the fluid is a perfect Newtonian gas,
- the physical properties of the fluid are constant,
- volume forces and their work, mass and energy sources as well as radiation are neglected.

* Graduate student, Department of Mechanical Engineering
** Professor, Department of Civil Engineering
*** Professor, Department of Mechanical Engineering

The dimensionless form is obtained using the following reference quantities :

- all lengths are scaled by the chord length c,
- ρ, T, $\underline{v}$, λ, μ, cp by their free stream values,
- p by ρv^2.

The flow is governed by the conservation laws of mass, momentum and energy, written in non-dimensional form and in a cartesian coordinate system as follows :

$$\frac{\partial \rho}{\partial t} + \mathrm{div}(\rho \underline{v}) = 0 \tag{1}$$

$$\rho \frac{D\underline{v}}{Dt} = -\frac{1}{E_0}\left(\rho R \,\mathrm{grad}(T) + T\,\mathrm{grad}(\rho R)\right) + \frac{1}{Re_0}\mathrm{div}(\tau) \tag{2}$$

$$\rho(cp - R)\frac{DT}{Dt} + \rho RT\,\mathrm{div}(\underline{v}) = \frac{1}{Re_0 Pr_0}\mathrm{div}(\lambda\,\mathrm{grad}\,T) + \frac{E_0}{Re_0}\tau:\mathrm{grad}(\underline{v}) \tag{3}$$

with

$$p = \frac{1}{E_0}\rho R T \;;\; Re_0 = \frac{\rho_0 V_0 c}{\mu_0} \;;\; Pr_0 = \frac{\mu_0 cp_0}{\lambda_0} \;;\; E_0 = (\gamma - 1)M_0{}^2.$$

The boundary conditions are :

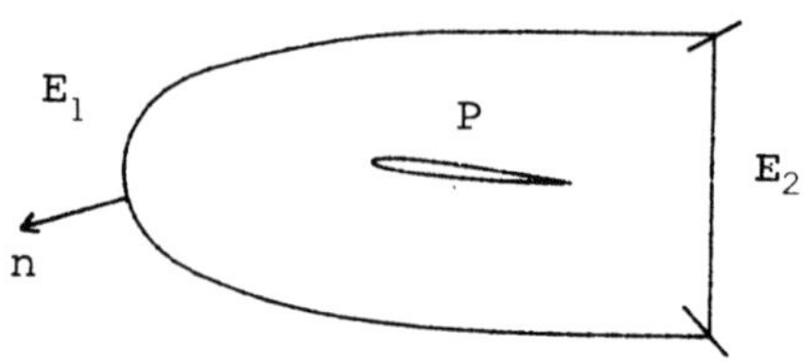

on E_1 : Dirichlet conditions : $u_0 = 1$, $v_0 = 0$, $\rho_0 = 1$, $T_0 = 1$,

on E_2 : von Neumann conditions : $\tau \cdot n = 0$, $\frac{\partial T}{\partial n} = 0$ n is the outward normal direction

on P : Dirichlet conditions : $u = 0$, $v = 0$, no-slip condition

on P : von Neumann conditions : $\frac{\partial T}{\partial n} = 0$, adiabatic wall.

The dependent unknowns used in the model are the primitive variables ρ, $\underline{v}$, T. The variational formulation underlying our finite element model is obtained by the use of the standard Galerkin method. Higher order derivative terms in the resulting integral form are integrated by parts using Green theorem. The piecewise approximations of velocity components and density-temperature are respectively biquadratic and bilinear in the standard reference element, which is a nine-node bilinear isoparametric element.

All integrals arising from the finite element approximation are evaluated using a 3x3 points Gaussian quadrature.

The finite element model, approximating equations (1-3), yields to solve a non-linear system of equations. Newton-Raphson's method coupled with the L.S. matrix decomposition have been adopted to solve these equations.

STRAGEGY OF RESOLUTION

The following strategy of resolution has been adopted :

Step 1 : Consider a starting mesh defined by the size "h_s".

Step 2 : The problem is solved in a mesh (with size "h_i") until we reach a solution free of oscillation for the required Mach number and for the maximum Re number allowed by the grid.

Step 3 : The solution obtained in the previous step is interpolated in a refined mesh ("h_i + i"). Thus, Step 2 serves to provide the initial conditions for this mesh. If the maximum Re number providing a stable solution is still less than the specified one, return to Step 2.

Remarks :
- The initial solution of the above process has been obtained by setting Dirichlet conditions on velocity at both E_1 and E_2 boundaries. For the subsequent solutions however, they have been relaxed at the outlet E_2 boundary; i.e. transformed into Neumann conditions.
- Long wave oscillations should be smoothen in the coarse mesh before any further treatment since they cannot be eliminated in any more refined mesh [Brandt, 1972].
- No upwinding procedure has been used.

RESULTS

We have solved problems A1, A2, A3, A5.

A bounded square domain around the profile of side nearly 18 times the chord length is decomposed into quadrilateral elements. A C-type grid is generated by using the algorithm of Sorenson and Steger.

The pressure distribution is simply deduced from the equation of state. The partial derivatives of the velocity components are computed directly from the finite element approximation. The shear stress at the profile is obtained by averaging the values given from the adjacent elements, since the tangential direction is discontinuous at corner nodes.

The presented results have been obtained on a final mesh of 5152 nodes and 1248 elements. For the A3 case, the grid has been adapted to the shock-form.

In general, convergence is achieved after 5-6 iterations (maximum 10) for a given load (Re and Ma Number). For flows conditions of Ma = 0.8,

Re = 500, α = 10.0 and with an initial solution at Ma = 0.6, Re = 500, α = 10.0, the convergence history is shown in Fig. 1. Global Cl, Cd, Cm for the 4 cases, as well as the separation point for case A2 are given in:

Table 1 Cl, Cd, Cm, separation point

	A1	A2	A3	A4
Cl	0.6127	0.4438	0.4059	0.0000
Cd	0.6377	0.2627	0.4791	0.2181
Cm	0.2453	0.1238	0.2044	0.0000
Sep.	--	0.37	--	--

OBSERVATIONS AND RECOMMENDATIONS

- Equal biquadratic interpolation for velocity and temperature leads to unstable solutions. Changing to a bilinear approximation for temperature eliminates this problem. This lead to think that a consistent stability conditions as in the case of the mixed finite element theory of Navier-Stokes equations (L.B.B. conditions) should exist. It is however not clear how one can handle theoretically the stability conditions for the present problem. Further theoretical works in this direction would be interesting.

- Considering the relatively small domain used in the computation, Dirichlet conditions for the velocity at the outlet generates oscillations. Natural conditions eliminates this problem.

- Our next study will be looking for saving storage requirement by developing methods where the equations are uncoupled.

- In order to refine the grid dynamically, a triangular element would be more appropriated. It would be interesting to test the mini-element of Arnold-Brezzi-Fortin.

HARDWARE FACILITIES, CPU-TIME, MEMORY REQUIREMENT

The computation has been done on the DEC VAX-785 computer of CRANI (Centre de Recherche sur les Applications Numériques en Ingénierie) with 8 Mbytes in-core memory, 25 Mbytes virtual memory and rated at 0,3 MFlops. The CPU-time per iteration and memory requirement for each grid used are summarized in Table 2.

Table 2 CPU-cost, Memory requirement

Grid	Element	Nodes	Deg. Freed	Mbytes	CPU-cost
1	100	440	1100	1.3	1' 44"
2	440	1850	4670	11.2	22' 30"
3	600	2500	6300	18.2	44' 50"
4	1248	5152	13000	56.0	6h 36'

For the final grid, we had to split the matrix on peripherical mass-data storage units. This operation increases drastically the CPU-Time. The total cost of a solution is hard to evaluate since the strategy of resolution has been adopted during the resolution.

REFERENCES

ARNOLD, D.N., BREZZI, F., FORTIN, M. : "A stable finite element for the Stokes equation", CALCOLO (1985).

BRANDT, A. : "Multi-level adaptive technique (MLAT) for fast numerical solution to boundary value problems", 3e Conf. on Num. Methods in Fluid Mech. (1972).

BRISTEAU, M.O., GLOWINSKI, R., DIMOYAT, B., PERIAUX, J., PERRIER, P., PIRONNEAU, O. : "Finite element methods for the compressible Navier-Stokes equations", AIAA 6th Computational Fluid Dynamics Conference (1983).

DHATT, G., TOUZOT, G. : "Une présentation de la méthode des éléments finis", P.U.L., Québec et Maloine S.A. Editeurs, Paris (1981).

FORTIN, M. and al : "Simple continuous pressure elements for incompressible flows" in preparation (1985).

SECRETAN, Y. : "Modélisation par éléments finis des équations de Navier-Stokes compressibles. Application à des écoulements externes". Thèse de maîtrise, Université Laval, Québec (1986).

SORENSON, STEGER : "Simplified clustering of nonorthogonal grids generated by elliptic partial differential equations", NASA TM 73252 (1977).

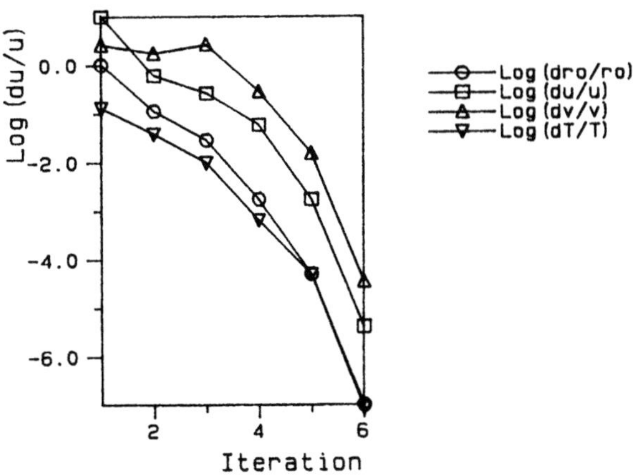

Fig.1 Convergence history

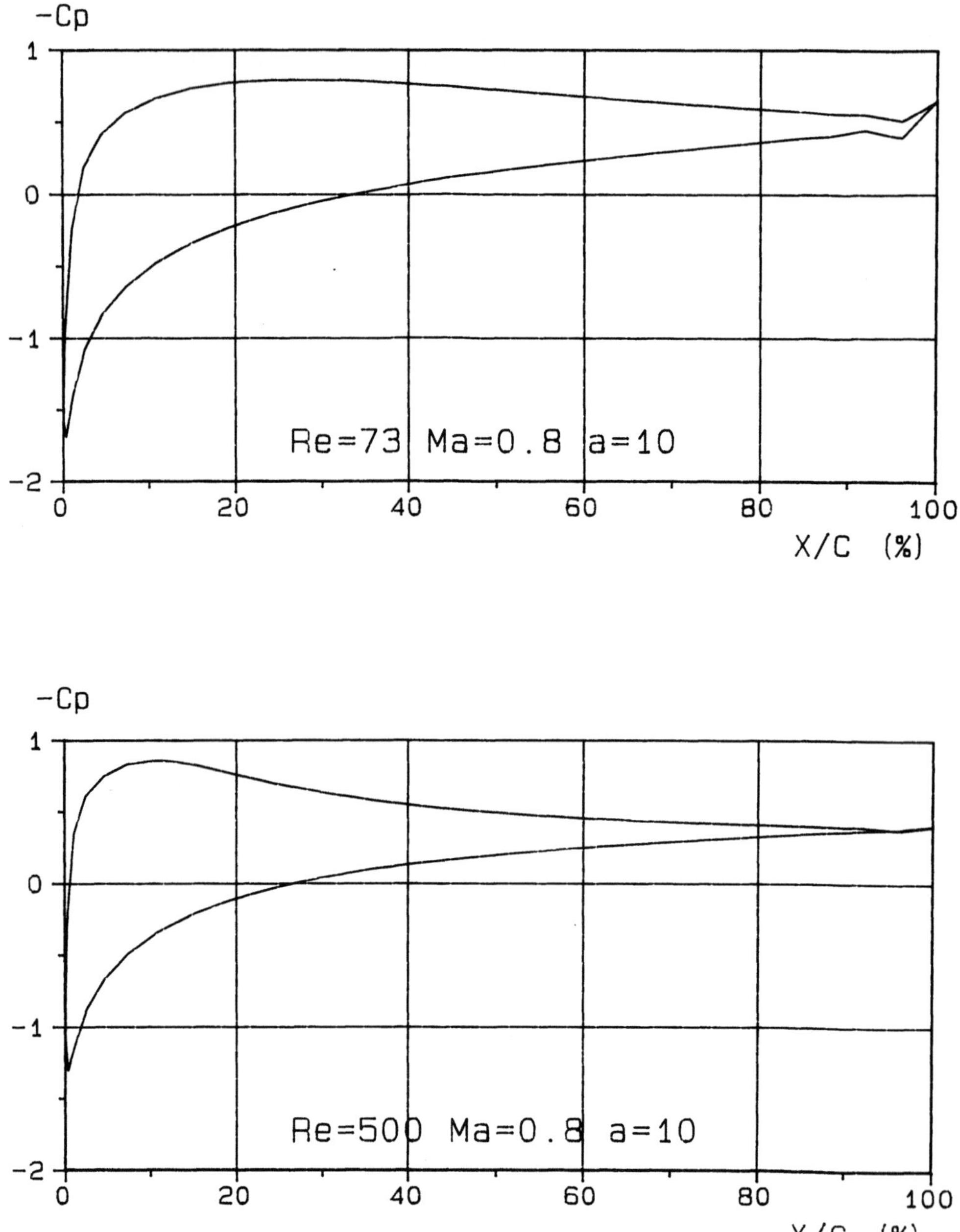

Fig 2.a Pressure coefficient

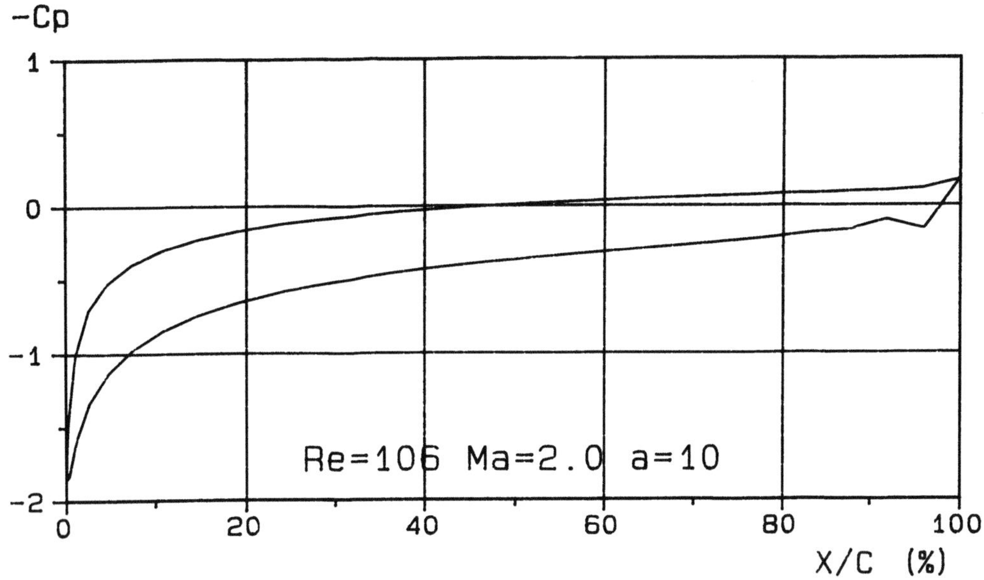

Fig 2.b Pressure coefficient

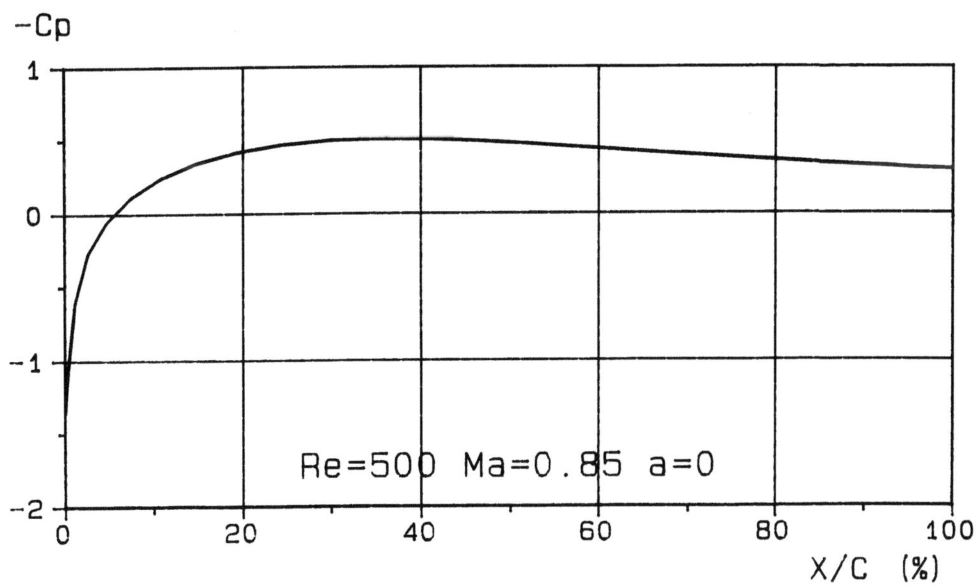

Fig 2.c Pressure coefficient

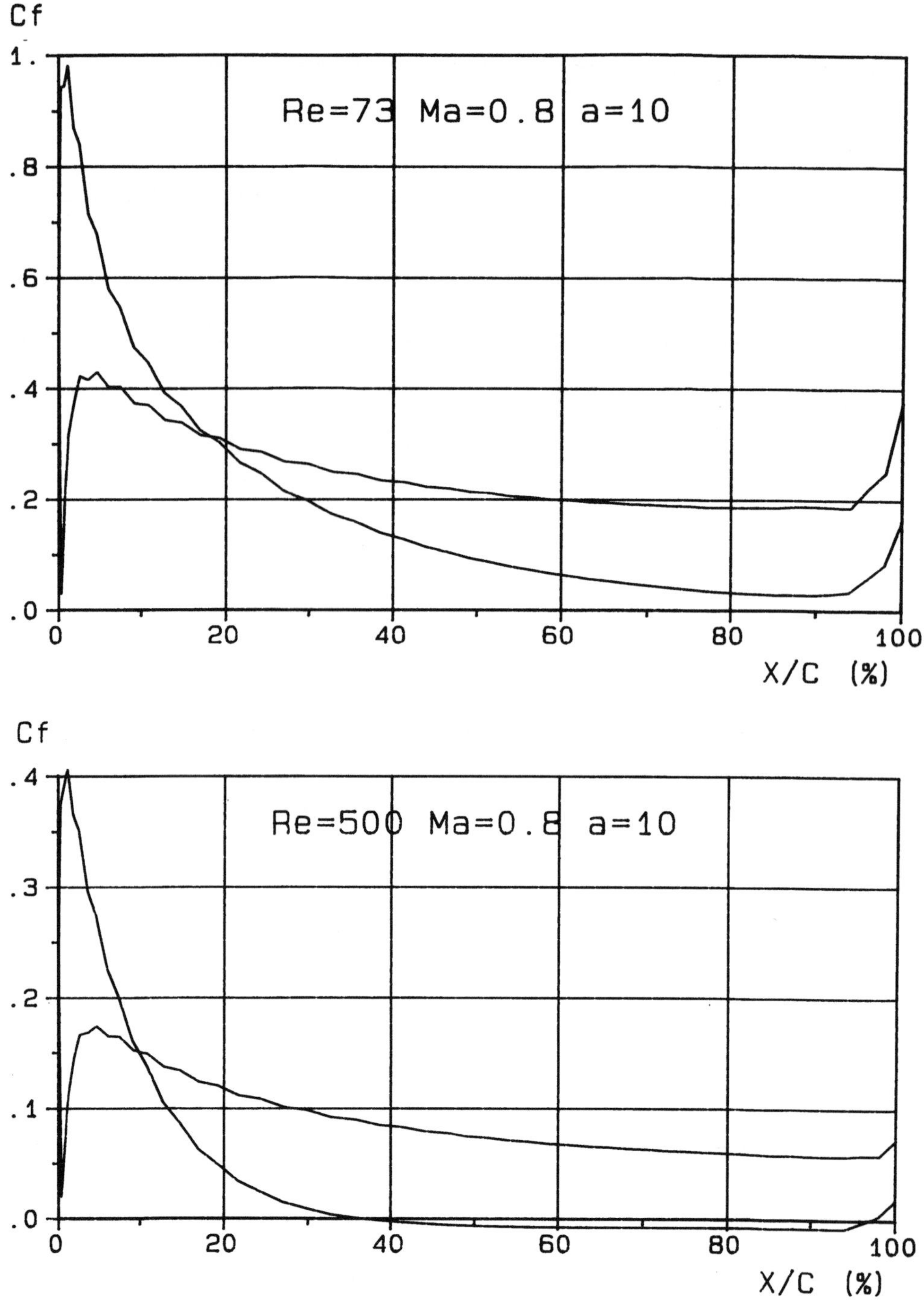

Fig 2.d Friction coefficient

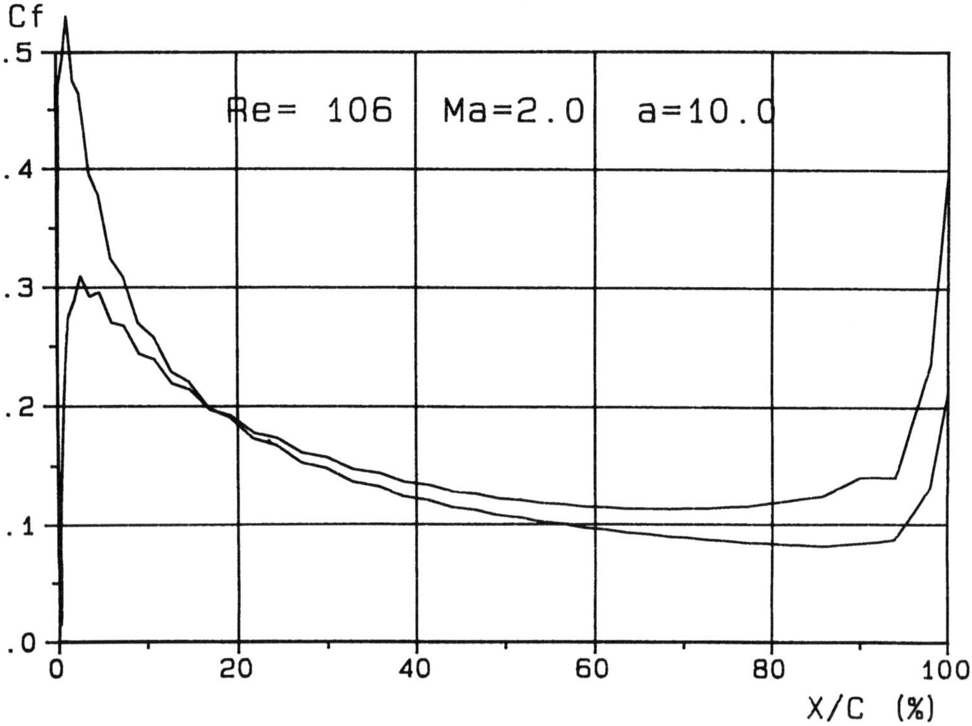

Fig 2.e Friction coefficient

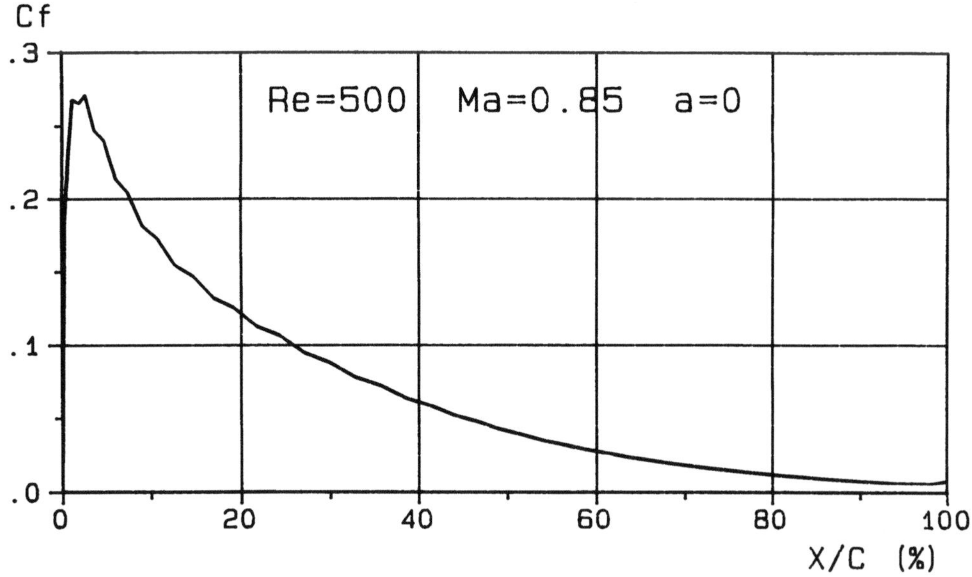

Fig 2.f Friction coefficient

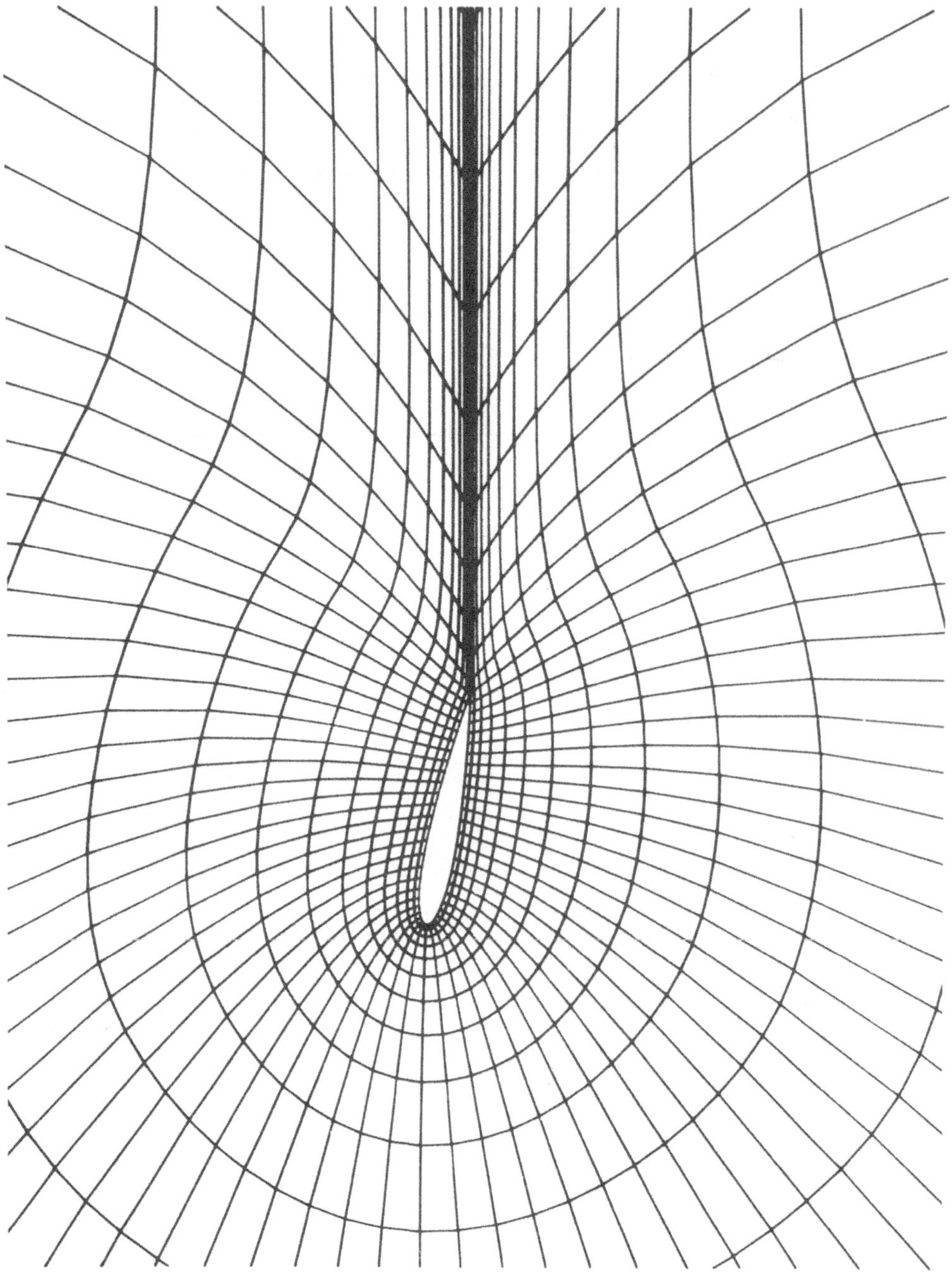

Fig. 3.a Enlargement of the final mesh : 5152 Nodes 1248 Elements

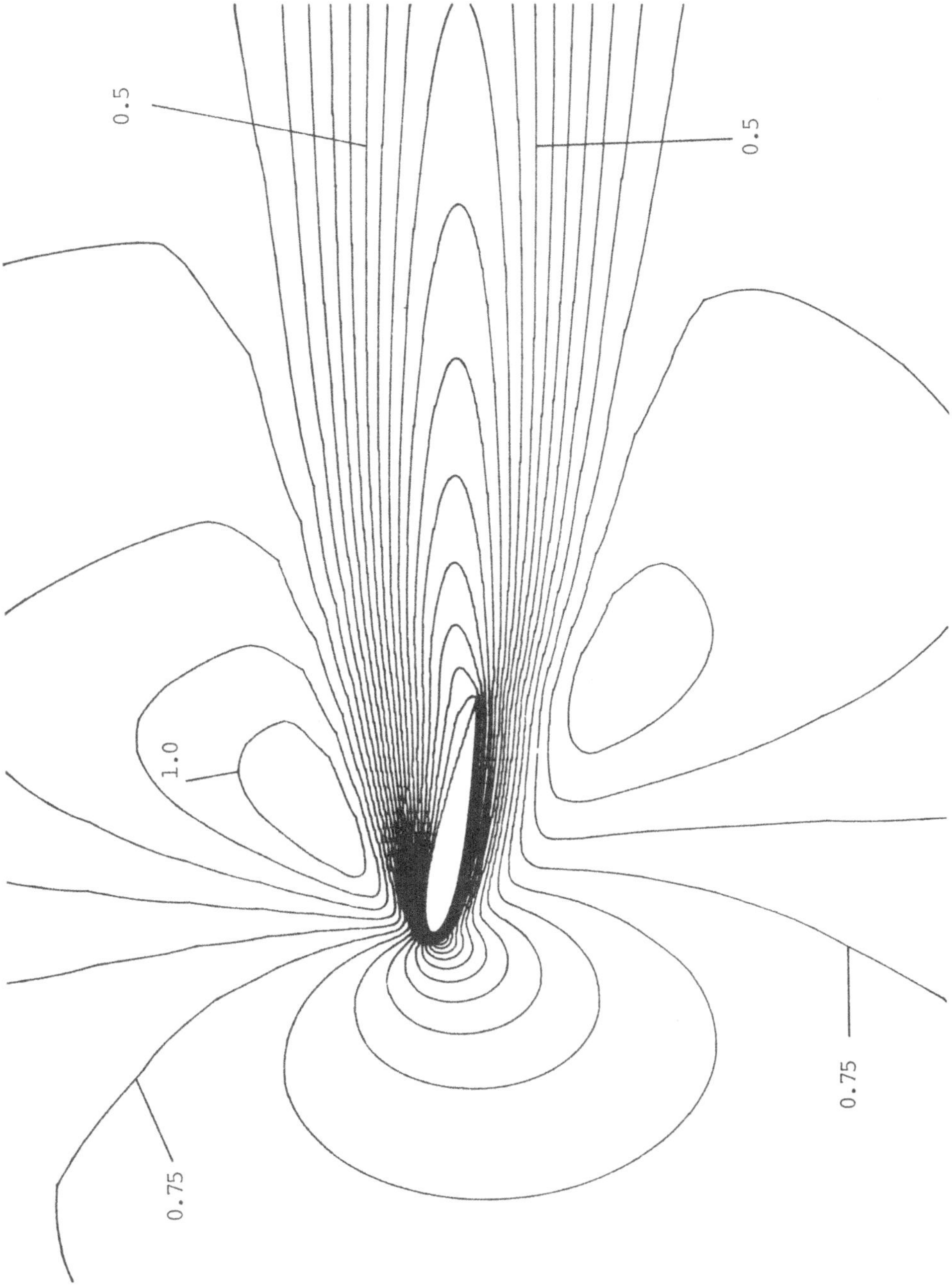

Fig. 3.b1 Case A1 Iso-Mach lines Re = 73 Ma = 0 8
α = 1.0 Δ = 0.05

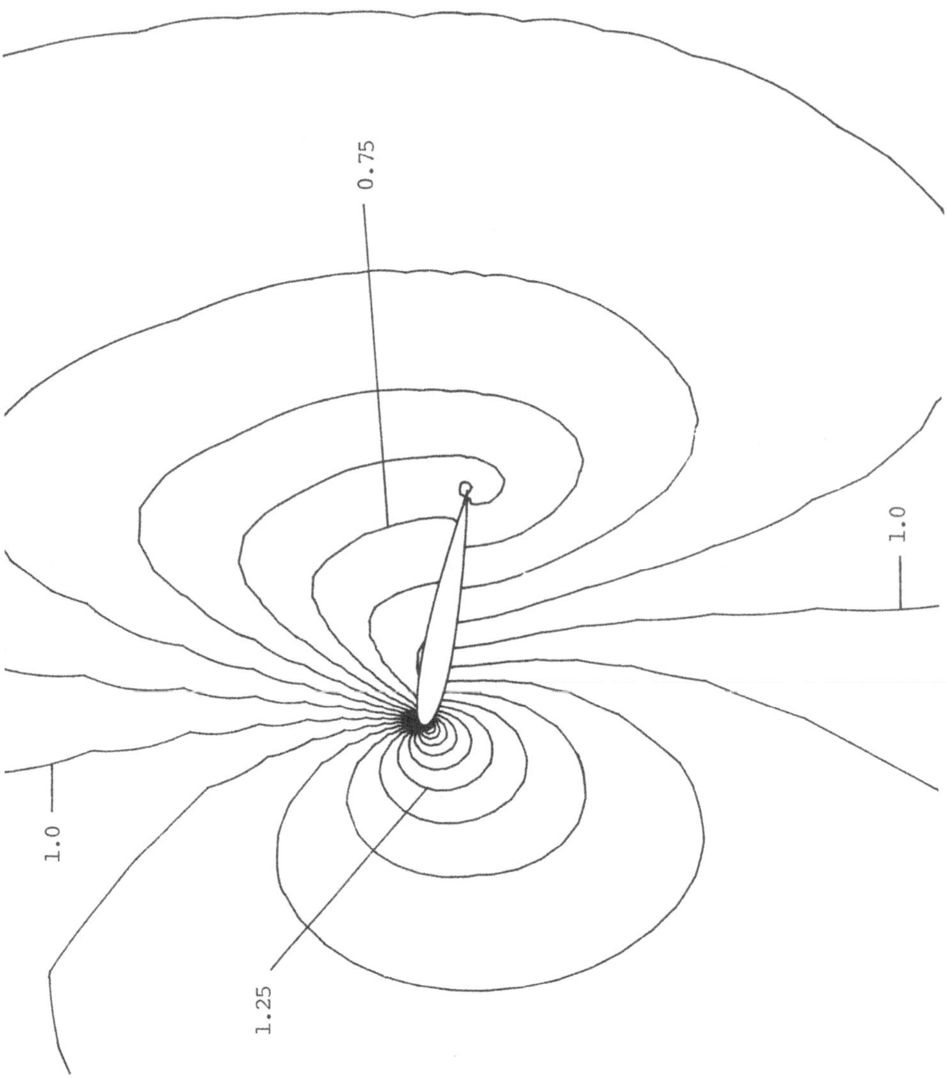

Fig. 3.b2 Case A1 Iso (ρ/ρ_{inf}) lines Re = 73 Ma = 0.8 α = 10 Δ = 0.05

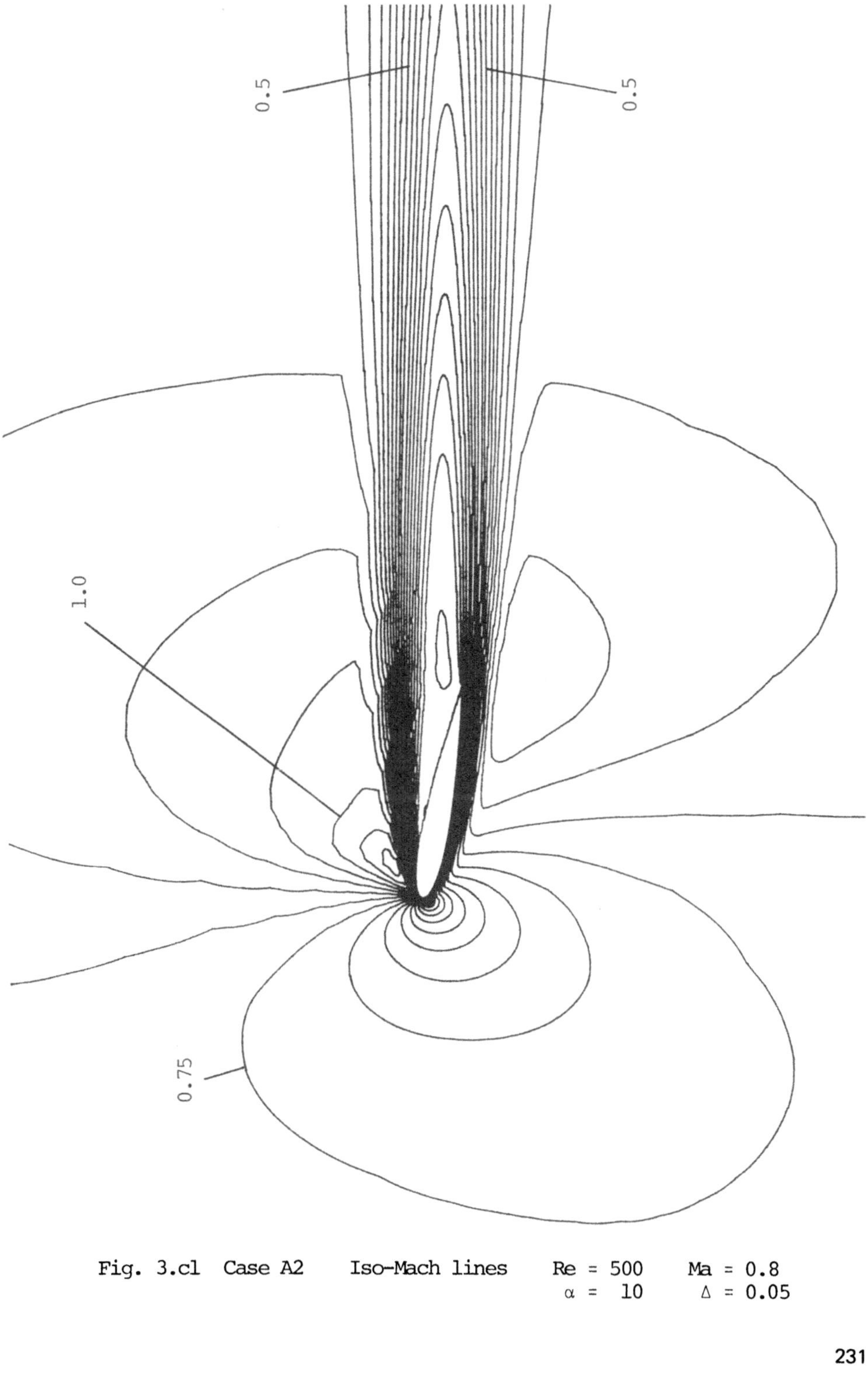

Fig. 3.cl Case A2 Iso-Mach lines Re = 500 Ma = 0.8
α = 10 Δ = 0.05

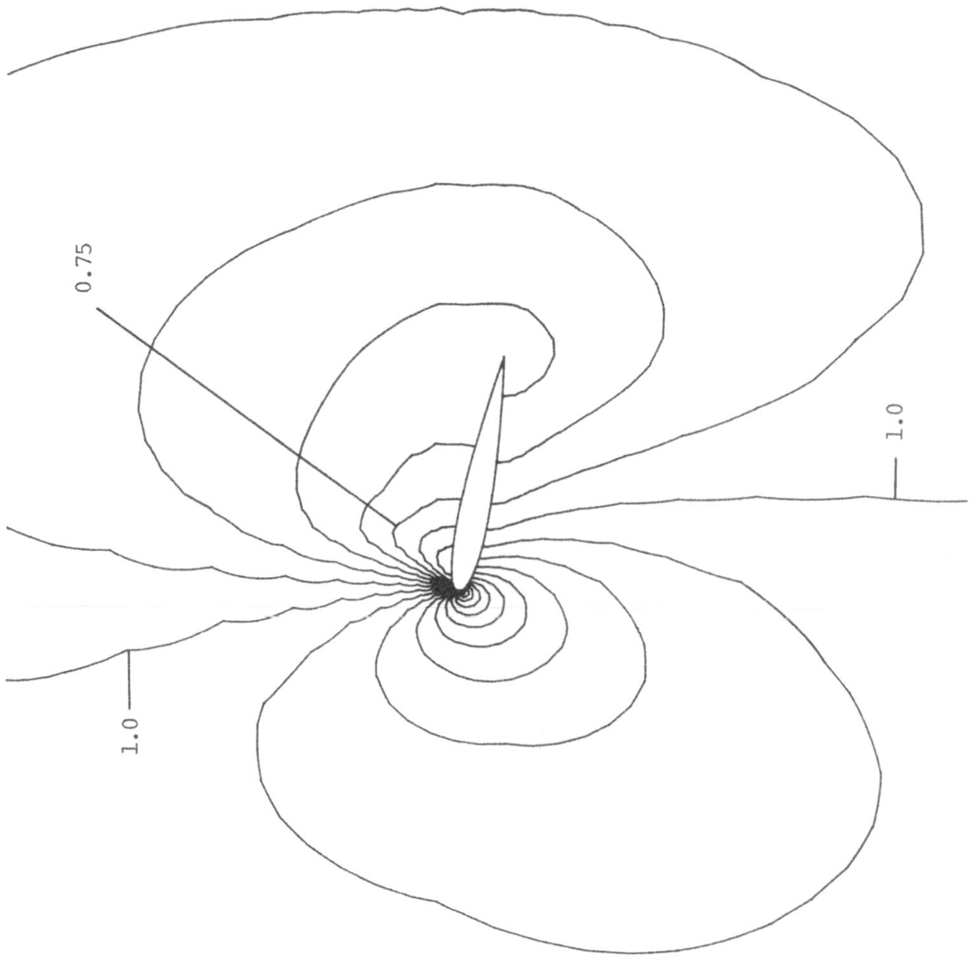

Fig. 3.c2 Case A2 Iso (ρ/ρ_{inf}) lines Re = 500 Ma = 0.8
α = 10 Δ = 0.05

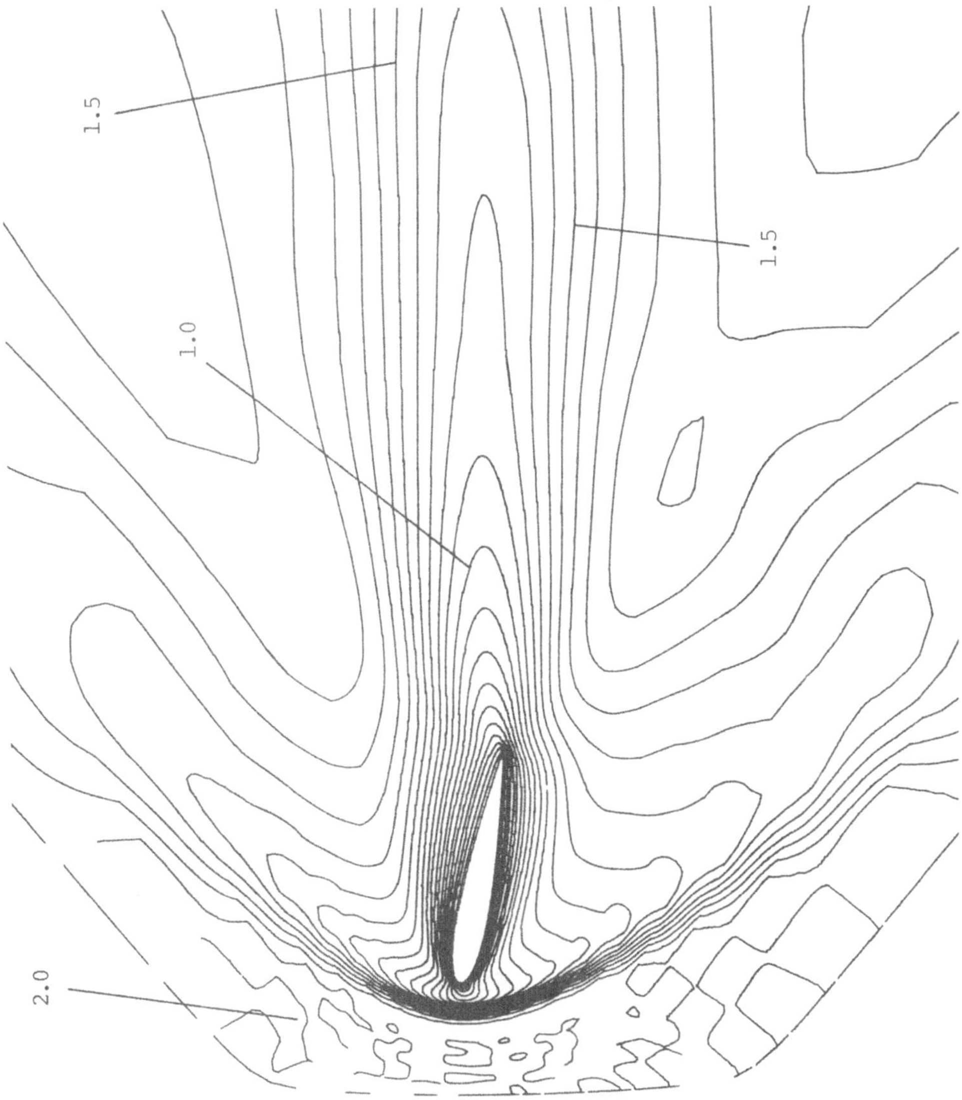

Fig. 3.dl Case A3 Iso-Mach lines Re = 106 Ma = 2.0
α = 10 Δ = 0.1

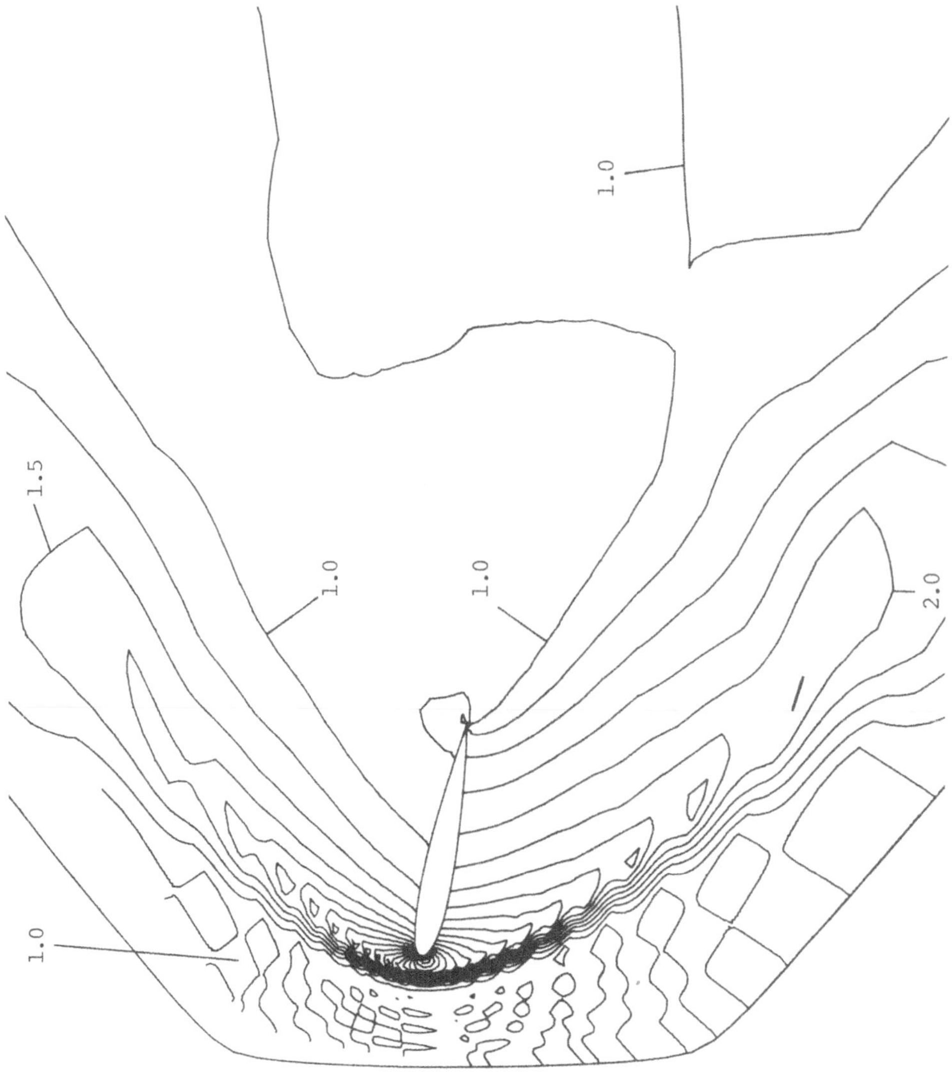

Fig. 3.d2 Case A3 Iso (ρ/ρ_{inf}) Re = 106 Ma = 2.0
α = 10 Δ = 0.25

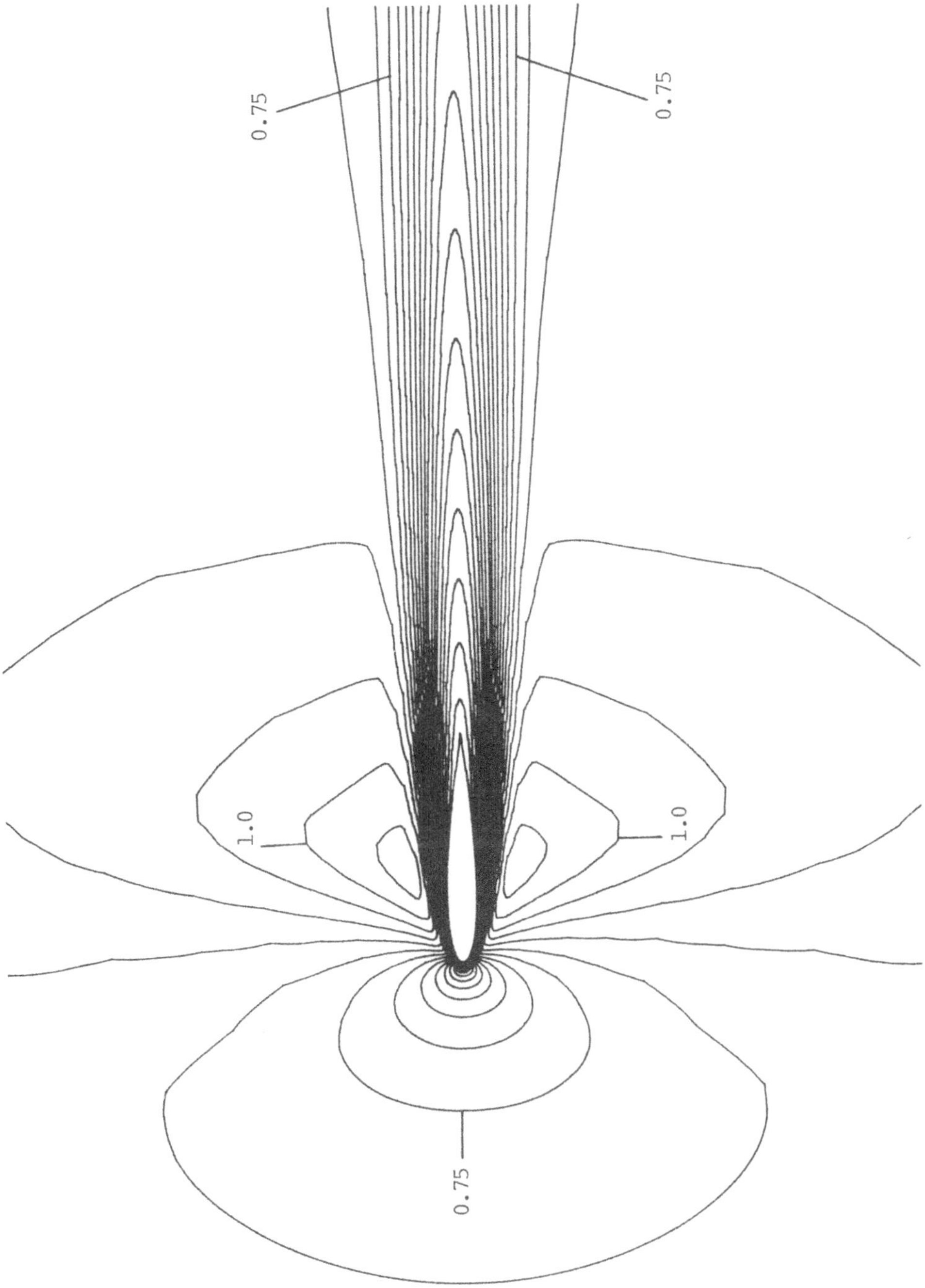

Fig. 3.e1 Case A5 Iso-Mach lines Re = 500 Ma = 0.85
α = 0 Δ = 0.05

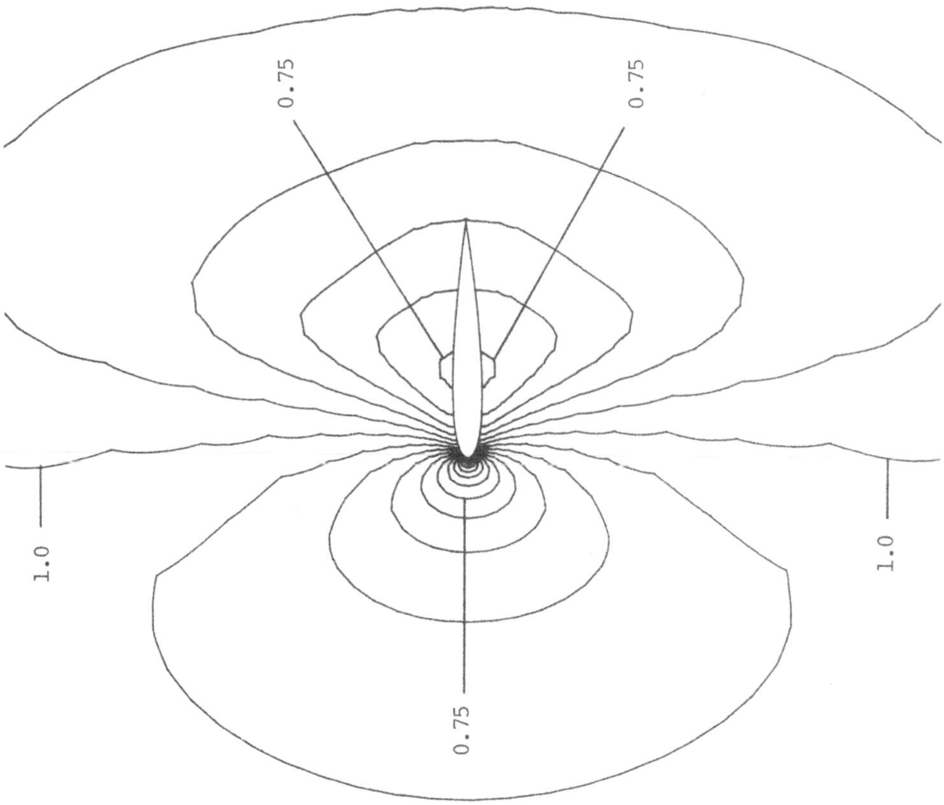

Fig. 3.e2 Case A5 Iso (ρ/ρ_{inf}) Re = 500 Ma = 0.85
α = 0 Δ = 0.05

SOLUTION OF THE COMPRESSIBLE NAVIER-STOKES EQUATIONS

FOR A DOUBLE THROAT NOZZLE

F. Bassi
Dip. di Energetica, P.zza L. da Vinci 32 Milano Italy 20133

F. Grasso
Ist. di Gasdinamica, P.le Tecchio 80, Napoli, Italy 80125

A. Jameson, L. Martinelli
Dept. of Mech. and Aero. Eng., Princeton, N.J. 08544

M. Savini
Ist. Ric. Prop. e Energetica, CNR, V.le Baracca 69, Peschiera Borromeo (Milano), Italy 20068

SUMMARY

An explicit multistage finite volume method for the solution of the compressible Navier Stokes equations has been applied to resolve the transonic flow through a double throat nozzle. The accuracy and robustness of the method are shown by the accurate predictions of: extent of recirculation regions; effects of the Reynolds number on the shock structure; the viscous-inviscid interaction. The dependency of the solution upon grid refinement and wall boundary condition is also studied.

INTRODUCTION

The numerical solution of the compressible Navier Stokes equations (CNS) is a severe test for most numerical algorithms as far as accuracy and robustness of the methods are concerned, on account of the complexities of the flows to be simulated.

In the last decade several implicit or semimplicit algorithms have been developed [1]-[5]. The progress in high speed computers and in the development of efficient numerical methods for the solutions of the Euler equations [6]-[7] have stimulated the application of explicit algorithms to CNS [8]-[10].

In the present work the viscous transonic flow through a double throat nozzle is analized by using an explicit multistage finite volume method. The objective of the work is to assess the accuracy and the robustness of the method in the presence of recirculation regions and strong viscous-inviscid interactions.

The numerical algorithm is an extension of that developed for the solution of the Euler equations by Jameson [11]. A cell centered formulation (i.e. the flow variables are defined at cell centers) is employed so as to use a fully conservative form of the Navier Stokes equations. However, as already pointed out in Ref. [10], such an approach requires some interpolation to satisfy Dirichlet boundary conditions for temperature, as requested in the test cases.

The computations have been performed on H-type meshes, best suited for the present applications. Two different grid refinements and two different nodal point distributions have been used to study the influence of the discretization on the solution.

The test cases require to compute the flow for three different values of the Reynolds number. The computed results show that the present method is able to predict the main features of the flow with good agreement with the results obtained by others.

In the next sections the governing equations are presented, then the numerical algorithm is discussed together with the treatment of the boundary conditions and the grid generation technique. Finally the results and some concluding remarks are given.

GOVERNING EQUATIONS

In the present approach a fully conservative formulation of the compressible Navier Stokes equations is employed. The dimensionless form of the equations is obtained by choosing the following reference quantities:

$$x_r, y_r = L \qquad t_r = \frac{L}{c_o} \qquad \rho_r = \rho_o \qquad p_r = \gamma p_o$$

$$u_r = c_o \qquad E_r = c_o^2 \qquad \mu_r = \frac{\rho_0 c_0 L}{Re} \qquad \lambda_r = \frac{c_p \mu_r}{Pr}$$

where L, c, Re and Pr are respectively the half throat nozzle height, the speed of sound, the Reynolds and the Prandtl numbers.

In a cartesian coordinate system the equations are:

$$\frac{\partial}{\partial t} \int_V w dV + \oint_{\partial V} (f n_x + g n_y) dS = \oint_{\partial V} (P n_x + Q n_y) dS \qquad (1)$$

where w, f, g, P and Q are respectively the vector unknown, the non dissipative flux components in x and y direction and the diffusive flux components in x and y, and they are defined as follows:

$$w = [\rho, \rho u, \rho v, \rho E]^T$$

$$f = [\rho u, \rho u^2 + p, \rho uv, \rho uE + up]^T$$

$$g = [\,\rho v,\ \rho uv,\ \rho v^2 + p,\ \rho vE + vp]^T$$

$$P = [0,\ \sigma_{xx},\ \sigma_{xy},\ u\sigma_{xx} + v\sigma_{xy} - q_x]^T$$

$$Q = [0,\ \sigma_{xy},\ \sigma_{yy},\ u\sigma_{xy} + v\sigma_{yy} - q_y]^T$$

where u and v are the velocity components in x and y; e, p and E are respectively the density, pressure and total energy.
Equation of state

$$p = (\gamma - 1)\,[E - 1/2(u^2 + v^2)]$$

Constitutive equations

$$\underline{\underline{\sigma}} = \frac{1}{Re}\,[(\nabla \underline{u} + \nabla \underline{u}^T) - 2/3\nabla\cdot\underline{u}\ \underline{\underline{U}}]$$

$$\underline{q} = -\frac{\gamma}{RePr}\,\nabla e$$

where e, γ and $\underline{\underline{U}}$ are respectively the specific internal energy, the ratio of the specific heat coefficients and the unit tensor.

NUMERICAL SOLUTION

The computational domain is partitioned into arbitrary quadrilateral cells. The discretized form of the governing equations is obtained by applying Eqn. (1) to each cell and assuming that the flow variables are uniform within each computational volume. The spatial and temporal terms are decoupled by using the method of lines [11]-[12], and the system of governing equations is reduced to a system of ordinary differential equations (ODE).

The volume integrals are approximated by means of the mean value theorem, and the mid-point rule is employed for the surface integrals. The following ODE is obtained:

$$\frac{d}{dt}(Vw)_{ij} = -\underset{\mathrm{I}}{\sum_{\beta=1,4}[(\hat{f}n_x + \hat{g}n_y)\Delta S]_\beta} + \underset{\mathrm{II}}{\sum_{\beta=1,4}[(\hat{P}n_x + \hat{Q}n_y)\,\Delta S]_\beta} \qquad (2)$$

where n_x and n_y are respectively the components of the positive unit normal in x and y.

The numerical flux $(\hat{f},\hat{g})$ at cell face β is evaluated as the average of the corresponding values at the two adjacent cells. This definition guarantees that $(\hat{f},\hat{g})$ be consistent with (f,g) in the following sense:

$$\hat{f}(w,w) = f(w)$$

$$\hat{g}(w,w) = g(w)$$

moreover it enhances the computational efficiency and it yields second order accuracy on a smooth grid. Observe that term I coincides with the Euler-type flux contribution (R_E).

The discretized counterpart of the diffusion fluxes ($\hat{P},\hat{Q}$) are evaluated by applying Gauss theorem to a computational volume centered around a grid node and whose vertices are the centers of the four adjacent cells, so as to calculate the velocity and temperature gradients [10], [12].

Adaptive dissipative terms, proportional to the second and fourth differences of the variables, are added to the discretized governing equations to prevent even/odd decoupling and to inhibit expansion shocks [11], when solving the Euler equations. The Navier Stokes equations contain physical dissipative terms. However some adaptive dissipation is still added to guarantee the proper behaviour of the scheme in regions where the convection and diffusion contributions are not of the same order. Hence Eqn. (2) is augmented by adding adaptive dissipation terms similar to the ones employed in Euler calculations [10]-[11], yielding:

$$\frac{d}{dt}(Vw)_{ij} = - R_E(w) - D(w) - AD(w) \tag{3}$$

where D represents term II of Eqn. (2), and AD is the adaptive dissipation.

The time integration of Eqn. (3) is obtained by employing an explicit 3-stage Runge-Kutta scheme as that used in ref. [10]. To enhance the computational efficiency both the physical and artificial dissipative terms are evaluated once per time step. The solution is then advanced in time as follows:

$$w^{(0)} = w^n$$

$$w^{(1)} = w^{(0)} - \alpha_1 \Delta t/V[R_E^{(0)} + D^{(0)} + AD^{(0)}]$$

$$w^{(2)} = w^{(0)} - \alpha_2 \Delta t/V[R_E^{(1)} + D^{(0)} + AD^{(0)}]$$

$$w^{n+1} = w^{(0)} - \Delta t/V[R_E^{(2)} + D^{(0)} + AD^{(0)}]$$

where $\alpha_1 = \alpha_2 = .6$.

The efficiency of the method is further increased by use of local time stepping and implicit smoothing.

Mesh Description

The mesh used in the computations is a body fitted H-mesh obtained by solving the Laplace equation for the stream function (ψ) on an equally spaced grid that covers the physical domain. The coordinates of the nodes are obtained by finding lines of constant Φ (the velocity potential) and constant ψ, yielding a system of orthogonal grid coordinate lines.

To improve viscous layer resolution the grid is refined near the wall. The refinement is accomplished by use of the following nodal point distribution (along ψ) [13]:

$$f(\psi) = m\psi + (1-m)[1 - \frac{\tanh D(1-\psi)}{\tanh \psi}]$$

where D measures the rate of deviation from a uniform linear distribution with slope m.

Boundary Conditions

At the wall no slip boundary conditions are imposed on the velocity components, i.e.

$$\underline{u} \cdot \underline{n} = 0$$

$$\underline{u} \cdot \underline{t} = 0 .$$

The temperature is set equal to the stagnation value as requested in the test cases.

The pressure is needed at the wall when evaluating the momentum balance equation for the cells adjacent to the solid boundary. For this purpose two different boundary conditions are investigated:

$$\nabla p \cdot \underline{n} = 0 \tag{4}$$

$$(-\nabla p + \nabla \cdot \underline{\underline{\sigma}}) \cdot \underline{n} = 0 . \tag{5}$$

Computations performed using either Eqn. (4) or (5) have shown negligible differences in the results.

Due to symmetry of the flow, symmetry boundary conditions are imposed along the centerline axis.

At the upstream boundary the viscous effects are neglected and the following boundary conditions are imposed (being $M_0 < 1$):

$$H = H_0 \tag{6}$$

$$s = s_0 \tag{7}$$

$$\frac{2}{\gamma-1} c + q_n = \frac{2}{\gamma-1} c_e + q_{n_e} \tag{8}$$

$$q_n n_y - q_t n_x = 0 \tag{9}$$

where n_x, n_y are the components of the normal to the surface, positive if pointing outward; q_n, q_t, H and s are respectively the normal and tangential velocity components, the total enthalpy and the entropy.

Eqn. (8) corresponds to a boundary condition for the outgoing Riemann invariant. Eqn. (9) implies that the flow be parallel to the centerline axis.

At the downstream boundary, numerical boundary conditions are obtained by extrapolation from the interior in the direction normal to the outflow surface, i.e.

$$\frac{\hat{\partial}^2}{\partial n^2}\left(\frac{2c}{\gamma-1}+q_n\right)=0 \qquad (10)$$

$$\frac{\hat{\partial}^2 s}{\hat{\partial} n^2}=0 \qquad (11)$$

$$\frac{\hat{\partial}^2 q_t}{\hat{\partial} n^2}=0 \qquad (12)$$

$$\frac{\hat{\partial}^2}{\partial n^2}\left(\frac{-2c}{\gamma-1}+q_n\right)=0 \qquad (13)$$

where ^ stands for numerical discretization.

Observe that Eqn. (13), valid only for locally supersonic flow, is used also in the subsonic case, as required in the specifications of the test cases.

RESULTS

The calculations for the three test cases at Re = 100; 400; 1600 have been performed on a 152x32 grid (Fig. 1) with the nodal point distribution of Fig. 2. The effects of mesh refinement and those due to the two different wall pressure boundary conditions (Eqn. (4)-(5)) have been studied at Re = 100.

All the computations have been performed on a GOULD 3267 (2 Mbytes CPU memory, 2 Mips). The computational times are given in table 1 and they correspond to a number of time steps N ≃ 300/hr, on a 152x32 grid without multigrid.

Re = 100 (CASE 1.A)

The flow variables have been initialized assuming uniform conditions corresponding to an inlet Mach number of .4. To drive the flow to a fully expanded situation all but Eqn. (13) are imposed at the outflow. Indeed the fourth boundary conditions has been set by imposing a pressure corresponding to an isentropic Mach number equal to 1.3 until a supersonic region develops. After that the outflow boundary conditions (Eqns. (10)-(13) are imposed.

Figs. 3-11 show the strong influence of the viscous effects and an extended recirculation region between the two throats as indicated by the wall skin friction distribution and the velocity profiles vs x.The flow separates at x_s=2.9788 and reattaches at x_R = 4.8343. Observe that for this case, strong recompression regions are absent .

Effects of Mesh Refinement (CASE 1.B-C)

Results obtained on a 152x32 grid with a high mesh refinement near the wall (see Figs. 30-35) indicate that, at least for such a low Reynolds number, it is not necessary to concentrate as many grid points near the wall. The level of convergence is worse than in the previous case as observed by comparing the mass flow rate along the channel for the two cases .

Comparison of the results obtained on the coarser mesh (102x22) show that the overall resolution of the flow field is adequate except through the compression region located downstream of the first throat (see Figs.36-41). Further differences are observed in the distribution of the heat flux coefficient vs x. However the differences seem due to the higher convergence reached on the 102x22 mesh.

Effects of the Wall Pressure Boundary Conditions (CASE 1.D)

Figs. 42-44 show the results obtained with the wall boundary conditions on pressure $\nabla p \cdot \underline{n} = 0$. The results indicate that the boundary layer approximation is adequate when using a highly refined mesh near the wall.

Re = 400 (CASE 2)

The flow variables have been initialized from the converged solution obtained at Re = 100. The isobar and isomach lines (Figs. 12-14), as well as the pressure and Mach distributions on the centerline (Figs. 15-20), show the formation of a rather strong shock. The interaction of the latter with the viscous zone is responsible for the delay in the flow reattachment as observed from Fig. 18 (x_S=3.0025; x_R=6.2829). Due to the viscous layer thickness reduction, the mass flow rate increases with respect to the one at Re = 100, as it appears from Fig. 19, which also shows the good convergence of the computation.

Re = 1600 (CASE 3)

The flow field has been initialized starting from the converged solution at Re = 400. At the Reynolds number 1600, apart from the viscous zone, the flow is supersonic from the first throat to the outlet section. The shock structure becomes oblique and the extent of the recirculation zone between the two throats is reduced: x_S=3.4006; x_R=4.7669 (see Figs. 21-29). A second recirculation bubble appears just downstream of the second throat due to the interaction between the impinging oblique shock and the viscous layer (x_S=8.0326; x_R=8.2560).

CONCLUSIONS

An explicit 3-stage finite volume method for the solution of the compressible Navier Stokes equations has been applied to resolve the transonic flow through a double throat nozzle at different Reynolds numbers. The efficiency of the method is increased by using local time stepping, implicit smoothing and a cell centered formulation. The main features of the flow (extent of recirculation regions, location of the points of separation and reattachment, effects of the Reynolds number on the shock structure, viscous-inviscid interaction) have been accurately predicted.

The effect of different mesh refinements and different nodal point distributions have been studied at Re = 100. At this Reynolds number, the computed results show no need for using very refined meshes. However an adaptive local mesh refinement should be used to improve the accuracy [14], [15] at higher Reynolds numbers, where the flow structure is such that large gradients of the variables arise in small regions.

To enhance the computational efficiency of the method a multigrid technique should also be used [10], [11].

REFERENCES

[1] Steger J.L., "Implicit Finite Difference Simulation of Flow About Arbitrary Geometries with Applications to Airfoils", AIAA paper no. 77-665, 1977.

[2] Beam, R. and Warming, R., "An implicit Factored Scheme for the Compressible Navier-Stokes Equations", AIAA J. Vol. 16, 1978.

[3] Briley, W.R. and Mc Donald, H., "Solution of the Multi-Dimensional Compressible Navier-Stokes Equations by a Generalized Implicit Method", J. of Comp. Physics, Vol. 24, 1977.

[4] Barth, J., Pulliam, T. and Bunning, P., "Navier Stokes Computations for Exotic Airfoils", AIAA paper no. 85-0109, 1985.

[5] Ong, C. and Knight, D., "A Comparative Study of the Hybrid MacCormack and Implicit Beam-Warming Algorithms for a Two-Dimensional Supersonic Compression Corner," AIAA paper no. 86-0204, 1986.

[6] Jameson, A., Schmidt, W. and Turkel, E., "Numerical Solutions of the Euler Equations by Finite Volume Methods Using Runge-Kutta Time-Stepping Schemes," AIAA paper no. 81-1259, 1981.

[7] Jameson, A., Baker, J. and Weatherhill, N., "Calculation of the Inviscid Transonic Flow over a Complete Aircraft," AIAA paper no. 86-0103, 1986.

[8] Agarwaal, R. and Deese, L., "Computations of Transonic Viscous Airfoil, Inlet and Wing Flowfields," AIAA paper no. 84-1551, 1984.

[9] Swanson, R.C. and Turkel, E., "A Multistage Time-Stepping Scheme for the Navier Stokes Equations," AIAA paper no. 85-0035, 1985.

[10] Martinelli, L., Jameson, A. and Grasso, F., "A Multigrid Method for the Navier Stokes-Equations," AIAA paper no. 86-0208, 1986.
[11] Jameson, A., "Transonic Flow Calculations," MAE Report no.1651, Princeton University, 1983.
[12] Peyret, R. and Taylor T.D., "Computational Methods for Fluid Flow," Springer Verlag 1982.
[13] Eiseman, P.R., "Viscous Transonic Cascade Analysis," J. Comp. Physics Vol. 26, 1978.
[14] Berger, M. and Jameson, A., 9th ICNMFD, Saclay, France, 1985.
[15] Dannenhoffer, J.F. and Baron, J.R., "Grid Adaptation for the 2-D Euler Equations," AIAA paper no. 85-0484, 1985.

ACKNOWLEDGEMENT

Research Supported by CNR PFE2 grant no. 84.02644.59.

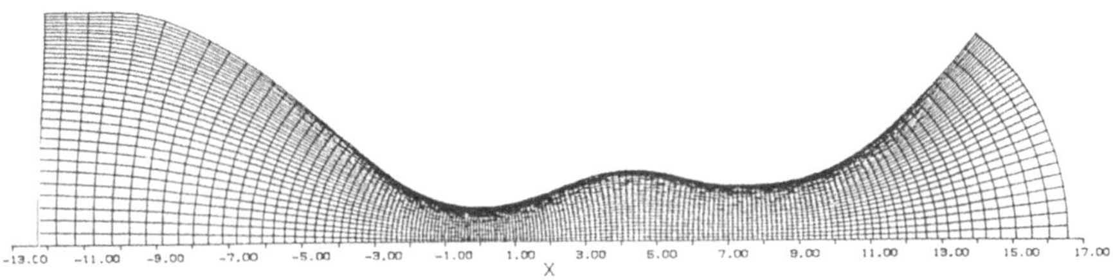

Fig. 1 - Computational Mesh (152x32)

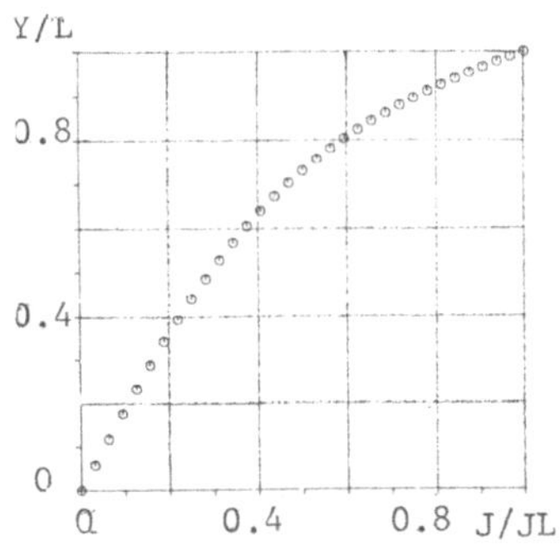

Fig. 2 - Mesh Point Distribution

Table 1 - CPU Time and Residual Values

Re	Grid	CPU Gould 3267	log RMS($\frac{d\rho}{dt}$)
100	152x32	1260	-3.9281
400	152x32	1000	-4.5302
1600	152x32	600	-4.5406

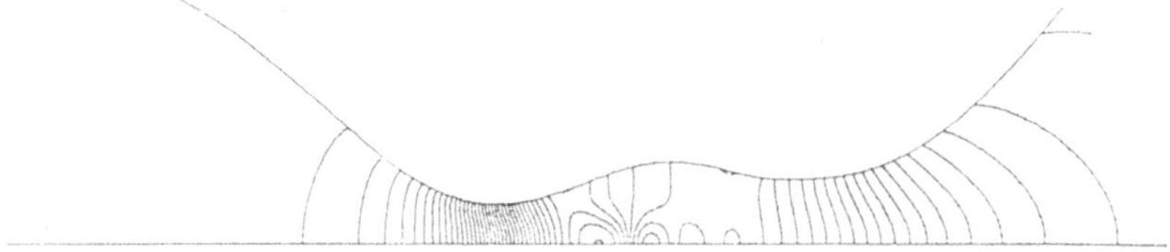

Fig. 3 - Pressure Contours (DP = .02)

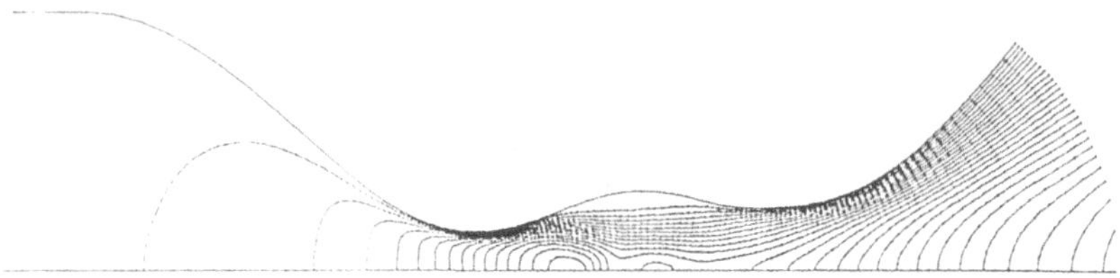

Fig. 4 - Mach Contours (DM = .1)

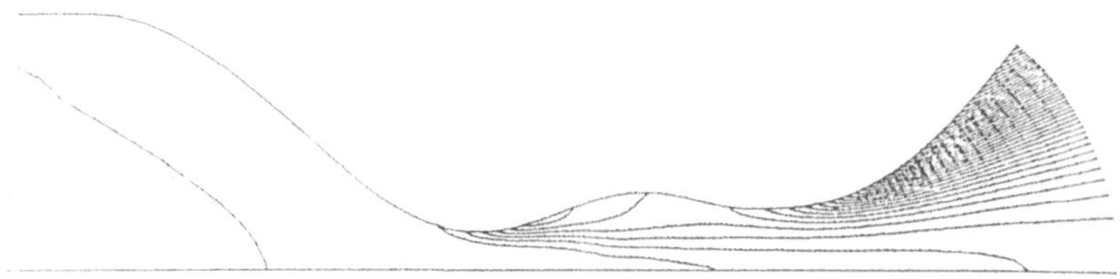

Fig. 5 - Entropy Contours (DS = .1)

Table 2 - RE = 100

	x	-10	-4	0	4	7	13
WALL VALUES	p/p_0	.99661	.98346	.54908	.35386	.33556	.023897
	c_f	.0011264	.0071064	.0095134	-.0025015	.041744	.022022
	c_h	.47032E-5	.69091E-4	.007057	.0011061	.0035174	.0037657
SYMMETRY AXIS VALUES	p/p_0	.99435	.97016	.59374	.40499	.34521	.070498
	M	.091126	.20912	.90042	1.14721	1.18759	2.16658
	S	.99995	1.00008	1.00041	1.05225	1.11369	1.18381

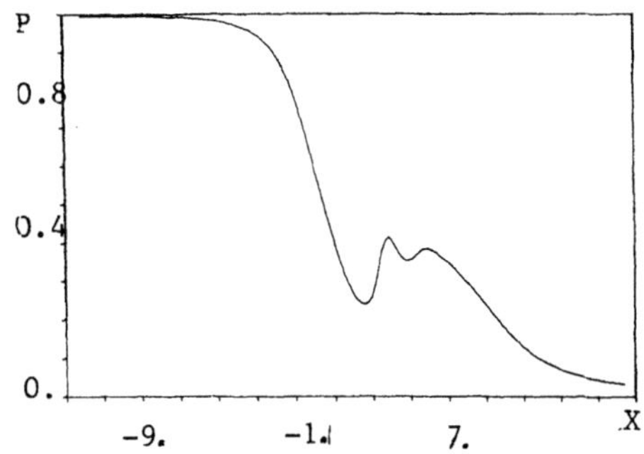

Fig. 6 - Centerline Pressure

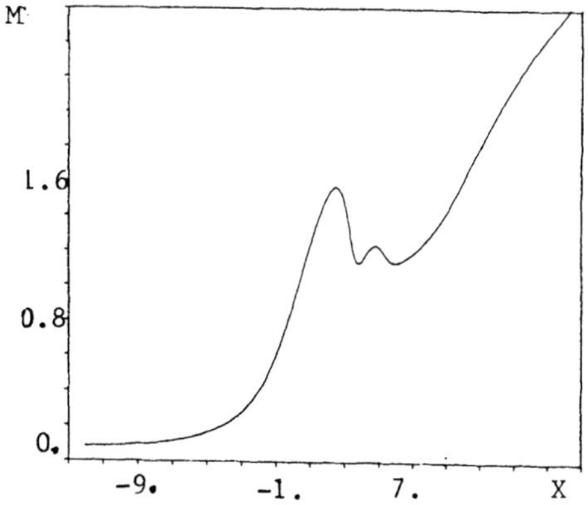

Fig. 7 - Centerline Mach Number

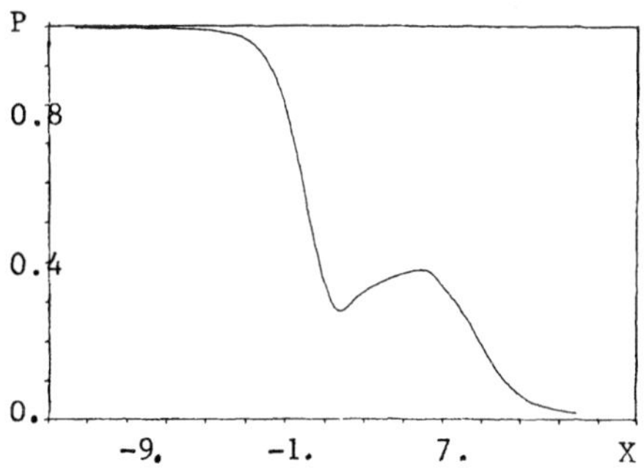

Fig. 8 - Wall Pressure

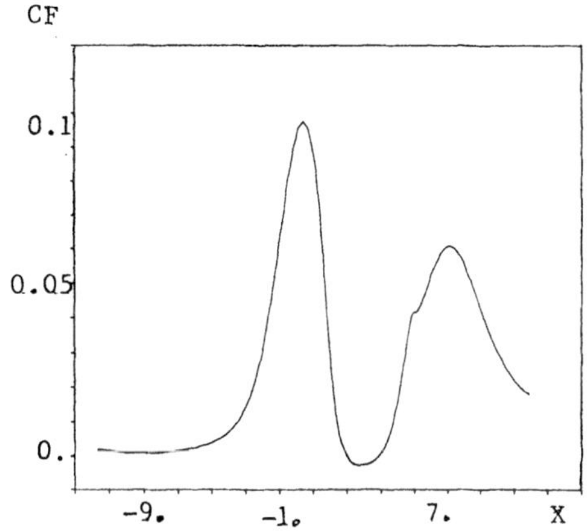

Fig. 9 - Wall Skin Friction

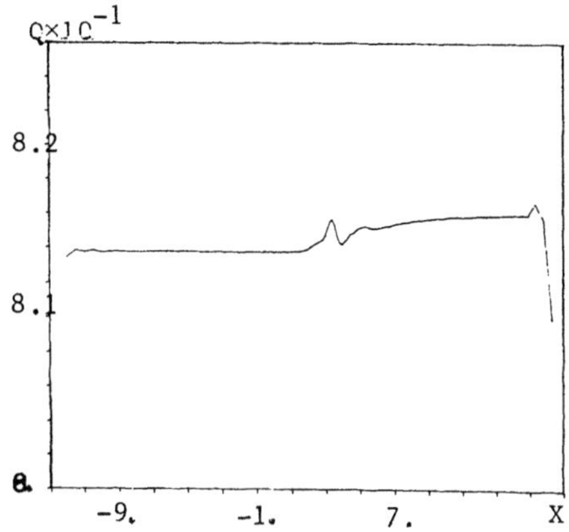

Fig. 10 - Mass Flow Rate

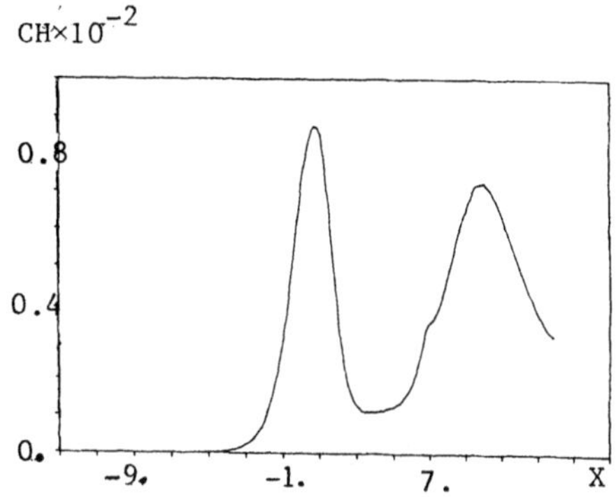

Fig. 11 - Wall Heat Flux

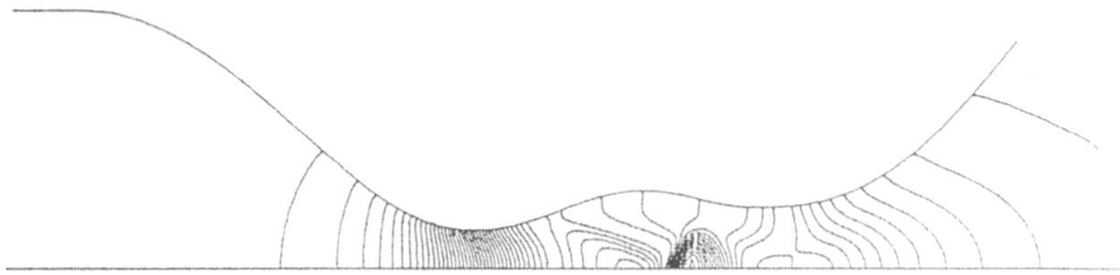

Fig. 12 - Pressure Contours (DP = .02)

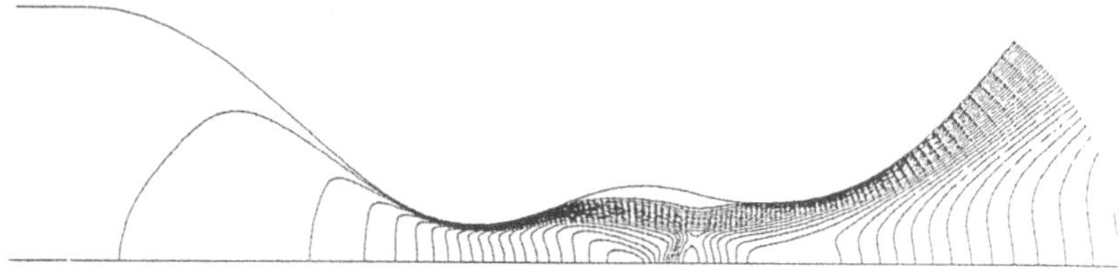

Fig. 13 - Mach Contours (DM = .1)

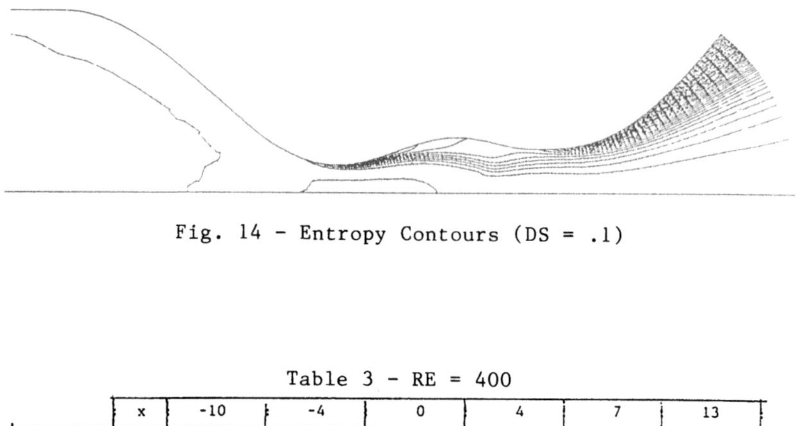

Fig. 14 - Entropy Contours (DS = .1)

Table 3 - RE = 400

	x	-10	-4	0	4	7	13
WALL VALUES	p/p_0	.99638	.98117	.512455	.25509	.28135	.18906
	C_f	.47717E-3	.0039085	.049432	-.00219	.015262	.0095716
	C_h	.23642E-5	.45368E-4	.0036684	.48878E-3	.001548	.0015795
SYMMETRY AXIS VALUES	p/p_0	.99357	.96778	.57570	.11183	.22272	.06108
	M	.096811	.21703	.92619	2.08007	1.5906	2.42434
	S	.99996	1.00002	.9995	1.00341	1.027	1.0324

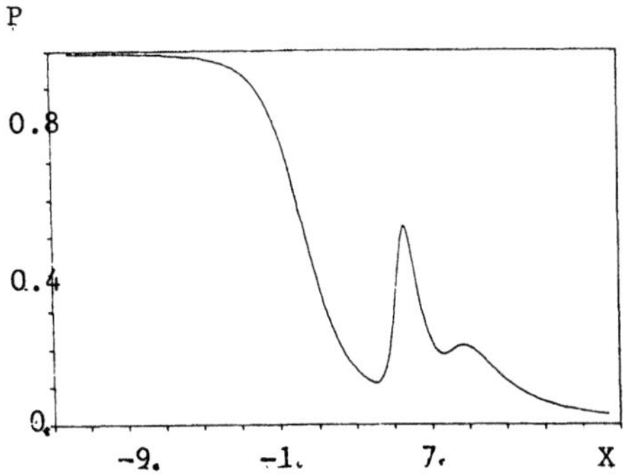

Fig. 15 - Centerline Pressure

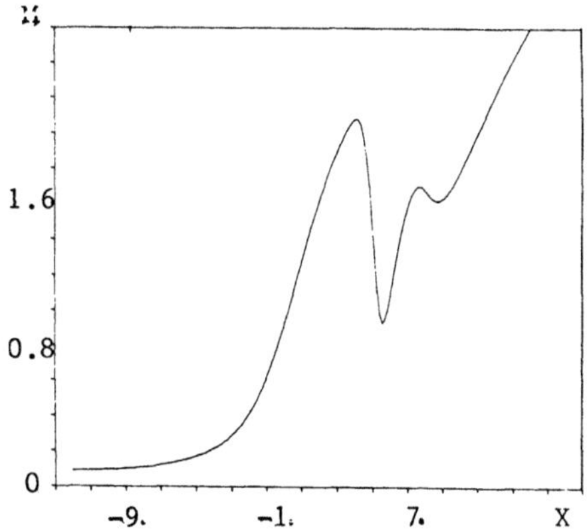

Fig. 16 - Centerline Mach Number

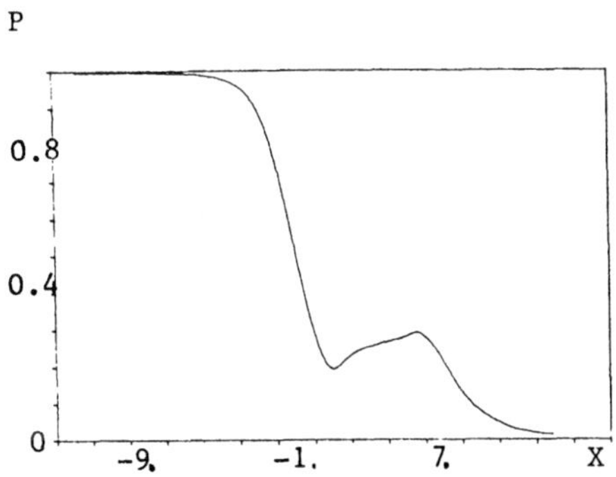

Fig. 17 - Wall Pressure

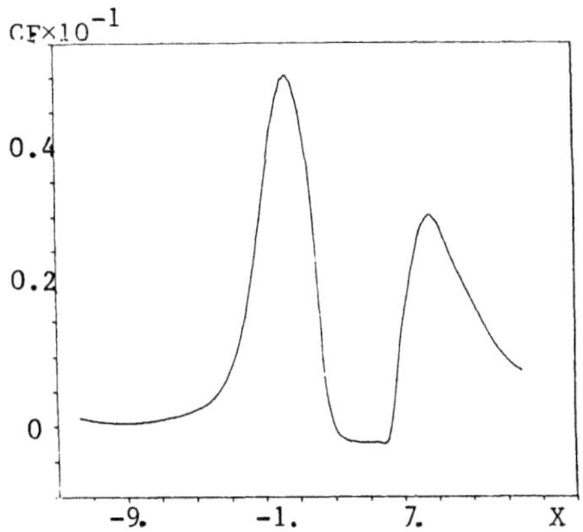

Fig. 18 - Wall Skin Friction

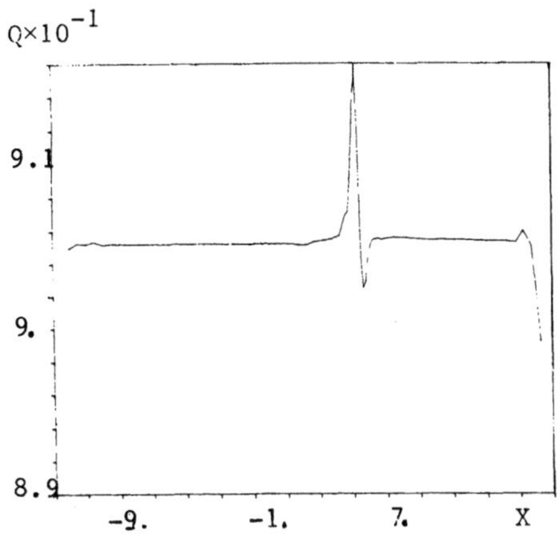

Fig. 19 - Mass Flow Rate

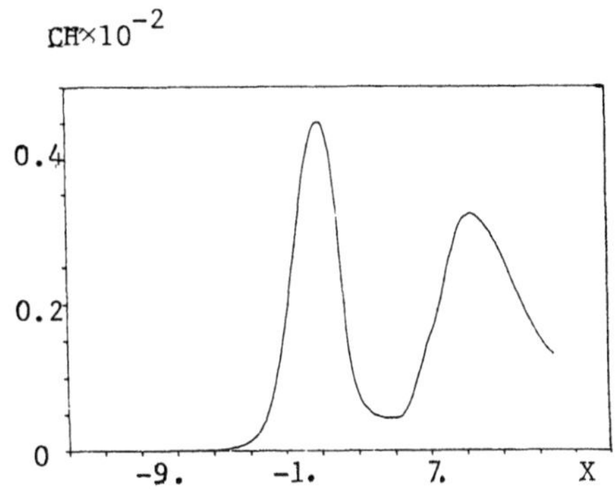

Fig. 20 - Wall Heat Flux

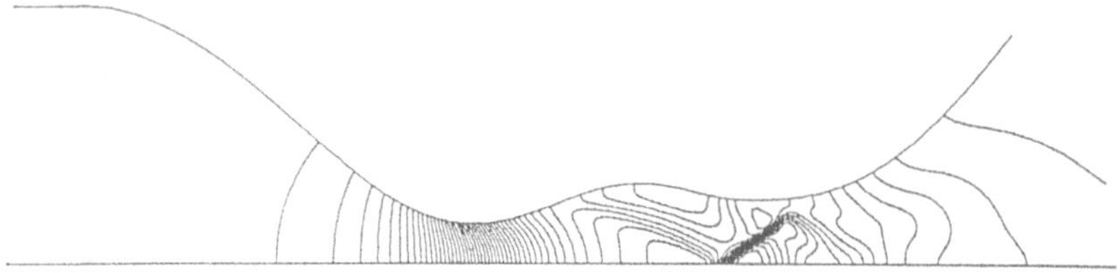

Fig. 21 - Pressure Contours (DP = .02)

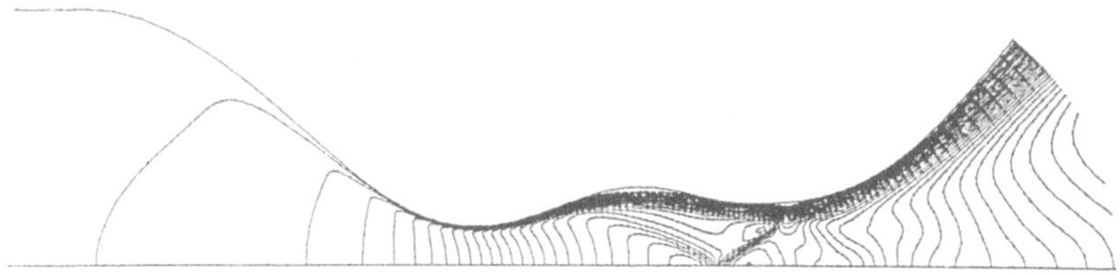

Fig. 22 - Mach Contours (DM = .1)

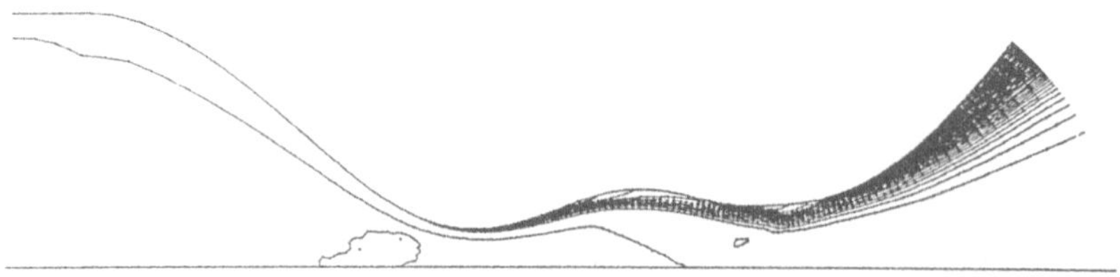

Fig. 23 - Entropy Contours (DS = .1)

Table 4 - RE = 1600

	x	-10	-4	0	4	7	13
WALL VALUES	p/p_0	.99625	.98005	.49071	.19565	.12673	.014435
	c_f	.21724E-3	.0021475	.02526	-.67258E-3	.013253	.004195
	c_h	.24654E-6	.25331E-4	.0019085	.35770E-3	.0014171	.68152E-3
SYMMETRY AXIS VALUES	p/p_0	.99317	.96640	.56657	.094517	.33586	.050162
	M	.099733	.22163	.93948	2.19385	1.30375	2.55742
	S	.99996	1.	.99976	.99964	1.0265	1.0254

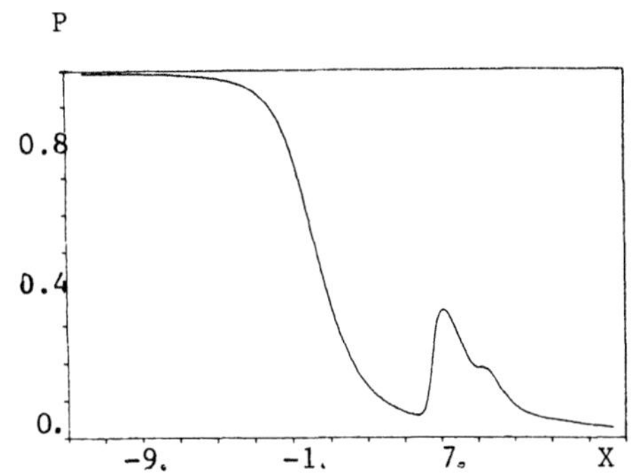

Fig. 24 - Centerline Pressure

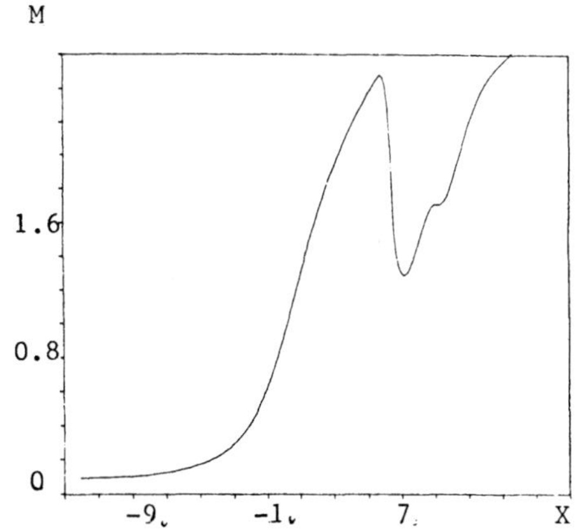

Fig. 25 - Centerline Mach Number

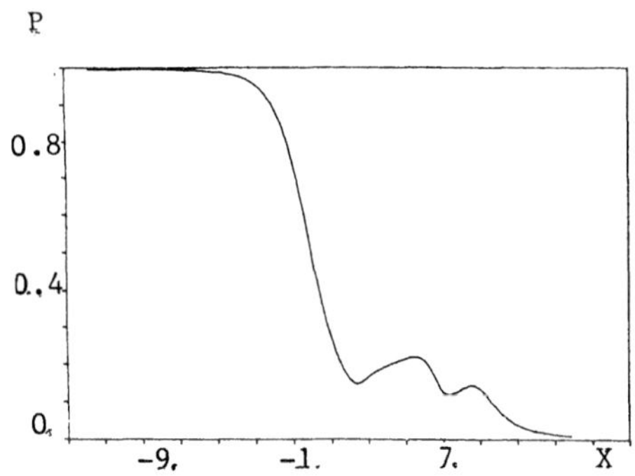

Fig. 26 - Wall Pressure

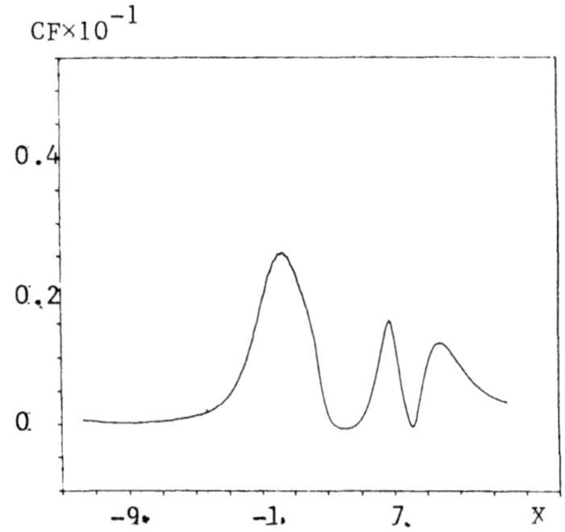

Fig. 27 - Wall Skin Friction

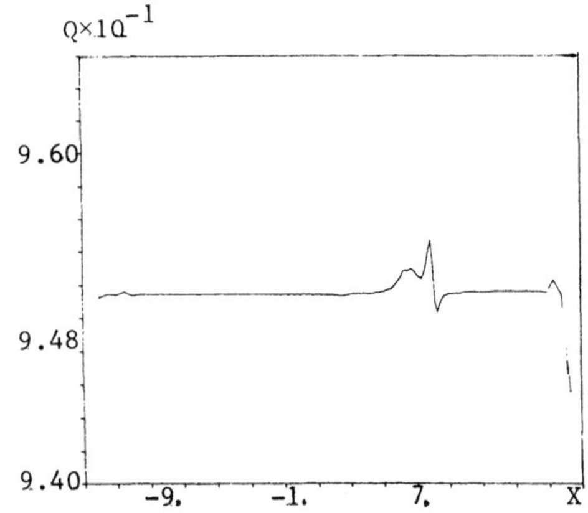

Fig. 28 - Mass Flow Rate

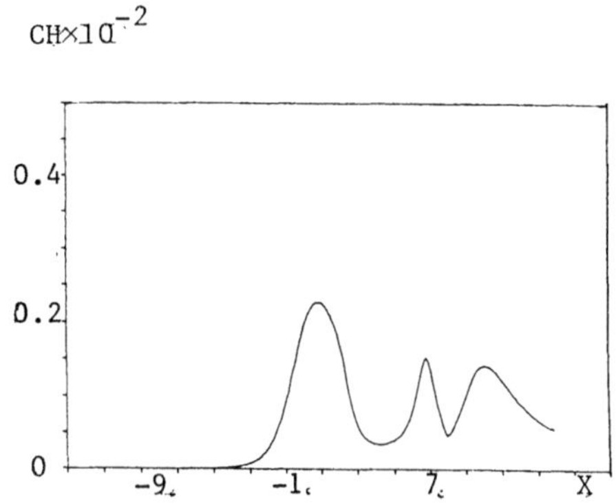

Fig. 29 - Wall Heat Flux

CASE 1.B

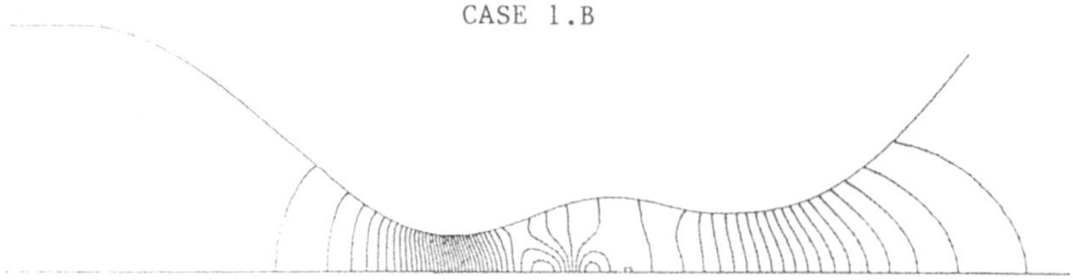

Fig. 30 - Pressure Contours (DP = .02)

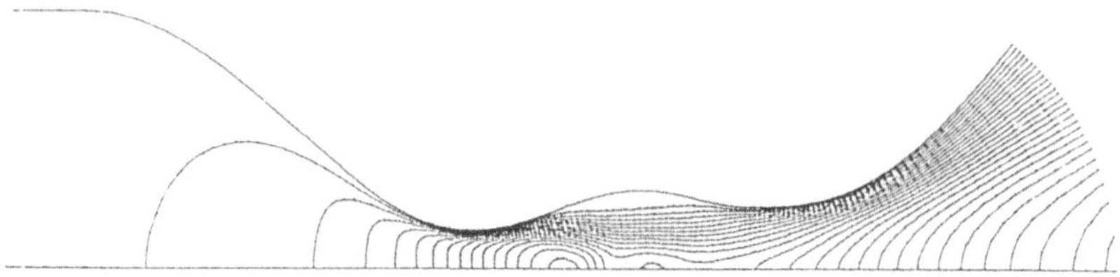

Fig. 31 - Mach Contours (DM = .1)

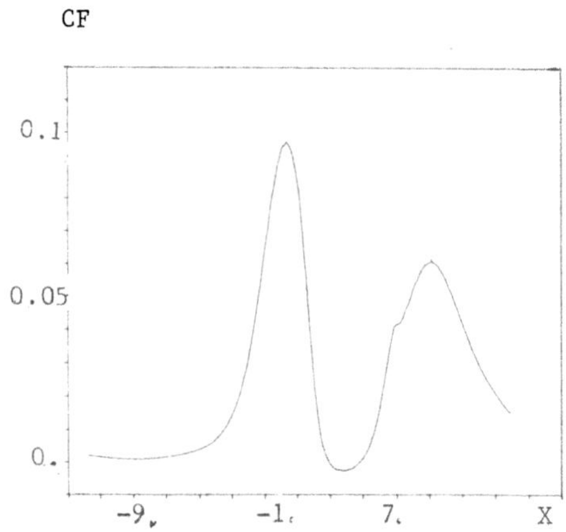

Fig. 32 - Wall Skin Friction

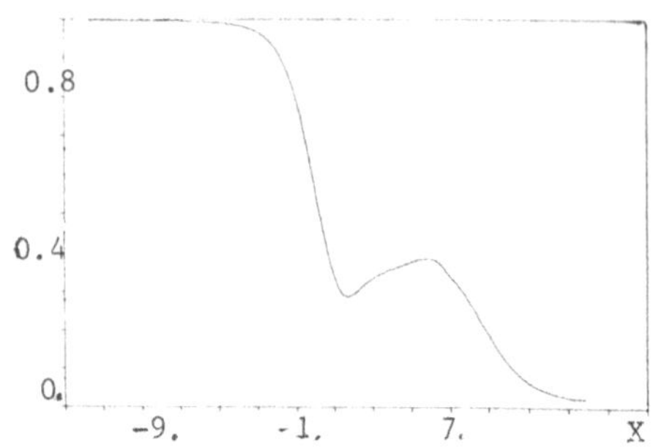

Fig. 33 - Wall Pressure

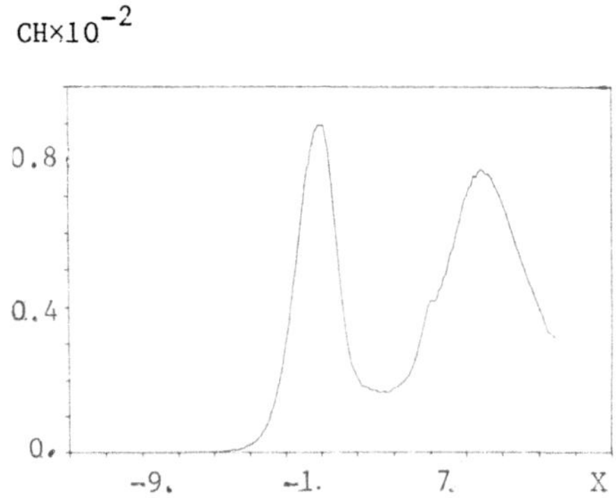

Fig. 34 - Wall Heat Flux

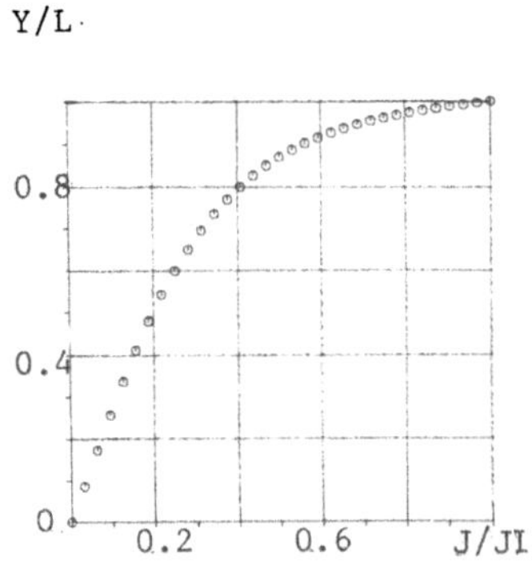

Fig. 35 - Mesh Point Distribution

CASE 1.C

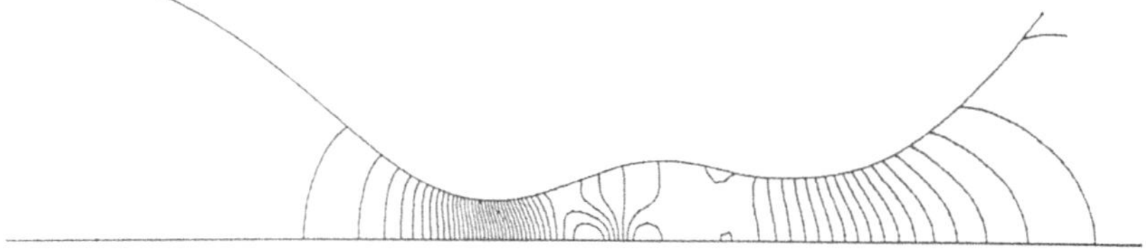

Fig. 36 - Pressure Contours (DP = .02)

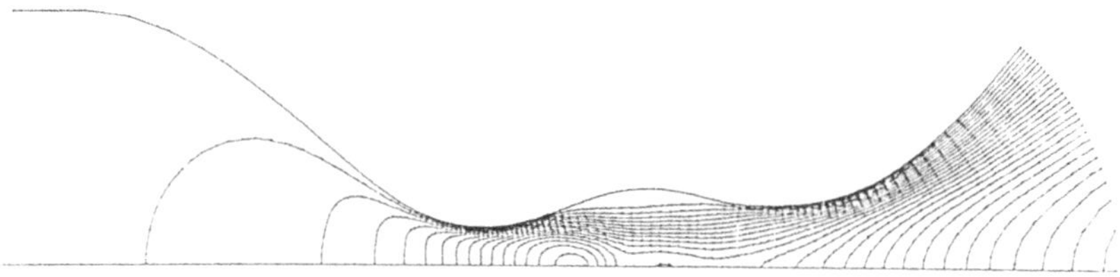

Fig. 37 - Mach Contours (DM = .1)

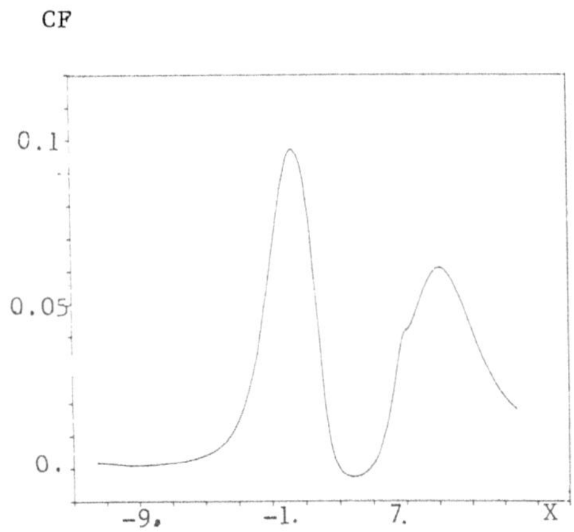

Fig. 38 - Wall Skin Friction

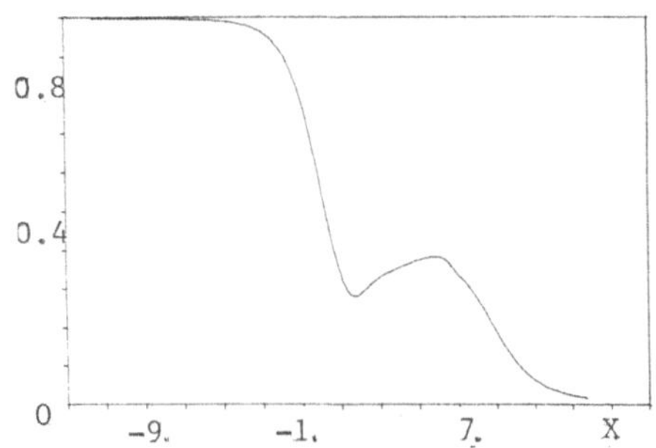

Fig. 39 - Wall Pressure

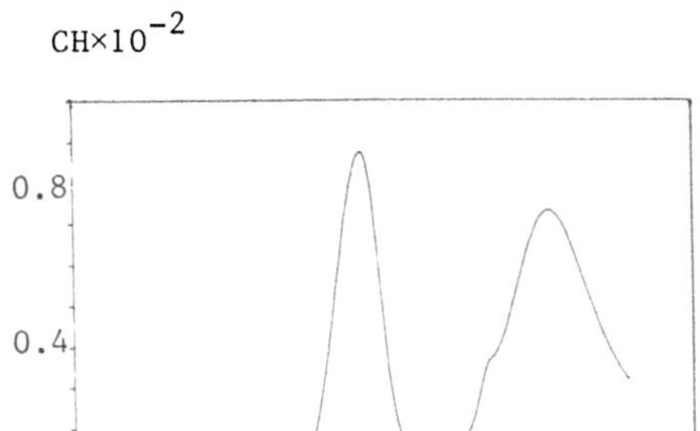

Fig. 40 - Wall Heat Flux

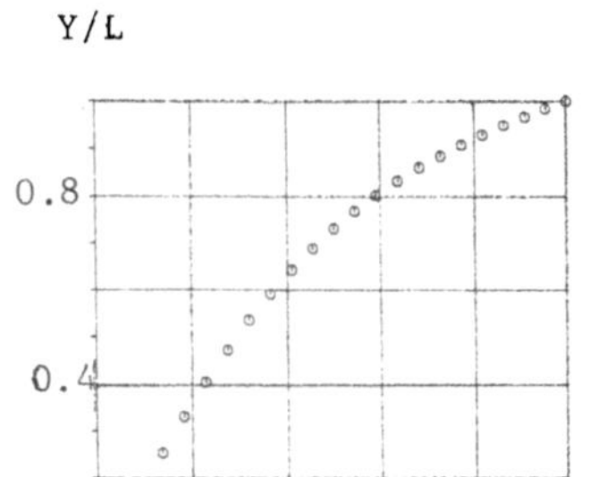

Fig. 41 - Mesh Point Distribution

CASE 1.D

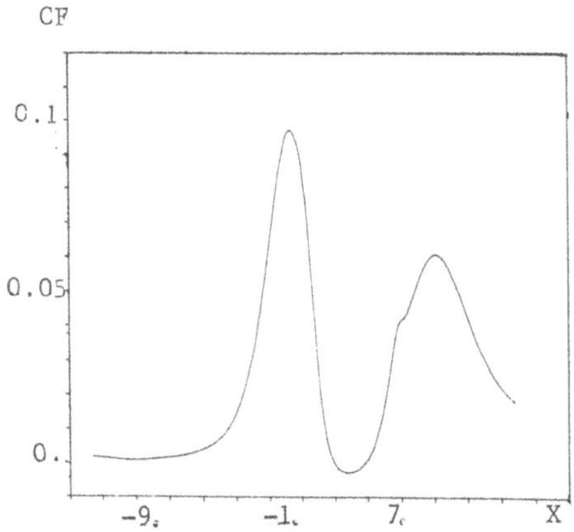

Fig. 42 - Wall Skin Friction

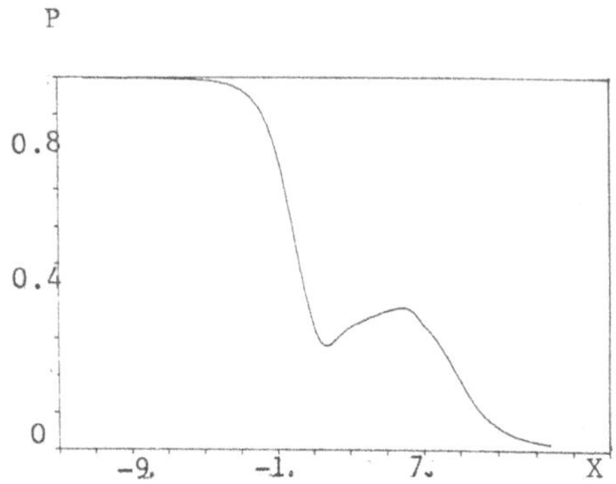

Fig. 43 - Wall Pressure

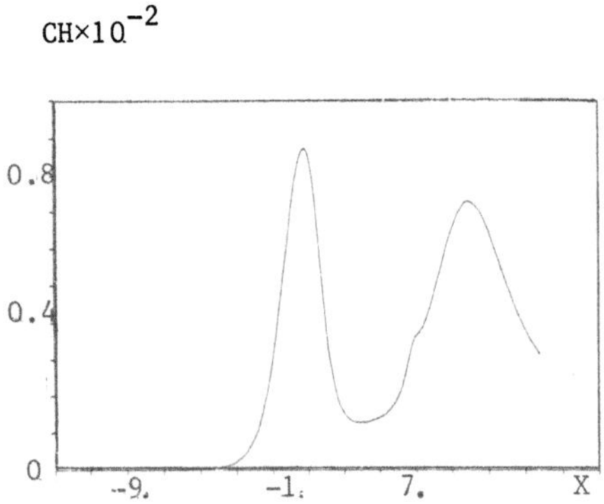

Fig. 44 - Wall Heat Flux

SOLUTION OF 2.D COMPRESSIBLE NAVIER-STOKES EQUATIONS FOR A DOUBLE THROAT NOZZLE BY MEANS OF IMPLICIT FINITE-VOLUME METHOD

O. Labbé
Office National d'Etudes et de Recherches Aérospatiales
BP 72 - 92322 Châtillon Cedex, France

In the work reported here, a numerical method is implemented for predicting the flow characteristics and performance of a 2.D nozzle. The approach is based on an implicit finite-volume method to solve the unsteady Navier-Stokes equations in a boundary conforming curvilinear coordinate system and to obtain the desired time asymptotic steady state solution. This method has already been successfully developed in 3.D and 2.D axisymmetric cases [1, 5].

I. GOVERNING EQUATIONS

The governing equations for compressible, viscous flow consist in conservation equations for mass, momentum and energy (plus state equations). They are here collectively termed the Navier-Stokes equations.

The equations in nondimensional conservation form can be written as :

$$\frac{\partial W}{\partial t} + \frac{\partial F(W)}{\partial x} + \frac{\partial G(W)}{\partial y} = 0 \qquad (1)$$

where $W = (\rho, \rho u, \rho v, \rho E)^T$

u, v are the velocity components in the coordinates directions x and y, ρ is the density, $E = e + (u^2 + v^2)/2$ where e is the specific internal energy and p is the pressure $p = (\gamma - 1)\, \rho e$.
F and G represent the fluxes in the two directions x and y.

We define the Reynolds number Re_o by $Re_o = \dfrac{\rho_o a_o L_o}{\mu_o}$ and the Prandtl $P_r = \dfrac{\mu C_p}{K}$.

The subscript o refers to the reservoir conditions. The viscosity coefficients λ and μ are related by $3\lambda + 2\mu = 0$.

The computational flow region is divided into cells and flow variables are calculated at cell centers. The finite volume method is applied to the integral form of equation (1).

II. SURVEY OF THE METHOD

This method is second-order accurate in time and in space, highly stable and dissipative in weakly viscous regions. At each time-step, a provisional value is calculated by means of an explicit centered predictor-corrector scheme, this value is then corrected by implicit operators factorized with an A.D.I technique. The present method is well-suited to the calculations at high Reynolds numbers since it extends an algorithm used for the Euler equations.

For the following 1.D scalar model equation :

$$W_t + \left(g\,(W,\, W_x)\right)_x = 0 \tag{2}$$

with

$$g(W,\, W_x) = f(W) - \nu(W)W_x \qquad \left(\nu(W) \geqslant 0\right) \tag{3}$$

The time scheme is given by :

$$\delta W^n + a\,\delta t\,\frac{\partial}{\partial x}\,(A\,\delta W)^n + b\,\frac{\delta t^2}{2}\,\frac{\partial}{\partial x}\left(A^2\,\frac{\partial}{\partial x}\,(\delta W)\right)^n - \frac{\delta t}{2}\,\frac{\partial^2}{\partial x^2}\,(\nu\,\delta W)^n = -\,\delta t\,\frac{\partial g}{\partial x}^n + (\,1 - 2a)\,\frac{\delta t^2}{2}\,\frac{\partial}{\partial x}\,\left(A\,\frac{\partial g}{\partial x}\right)^n \tag{4}$$

where $W^n = W\,(x,\, t_n = n\,\delta t)$, $\delta W^n = W^{n+1} - W^n$, $A = \frac{df}{dW}$

and a, b two parameters related by $b = 2a - 1$. For more details of the scheme, the reader is referred to [1].

The R.H.S. of eq. (4) is discretized by means of a centered predictor-corrector scheme which is a generalization of the Thommen scheme [2]. This explicit scheme provides the solution at steady state ($\delta W^n = 0$). The implicit part of the scheme (L.H.S of eq. (4)) includes a linear combination of the inviscid first order term of the Beam-Warming scheme [3] $\delta t\,\frac{\partial}{\partial x}\,(A\,\delta W)^n$ and the typical second-order term of the Lerat schemes [4] $\delta t^2\,\frac{\partial}{\partial x}\left(A^2\,\frac{\partial}{\partial x}\,(\delta W)\right)^n$. The last term $\frac{\delta t}{2}\,\frac{\partial^2}{\partial x^2}\,(\nu\,\delta)W^n$ arises from the Crank-Nicolson approximation of the viscous terms in eq. (2).

The spatial derivatives of the implicit operator are discretized by using centered finite difference formulas, so that the implicit step requires the inversion of a tridiagonal matrix. In the inviscid case, unconditional stability, dissipation and good conditioning of the algebraic systems are achieved for $0 \leqslant a < 1/2$. These properties are preserved in the viscous case, except unconditional stability. However, it has been found [1], that large time-steps can be used and the domain of stability increases as "a" increases from 0 to 1/2.

III. IMPLEMENTATION

The 2.D Navier-Stokes equations are solved by the scheme described above in curvilinear meshes with a finite-volume technique. At each time step, the discretization of the integral form of the equations provides a mean value of the solution inside each mesh cell. The evaluation of the viscous terms requires to approximate velocity and temperature gradients at the interface of two adjacent cells. This is done by finite-difference formulas assuming a central localization of the mean values. A suitable splitting of the fluxes in the implicit operator has been chosen for easy application of an A.D.I technique.

As initial condition, we use the solution of monodimensional isentropic flow and we add a parabolic velocity profile near the wall.

The calculations presented here were made using a local time step technique deduced from the approximate explicit criterion :

$$\delta t_{i,j} = \eta . \mathrm{Min}(\delta t_1, \delta t_2) \qquad (\eta = cst > 0) \tag{5}$$

with $\delta t_1 = \left(\frac{d}{|\underline{u}| + c}\right)_{i,j}$, $\delta t_2 = \left(\frac{\rho P_r Re_o}{2\gamma\mu h}\right)_{i,j}$ where $c^2 = \gamma P/\rho$

and $h = (\mathrm{grad}\ \xi_\beta / \delta\xi_\beta)^2$ $\quad\beta = 1, 2$ (ξ_1, ξ_2 being curvilinear coordinates) where d is the shortest distance between the center of the mesh cell (i,j) and the center of the 4 adjacent cells in the mesh directions i and j. This criterion has been used for the case at $Re_o = 100$, but for $Re_O = 400$ and $Re_o = 1600$, for a line i = cst, we used the minimum of the time step found in "j" direction, i.e. $\bar{\delta t}_i = \min_j (\delta t_{i,j})$.

An artificial viscosity of Lax-Wendroff's type is added to the finite-volume method in order to reduce spurious oscillations in the numerical shock profiles. This artificial viscosity only appears in the momentum equations and is not expensive in computing time. For example, the first momentum component $(\rho\hat{u})^{n+1}_{i,j}$ computed by finite-volume method in the cell (i,j) at time t^{n+1} is corrected as follows :

$$\begin{aligned}(\rho u)^{n+1}_{i,j} = (\rho\hat{u})^{n+1}_{i,j} - \delta t \frac{\cdot\chi}{4} NI \Big\{ &(\rho_{i,j} + \rho_{i+1,j}) |\delta_i u_{i,j}| \delta_i u_{i,j} \\ &- (\rho_{i-1,j} + \rho_{i,j}) |\delta_{i-1} u_{i-1,j}| \delta_{i-1} u_{i-1,j} \Big\}\end{aligned} \tag{6}$$

where NI is the number of cells in the i-direction,
$\delta_i u_{i,j} = u_{i+1,j} - u_{i,j}$ and χ is a dimensionless positive coefficient. This coefficient will be specified in table I according to mesh and Reynolds number. Notice that the artificial viscosity has been added only in the i-direction.

IV. BOUNDARY CONDITIONS

Due to symmetry, the computational domain is restricted to its upper half part. In addition to symmetry conditions, we consider the following treatment of the other boundaries :

- At the wall, the fluxes are evaluated taking account of the no-slip condition, and temperature is equal to the reservoir temperature. At the implicit step, boundary conditions are obtained by :

$$\delta \underline{u} = \underline{0}, \quad \frac{\partial \delta_e}{\partial \xi_2} = 0 \qquad \text{and} \quad \frac{\partial \delta_p}{\partial \xi_2} = 0$$

where ξ_2-derivative is approximated with a first order formula.
Then the correct values of variables are obtained by using a non-centered differencing formula of the normal momentum equation.

- At the inflow boundary, the flow is assumed to be inviscid and the viscous terms are neglected in the governing equations. Since the normal velocity is subsonic at the inflow boundary, only three conditions are needed. In the present calculations, the values of entropy, total enthalpy and flow direction are given. The additional condition is provided by the compatibility relation associated with the outgoing characteristic :

$$p^{n+1} - p_s^{n+1} + (\rho c)^n \, (\underline{u}^{n+1} - \underline{u}_s^{n+1}).\underline{N} = 0$$

where $c^2 = \gamma p/\rho$, N is the outward unit vector normal to the boundary. The subscript s refers to the values at the inflow boundary obtained by extrapolation from the solution at the inner cells ;

- At last, in the outflow boundary cells, the unknowns are extrapolated from the inner cells.

V. RESULTS

Computations were performed on a CRAY 1.S computer. To study the influence of the mesh on the solution we used three meshes. The coarse mesh (104 x 27) needed the storage of 427,000 words. For the medium mesh (208 x 27) the memory required was 741,000 words while the memory required for the fine mesh (208 x 55) was 1.290,000 words. However, optimization of the storage of some matrices occurring in the implicit part of the scheme can reduce the memory up to 40%. In the i-direction, the points are equidistant and we use a geometric progression to mesh to domain in the j direction. For these three meshes the computing rate was about 12. 10^{-5} sec/iteration/grid point. Notice that a mesh of 105 x 30 geometrical points corresponds to 104 x 29 cells and only 104 x 27 computational cells because of the use of phantom points to satisfy boundary conditions in the y-direction.

The table I shows the influence of the mesh on the convergence. The comparison is done for 1000 iterations. The values of the parameters η of eq. (5) and χ of eq. (6) are given for each mesh and each Reynolds number.

Figure 1 shows an enlargement of the iso-Mach lines for the three meshes at $Re_{o,h}$ = 1600. We can notice that in the medium mesh, the oblique shock is less spread out than in the coarse mesh, but it is almost unchanged in the fine mesh. Figure 2 represents the convergence history for the maximal residues at $Re_{o,h}$ = 1600 in the medium mesh. To reach the same rate of convergence in the fine mesh for the same Reynolds number, we need about 4,000 iterations. In figure 3, the pressure distribution on the symmetry axis is plotted for the three Reynolds numbers in the medium mesh, in addition, we plotted the Euler solution obtained by J.P. Veuillot (private communication). For the three Reynolds numbers, the mass flow rate is more constant in the medium mesh than is the coarse one, but it does not improve in the fine mesh. In general, the results are more accurate in the medium mesh than in the coarse one, but they are unchanged in the fine mesh. The speed of convergence is better in the second mesh than in the third one, due to the size of the time step related to the size of the mesh. It seems that the fine mesh does not improve the results in this case, hence we present here all required results only in the medium mesh (Figs. 4 to 23).

REFERENCES

[1] Hollanders, H., Lerat, A., Peyret, R., "3-D Calculation of Transonic Viscous Flows by an Implicit Method," AIAA Journal, Vol. 23, No. 11, Nov. 1985.

[2] Thommen, H.U., "Numerical Integration of the Navier-Stokes Equations", Z.A.M.P., Vol. 17, No. 3, 1966, pp. 369-384.

[3] Beam, R.M. and Warming, R.F., "An Implicit Factored Scheme for the Compressible Navier-Stokes Equations," AIAA Journal, Vol. 16, 1978, pp. 393-402.

[4] Lerat, A., "Implicit Methods of Second-Order Accuracy for the Euler Equations," AIAA Journal, Vol. 23, No. 1, January 1985.

[5] Hollanders, H., Ravalason, W., "Résolution des équations de Navier-Stokes en fluide compressible par une méthode implicite," AAAF, Lille, Nov. 1985.

Table I

Reynolds number	MESH	η	χ	Maximal residues 1000 it	Mean residues 1000 it
100	104 x 27 1	10	0	10^{-8}, 10^{-9}	10^{-10}
100	208 x 27 2	10	0	10^{-7}, 10^{-8}	10^{-8}
100	208 x 55 3	10	0	10^{-3}, 10^{-4}	10^{-4}, 10^{-5}
400	1	10	0.006	10^{-9}, 10^{-10}	10^{-10}, 10^{-11}
400	2	8	0.005	10^{-4}, 10^{-5}	10^{-5}, 10^{-6}
400	3	7	0.005	10^{-2}	10^{-3}
1600	1	10	0.01	10^{-10}, 10^{-11}	10^{-11}
1600	2	7	0.005	10^{-3}	10^{-3}, 10^{-4}
1600	3	7	0.005	10^{-1}	10^{-2}, 10^{-3}

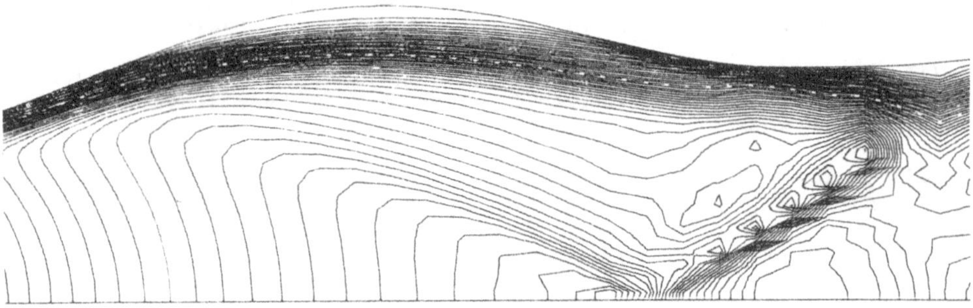

MESH 1 : 104 x 27

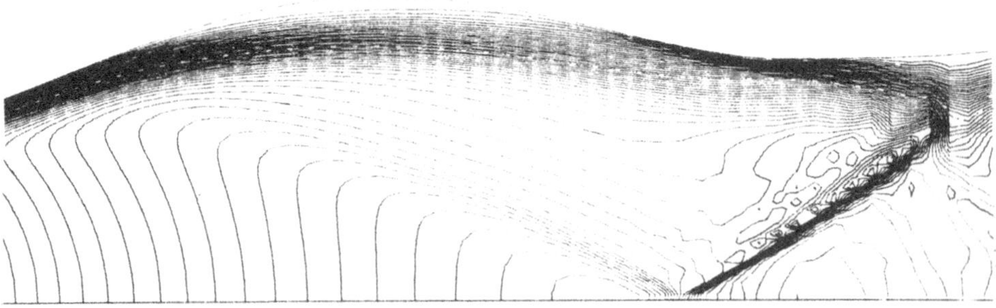

MESH 2 : 208 x 27

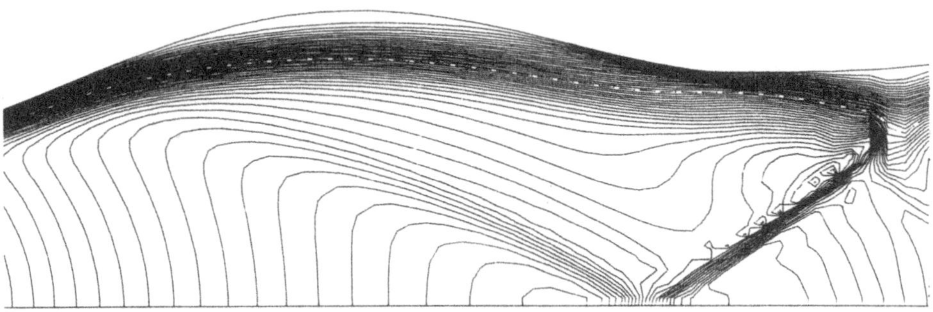

MESH 3 : 208 x 55

Fig. 1 Iso-Mach Lines $Re_{o,h}$ = 1600.

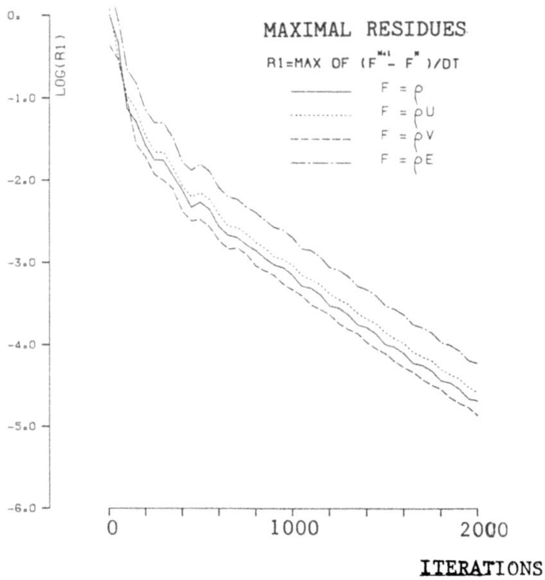

Fig. 2 Convergence history $Re_{o,h}$ = 1600.

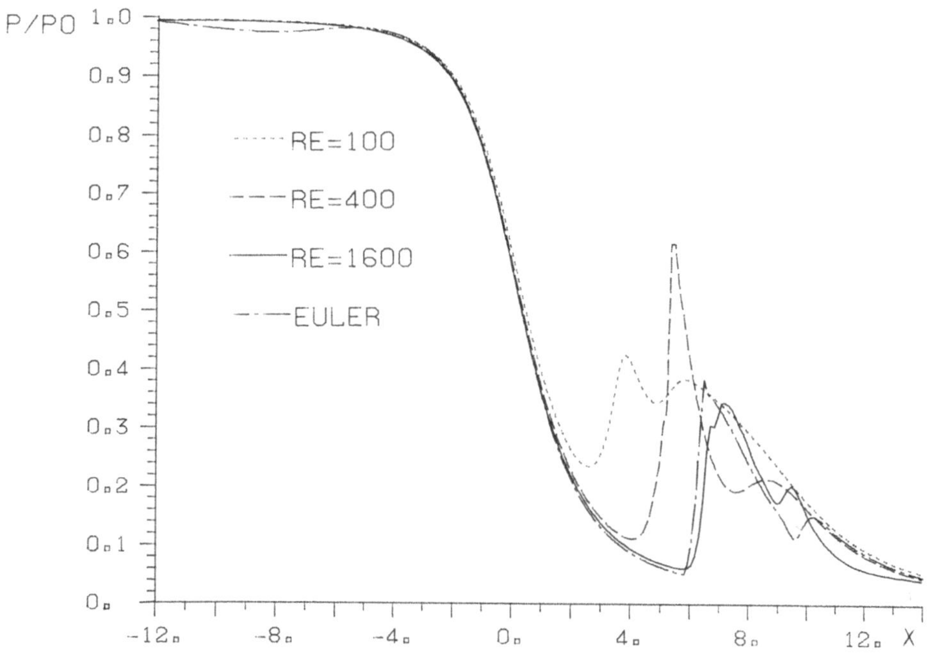

Fig. 3 Pressure on symmetry axis.

			-10	-4	0	4	7	13
$Re_o=100$	WALL VALUES	P/PO	0.9967557E+00	0.9834499E+00	0.5487110E+00	0.3525627E+00	0.3340262E+00	0.2609058E-01
		CF	0.1150429E-02	0.7090371E-02	0.9494783E-01	-.2555593E-02	0.4185350E-01	0.1900128E-01
		CH	0.4685613E-05	0.7122323E-04	0.7024956E-02	0.9739640E-03	0.3425836E-02	0.3478045E-02
	SYMMETRY AXIS VALUES	P/PO	0.9941810E+00	0.9700756E+00	0.5935426E+00	0.4115921E+00	0.3438882E+00	0.7056674E-01
		MACH	0.9137335E-01	0.2085790E+00	0.9000756E+00	0.1133387E+01	0.1189893E+01	0.2162723E+01
		S	0.1000002E+01	0.1000191E+01	0.1000696E+01	0.1052533E+01	0.1114197E+01	0.1186764E+01
$Re_o=400$	WALL VALUES	P/PO	0.9965926E+00	0.9811712E+00	0.5124994E+00	0.2552250E+00	0.2781365E+00	0.1909458E-01
		CF	0.4580081E-03	0.3885570E-02	0.4931015E-01	-.2205347E-02	0.1635132E-01	0.9245246E-02
		CH	0.2293559E-05	0.4408691E-04	0.3625076E-02	0.4322974E-03	0.1585904E-02	0.1521938E-02
	SYMMETRY AXIS VALUES	P/PO	0.9934476E+00	0.9677905E+00	0.5757960E+00	0.1119494E+00	0.2176694E+00	0.6104464E-01
		MACH	0.9737456E-01	0.2166331E+00	0.9257664E+00	0.2079215E+01	0.1613426E+01	0.2425868E+01
		S	0.9999893E+00	0.1000061E+01	0.9997113E+00	0.1003410E+01	0.1023185E+01	0.1032739E+01
$Re_o=1600$	WALL VALUES	P/PO	0.9964964E+00	0.9801249E+00	0.4912099E+00	0.1956161E+00	0.1263494E+00	0.1463984E-01
		CF	0.1249225E-03	0.2036083E-02	0.2521591E-01	-.5652315E-03	0.1339869E-01	0.4113313E-02
		CH	0.6069353E-06	0.2584686E-04	0.1869747E-02	0.3308563E-03	0.1429957E-02	0.6508128E-03
	SYMMETRY AXIS VALUES	P/PO	0.9930753E+00	0.9665343E+00	0.5670622E+00	0.9480143E-01	0.3321428E+00	0.4922419E-01
		MACH	0.1004152E+00	0.2210772E+00	0.9385596E+00	0.2191109E+01	0.1329492E+01	0.2582426E+01
		S	0.9999890E+00	0.1000028E+01	0.9999192E+00	0.1000175E+01	0.1016299E+01	0.1017164E+01

Tabulated wall values and symmetry axis values.

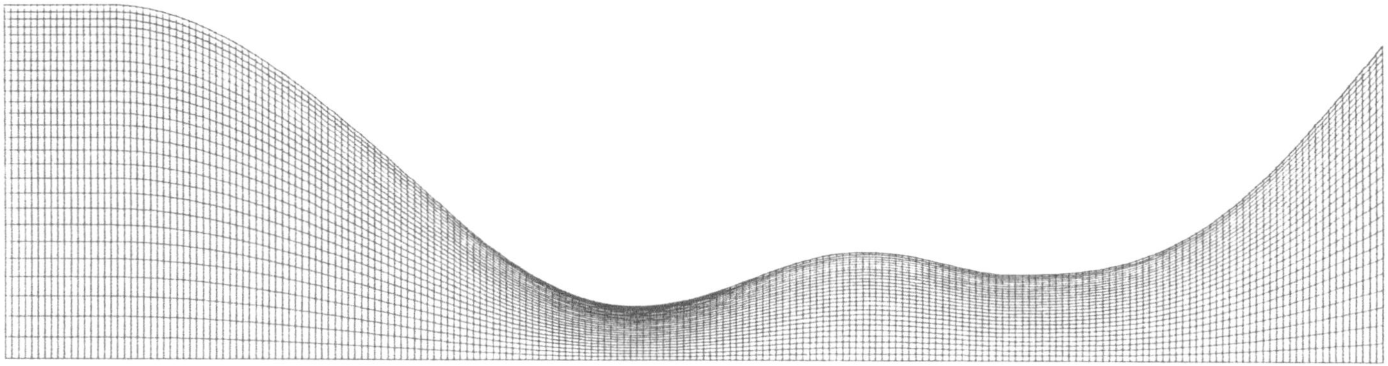

Fig 4 MESH : 209 x 28.

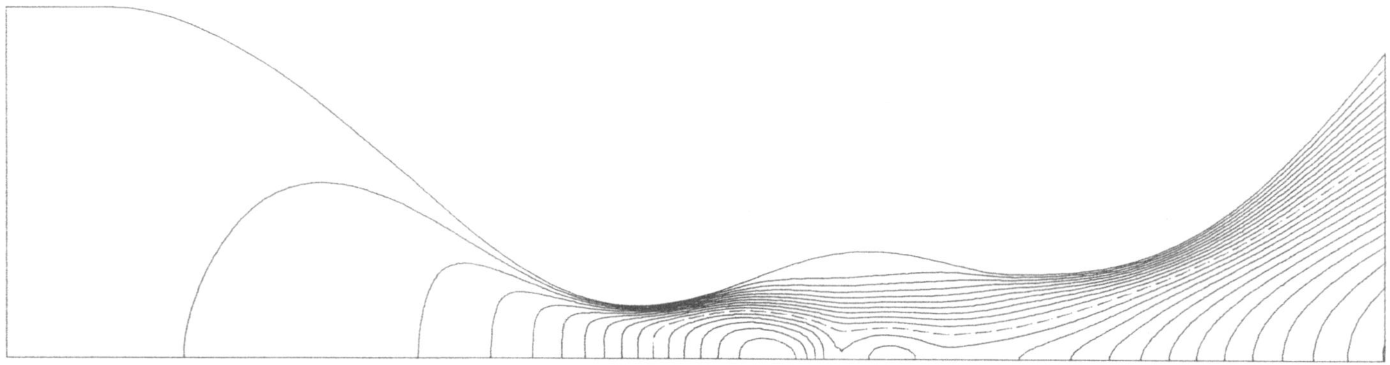

Fig. 5 Iso-Mach lines $Re_{o,h}$ = 100, ΔM = .1.

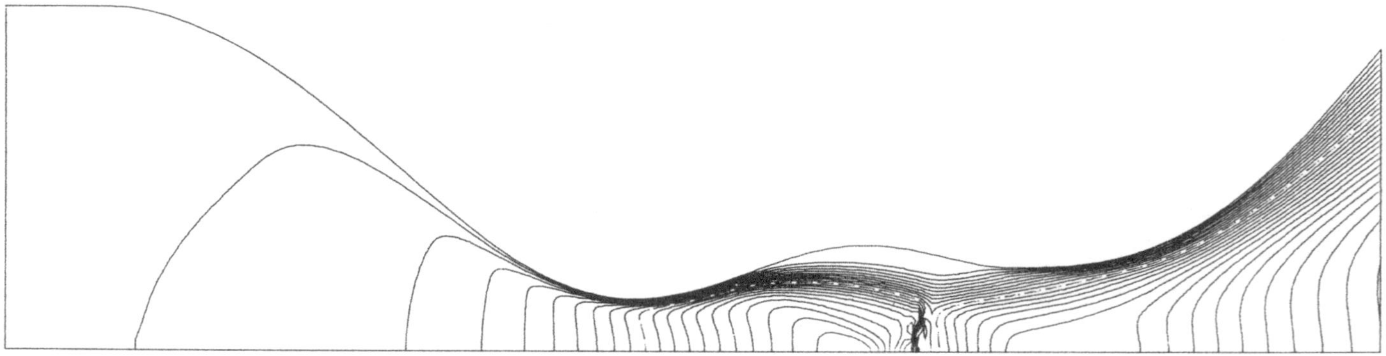

Fig. 6 Iso-Mach lines $Re_{o,h} = 400$, $\Delta M = .1$.

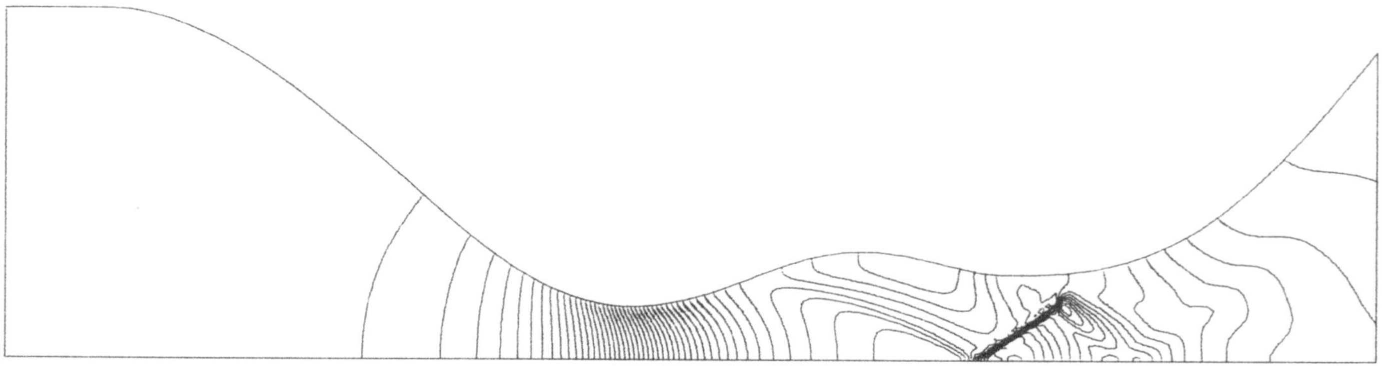

Fig. 7 Isobars $Re_{o,h} = 1600$, $\Delta P/P_o = .02$.

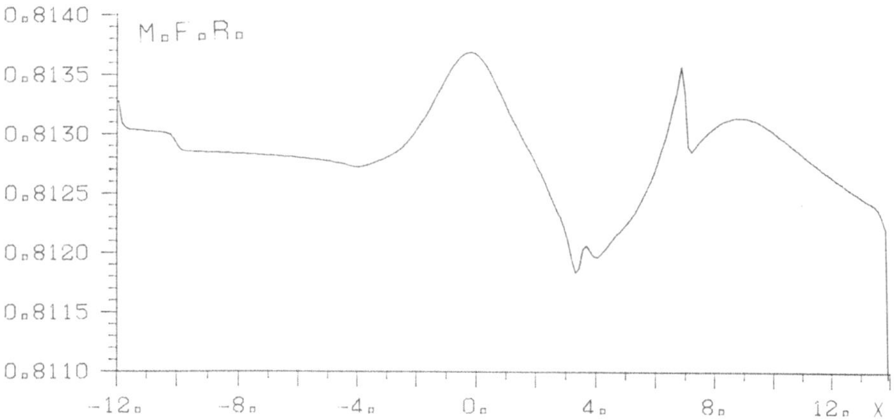

Fig. 8 Mass flow rate $Re_{o,h}$ = 100.

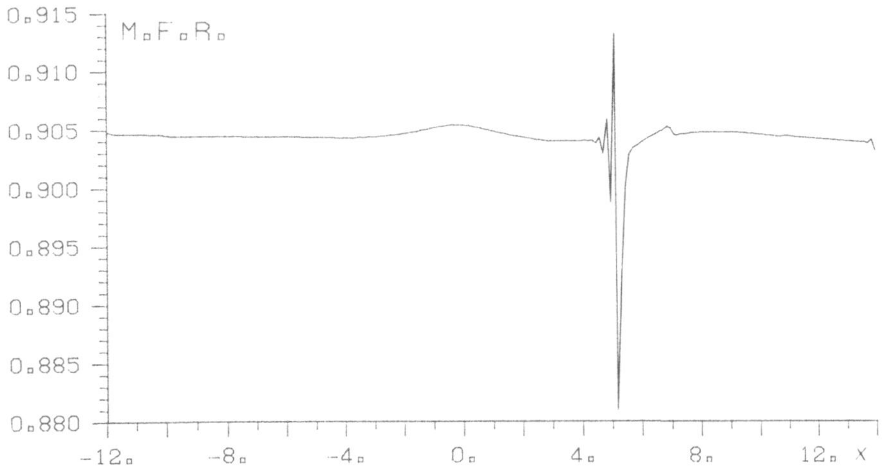

Fig. 9 Mass flow rate $Re_{o,h}$ = 400.

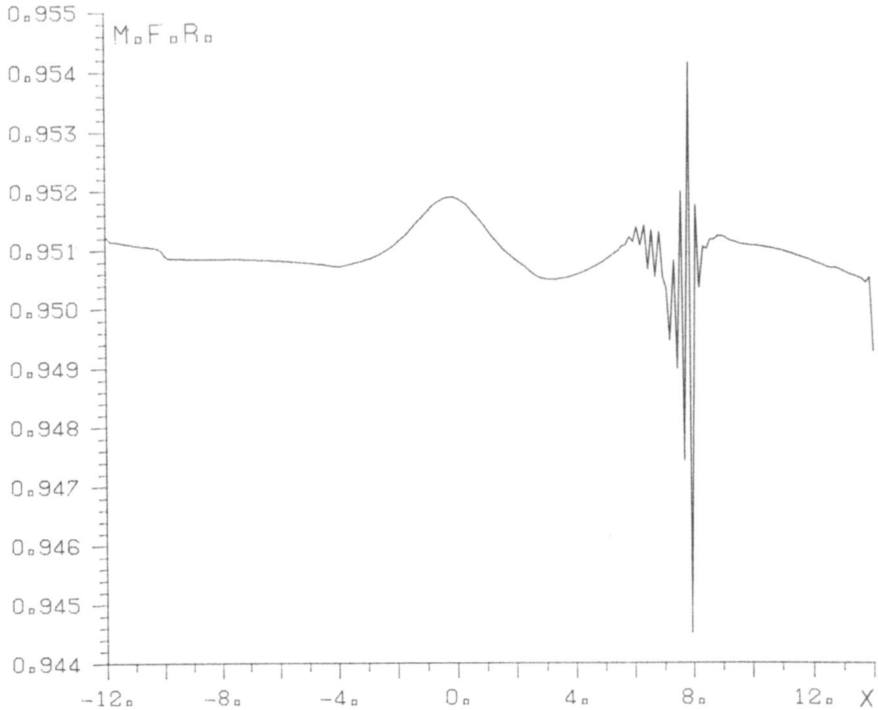

Fig. 10 Mass flow rate $Re_{o,h}$ = 1600.

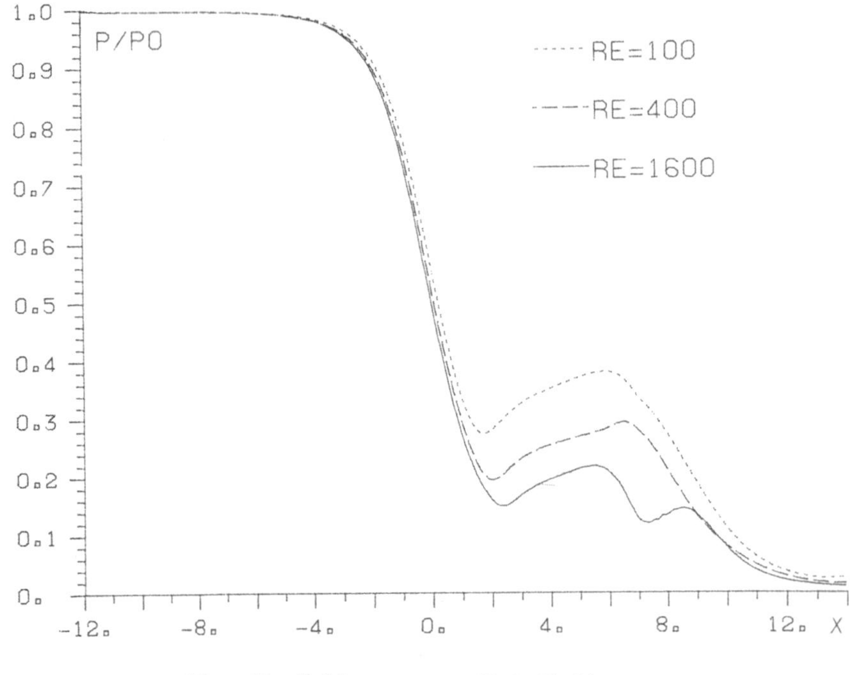

Fig. 11 Wall pressure distribution.

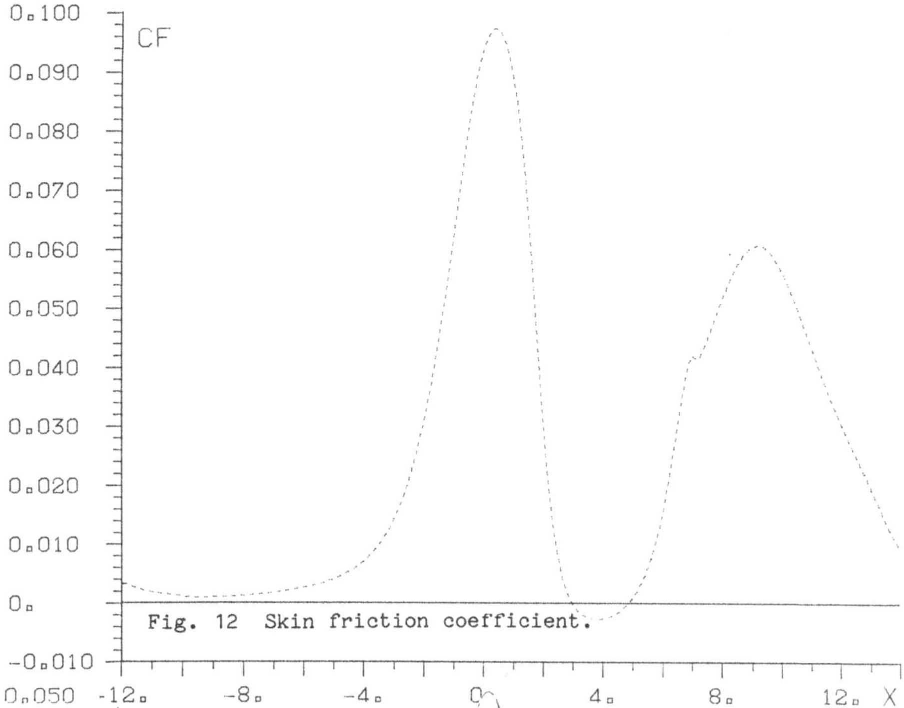

Fig. 12 Skin friction coefficient.

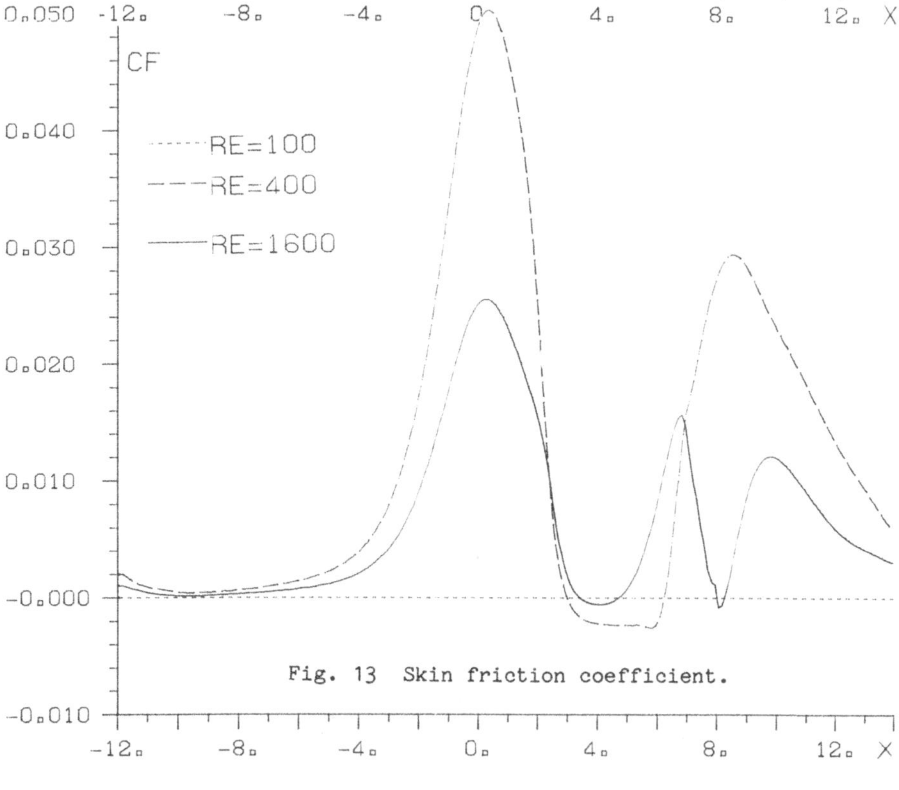

Fig. 13 Skin friction coefficient.

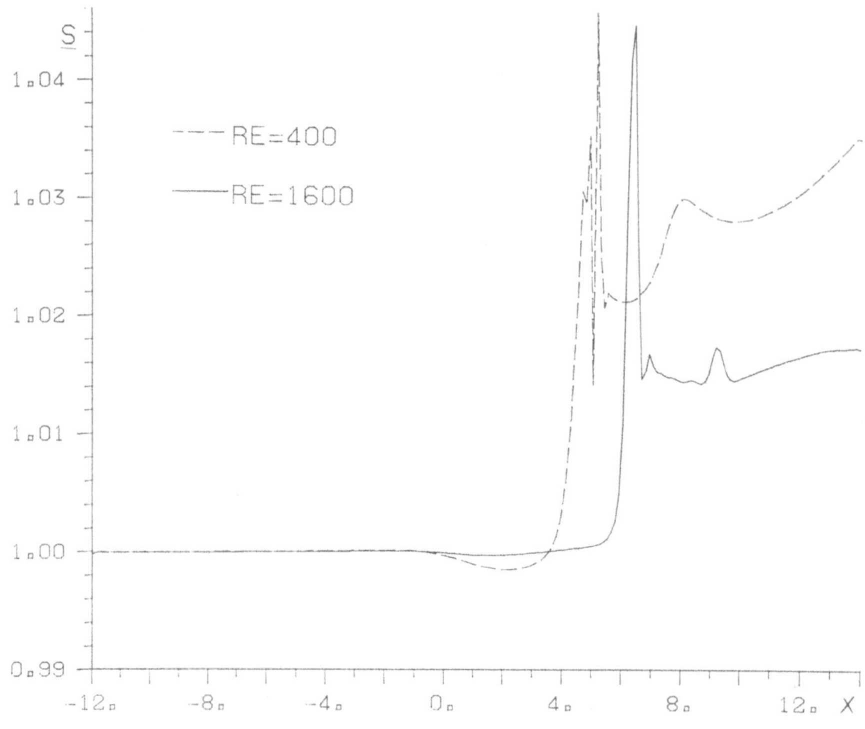

Fig. 14 Quantity related to entropy.

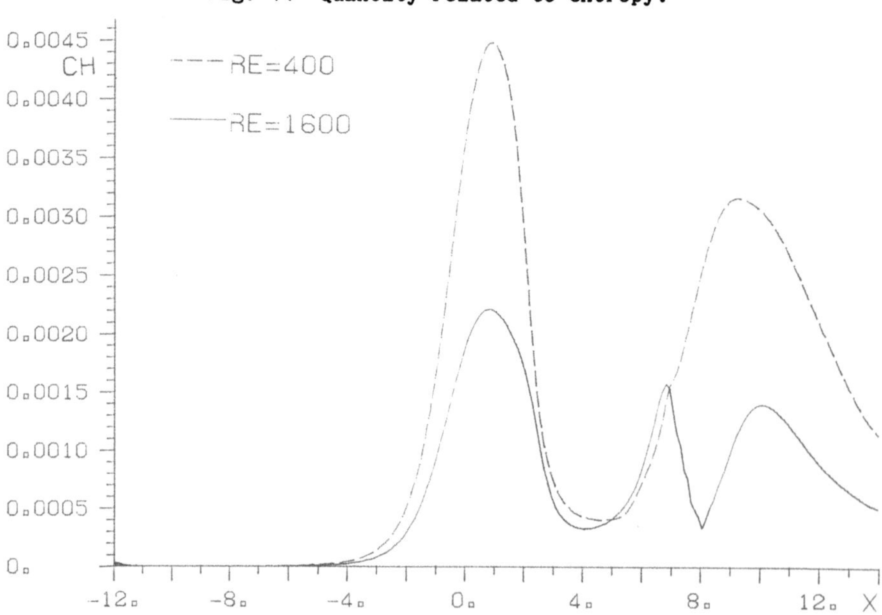

Fig. 15 Heat flux coefficient.

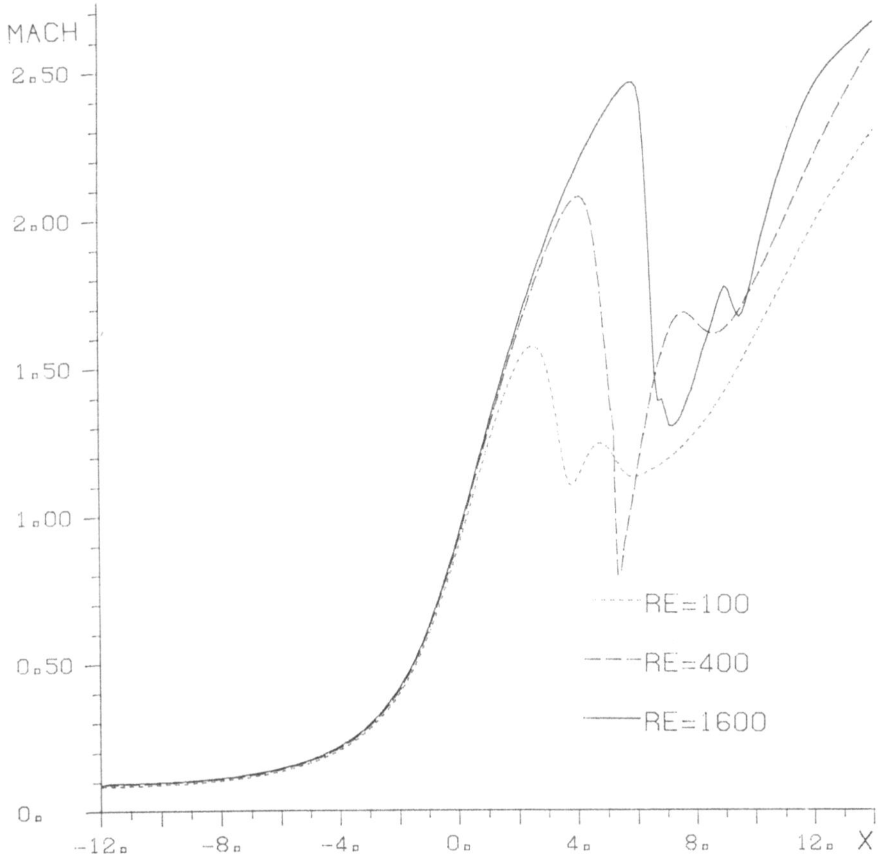

Fig. 16 Mach number on symmetry axis.

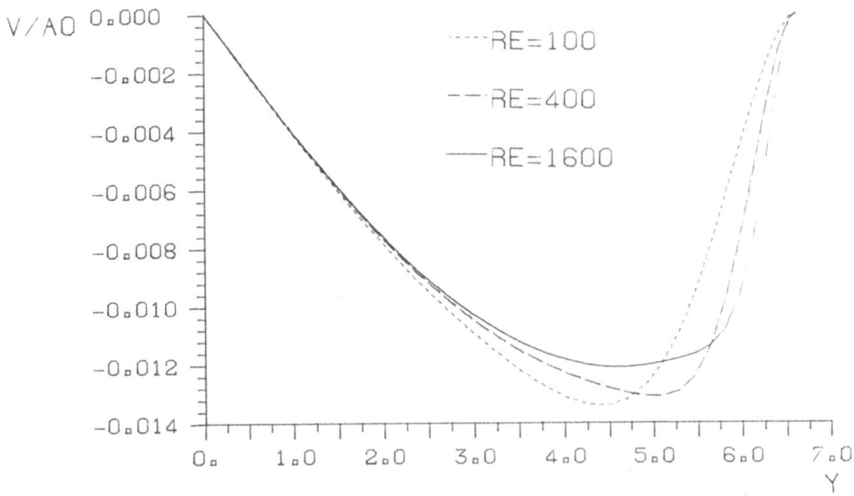

Fig. 17 Velocity profile at X = -10 (Y-component).

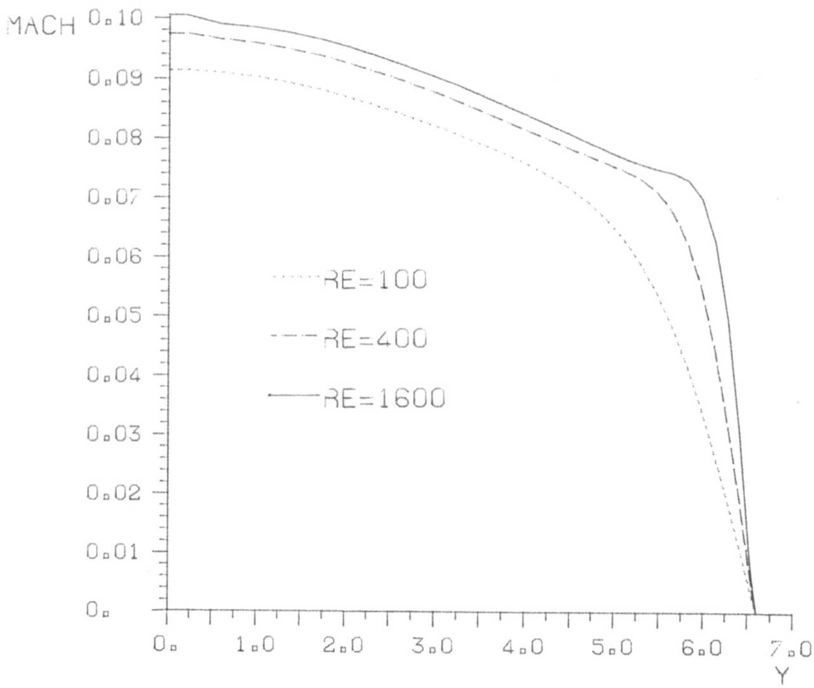

Fig. 18 Transverse Mach number at X = -10.

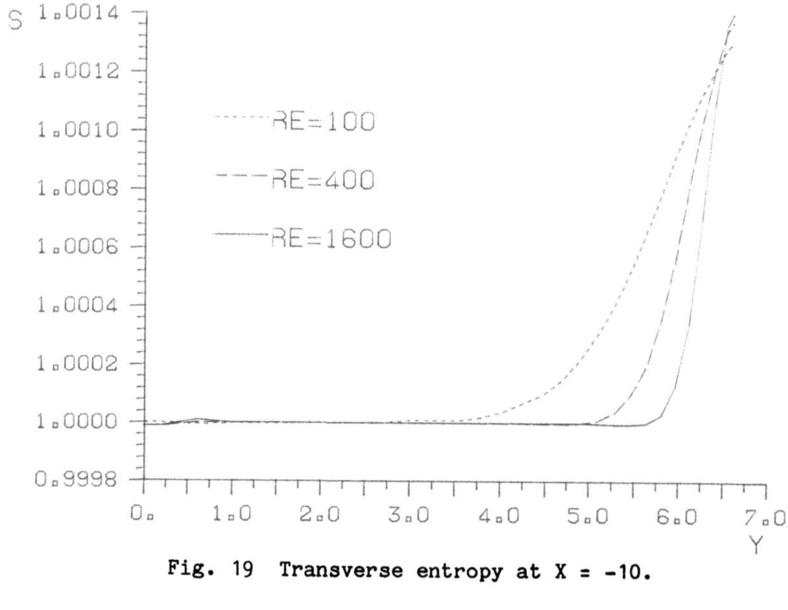

Fig. 19 Transverse entropy at X = -10.

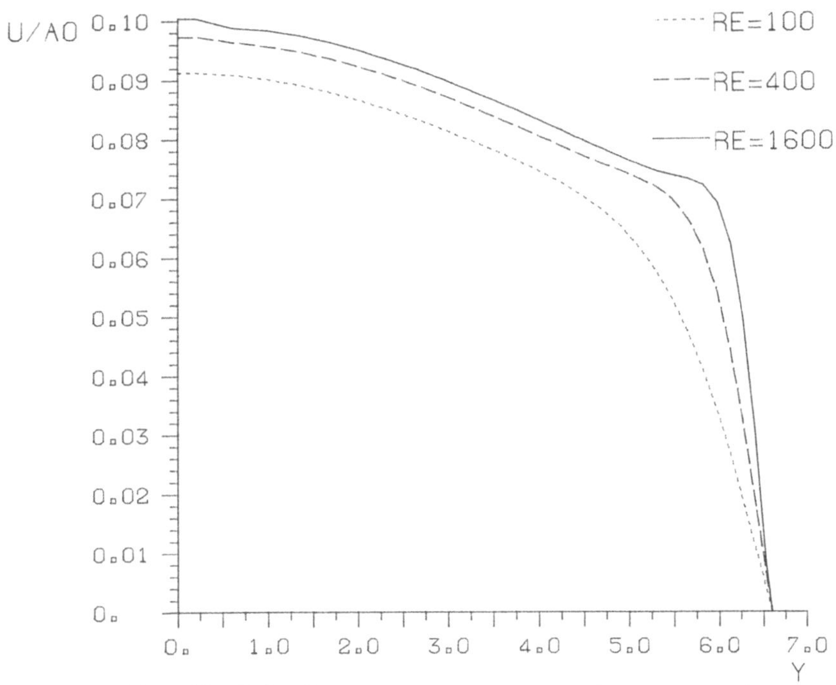

Fig. 20 Velocity profile at X = -10 (X-component).

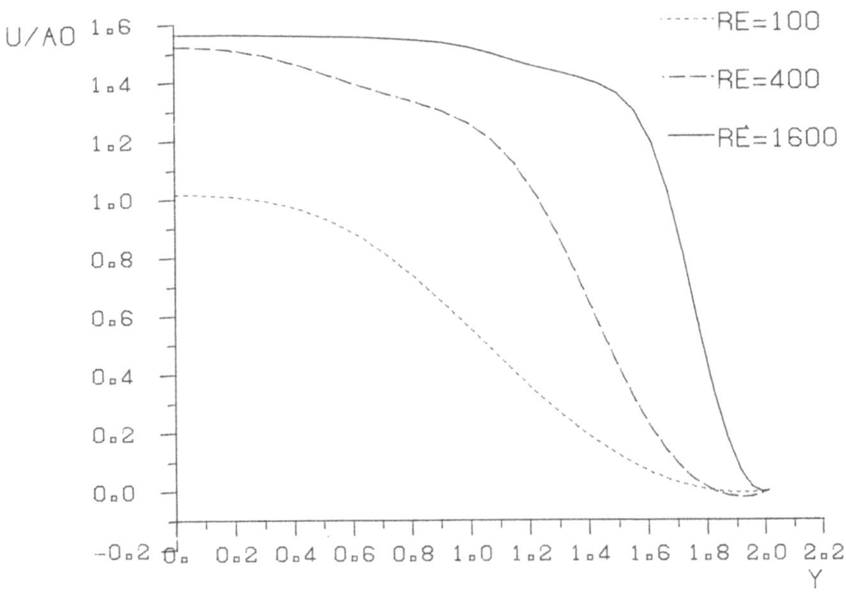

Fig. 21 Velocity profile at X = +4 (X-component).

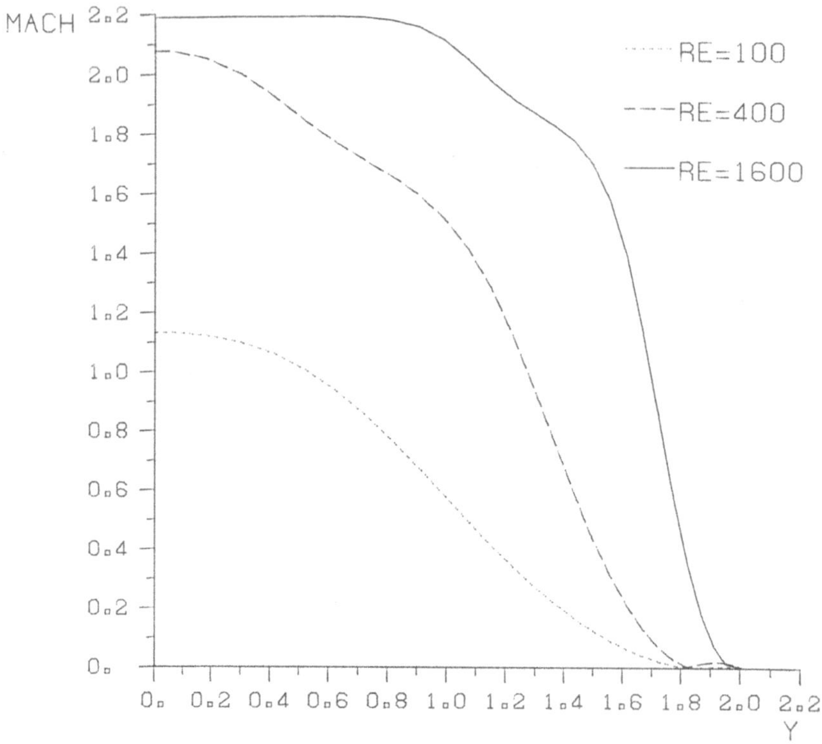

Fig. 22 Transverse Mach number at X = +4.

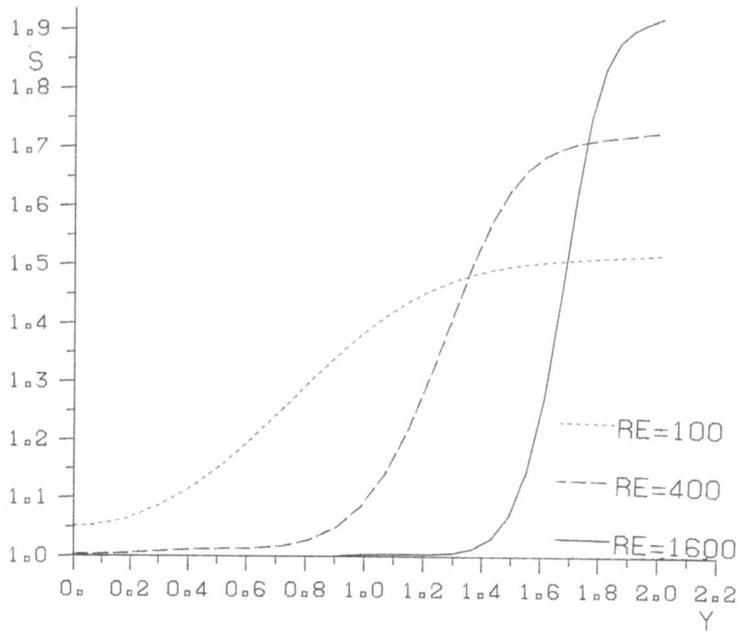

Fig. 23 Transverse entropy at X = +4.

COMPUTATIONS FOR VISCOUS COMPRESSIBLE FLOWS IN A DOUBLE THROAT NOZZLE

Y. MARX
Computational Fluid Dynamics Group
E. N. S. M.
1, rue de la Noë - 44072 Nantes, France

INTRODUCTION

The simulation of the flow field through a double throat nozzle is performed by solving the full Navier-Stokes equations with the help of a finite volume method.

GOVERNING EQUATIONS

The equations to be solved are the Navier-Stokes equations written in their conservative form, that is, using the vector notation

$$\frac{\partial U}{\partial t} + \frac{\partial (F+F')}{\partial x} + \frac{\partial (G+G')}{\partial y} = 0.$$

with

$$U = \begin{bmatrix} \rho \\ \rho u \\ \rho v \\ e \end{bmatrix} \quad F = \begin{bmatrix} \rho u \\ \rho u^2 + P \\ \rho u v \\ u(e+P) \end{bmatrix} \quad F' = \begin{bmatrix} 0 \\ -\lambda\left(\frac{\partial u}{\partial x} + \frac{\partial v}{\partial y}\right) - 2\mu \frac{\partial u}{\partial x} \\ -\mu\left(\frac{\partial u}{\partial y} + \frac{\partial v}{\partial x}\right) \\ -K\frac{\partial T}{\partial x} + u\left[-\lambda\left(\frac{\partial u}{\partial x} + \frac{\partial v}{\partial y}\right) - 2\mu \frac{\partial u}{\partial x}\right] + v\left[-\mu\left(\frac{\partial u}{\partial y} + \frac{\partial v}{\partial x}\right)\right] \end{bmatrix}$$

$$G = \begin{bmatrix} \rho v \\ \rho u v \\ \rho v^2 + P \\ v(e+P) \end{bmatrix} \quad G' = \begin{bmatrix} 0 \\ -\mu\left(\frac{\partial u}{\partial y} + \frac{\partial v}{\partial x}\right) \\ -\lambda\left(\frac{\partial u}{\partial x} + \frac{\partial v}{\partial y}\right) - 2\mu \frac{\partial v}{\partial y} \\ -K\frac{\partial T}{\partial y} + u\left[-\mu\left(\frac{\partial u}{\partial y} + \frac{\partial v}{\partial x}\right)\right] + v\left[-\lambda\left(\frac{\partial u}{\partial x} + \frac{\partial v}{\partial y}\right) - 2\mu \frac{\partial v}{\partial y}\right] \end{bmatrix}$$

NUMERICAL PROCEDURE

The method employed is an explicit finite volume method. The distribution of the vector U within each cell is taken to be linear. This distribution is obtained from the mean values by application of the Harten-Osher [1] reconstruction.

In the x direction the discontinuities that might appear at the cell interfaces are used to simulate upwinding.

For instance, if the one dimensional flow equation is considered

$$\frac{\partial U}{\partial t} + \frac{\partial F}{\partial x} = 0$$

the approximation yields

$$\bar{U}_i^{n+1} = \bar{U}_i^{n} - \Delta t \left[\frac{F^+(U^{i}_{i+\frac{1}{2}}) - F^+(U^{i-1}_{i-\frac{1}{2}})}{\Delta x_i} + \frac{F^-(U^{i+1}_{i+\frac{1}{2}}) - F^-(U^{i}_{i-\frac{1}{2}})}{\Delta x_i} \right]$$

where : $\bar{U}$ indicates the mean value of the vector U within the cell.

The inviscid fluxes F are splitted into forward F^+ and backward F^- contribution according to the Van Leer [2] flux splitting technique.

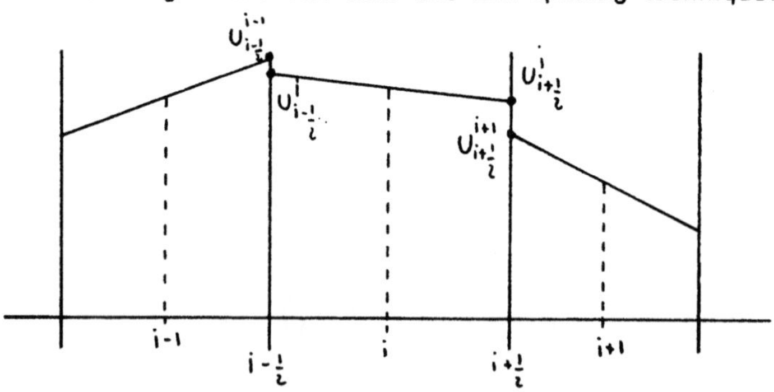

Across the bottom and the top part of the cell, an approximate Riemann solver is employed, that is : $\partial G/\partial y$ is approximated by

$$\frac{\partial G}{\partial y}_j = \frac{G(U_{j+\frac{1}{2}}) - G(U_{j-\frac{1}{2}})}{\Delta y_j}$$

the $U_{j+1/2}$ components are obtained from

$$\begin{cases} P_{j+\frac{1}{2}} = \left(\dfrac{P^{j}_{j+\frac{1}{2}} + P^{j+1}_{j+\frac{1}{2}}}{2}\right) - \widetilde{\rho a}\left(\dfrac{v^{j+1}_{j+\frac{1}{2}} - v^{j}_{j+\frac{1}{2}}}{2}\right) \\ v_{j+\frac{1}{2}} = \left(\dfrac{v^{j}_{j+\frac{1}{2}} + v^{j+1}_{j+\frac{1}{2}}}{2}\right) - \left(\dfrac{P^{j+1}_{j+\frac{1}{2}} - P^{j}_{j+\frac{1}{2}}}{2\,\widetilde{\rho a}}\right) \\ u_{j+\frac{1}{2}} = \tilde{u} \\ \rho_{j+\frac{1}{2}} = \tilde{\rho} + \left(\dfrac{P_{j+\frac{1}{2}} - \tilde{P}}{\tilde{a}^2}\right) \end{cases}$$

the symbol $(\tilde{\ })$ indicates a mean value at the interface which is obtained from the Roe [3] average.

All the viscous terms are calculated as differences across cell interfaces of central differences for first derivatives.

BOUNDARY CONDITIONS

The boundary conditions used for this configuration are :

. at the symmetry axis

$$\frac{\partial \theta}{\partial y} = 0. \qquad \theta = \rho\,,\ \rho u\,,\ e.$$

$$v = 0.$$

. at the exit .

All the variables are extrapolated.

. at the wall the pressure is calculated from a characteristic relation in the x = cte reference plane.

$$P_{wall} = P_A + \overline{\rho a}(v_A - tg\alpha\ u_A)$$

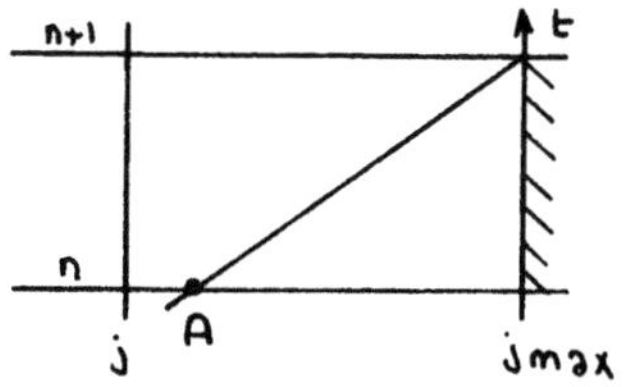

where $tg\alpha$ is the local tangent of the duct.

. at the inlet

$$v = 0$$

$$P = P_0 / \left(1 + \frac{\gamma-1}{2}\frac{u^2}{a^2}\right)^{\frac{\gamma}{\gamma-1}}$$

$$T = T_0 / \left(1 + \frac{\gamma-1}{2}\frac{u^2}{a^2}\right)$$

These relations give $\Delta P = -\rho u \Delta u$ at the inlet and allowed the elimination of the incoming information $\partial^-(.)/\partial x$ in the equations

$$\frac{\partial u}{\partial t} + \left(\frac{u+a}{2}\right)\frac{\partial^- u}{\partial x} + \left(\frac{u+a}{2\rho a}\right)\frac{\partial^- P}{\partial x} + \left(\frac{u-a}{2}\right)\frac{\partial^+ u}{\partial x} - \left(\frac{u-a}{2\rho a}\right)\frac{\partial^+ P}{\partial x} = 0 .$$

$$\frac{\partial P}{\partial t} + \rho a\left(\frac{u+a}{2}\right)\frac{\partial^- u}{\partial x} + \left(\frac{u+a}{2}\right)\frac{\partial^- P}{\partial x} - \rho a\left(\frac{u-a}{2}\right)\frac{\partial^+ u}{\partial x} + \left(\frac{u-a}{2}\right)\frac{\partial^+ P}{\partial x} = 0 .$$

therefore

$$\frac{\partial u}{\partial t} = -a\left(\frac{u-a}{u+a}\right)\left[\frac{\partial^+ u}{\partial x} - \frac{1}{\rho a}\frac{\partial^+ P}{\partial x}\right]$$

where $\partial^+/\partial x$ means downstream decentration.

COMPUTING DETAILS

The method has enough diffusion so that there is no need of additional artificial viscosity. Furthermore no special procedure is required for the computation of shocks.

The computation were performed on a Cray 1S and the computing time is of 2.8 10^{-5} second per grid point and time step. The storage for the 28 x 209 grid is 240 640 words, the storage requirements of the code are not yet optimized.

The actual method is time accurate, no acceleration procedure is employed. These methods such as local time step or implicit scheme are presently underway.

Acknowledgements

Computer facilities have been provided by the Scientific Committee of "Centre de calcul vectoriel pour la Recherche". Financial support of DRET through contract 85/005 is also gratefully acknowledged.

REFERENCES

[1] Harten, A. & Osher, S. - Uniformy High-Order Accurate Non Oscillatory schemes I, NASA CR 175768 Jun 85.

[2] Van Leer, B. - Flux vector splitting for the Euler Equations 8th ICMFD Lect. Notes in Physics 170, 1982.

[3] Roe, P. L . - Approximate Riemann solvers, parameter vectors, and difference schemes, J. Comp. Phys. 43, 1981.

x		-10	-4	0	4	7	13
WALL VALUES	P/PO	.9970	.9839	.5535	.3502	.3356	2.815E-2
	CF	8.674E-4	5.285E-3	9.460E-2	-3.456E-3	4.134E-2	1.197E-2
	CH	2.617E-5	1.624E-4	6.566E-3	1.050E-3	3.100E-3	3.499E-3
SYMMETRY AXIS VALUES	P/PO	.9935	.9699	.5917	.4063	.3447	6.900E-3
	MACH	9.888E-2	0.2091	.9026	1.143	1.186	2.175
	S	.9998	1.000	1.001	1.054	1.116	1.190

Re=100

x		-10	-4	0	4	7	13
WALL VALUES	P/PO	.9955	.9817	.5163	.2534	.2800	1.980E-2
	CF	4.007E-4	2.874E-3	4.919E-2	-2.232E-3	1.592E-2	6.154E-3
	CH	8.909E-5	6.374E-3	3.602E-3	4.745E-4	1.375E-3	1.557E-3
SYMMETRY AXIS VALUES	P/PO	.9953	.9677	.5759	.1117	.2192	6.035E-2
	MACH	.1024	.2173	.9262	2.082	1.602	2.428
	S	1.000	1.000	.9998	1.003	1.029	1.035

Re=400

x		-10	-4	0	4	7	13
WALL VALUES	P/PO	.9956	.9817	.5065	.1941	.1273	1.470E-2
	CF	1.974E-4	1.415E-3	2.471E-2	-4.794E-4	1.345E-2	2.753E-3
	CH	2.227E-5	1.593E-5	1.504E-3	4.389E-4	1.559E-3	6.644E-4
SYMMETRY AXIS VALUES	P/PO	.9954	.9677	.5709	9.440E-2	.3258	5.002E-2
	MACH	.1027	.2180	.9356	2.192	1.333	2.567
	S	1.000	1.000	.9999	.9998	1.025	1.027

Re=1600

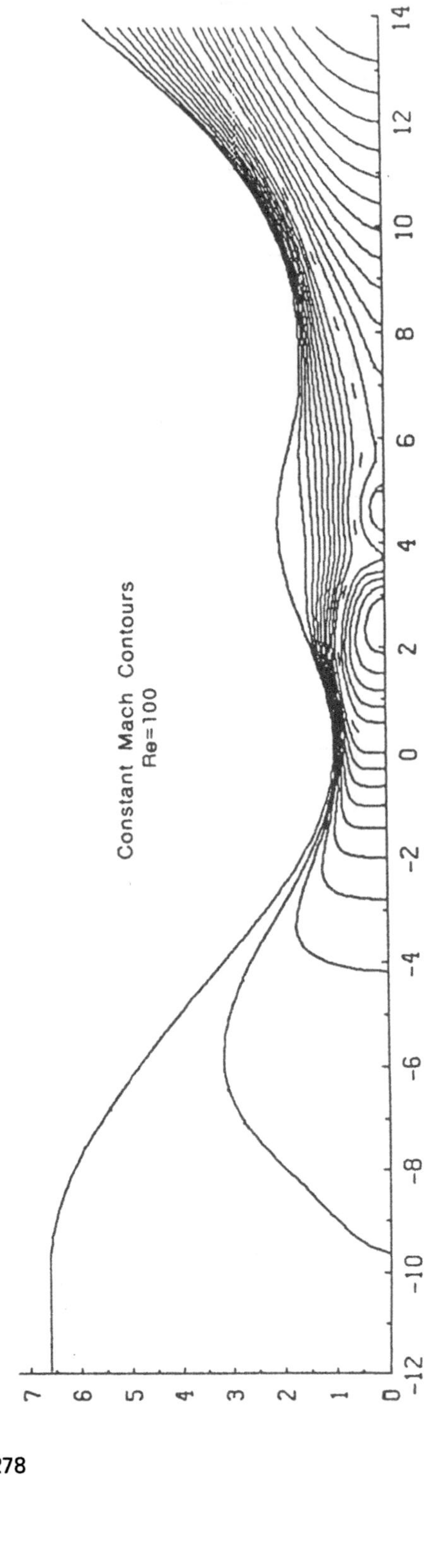
Constant Mach Contours
Re=100
-12 -10 -8 -6 -4 -2 0 2 4 6 8 10 12 14
0 1 2 3 4 5 6 7

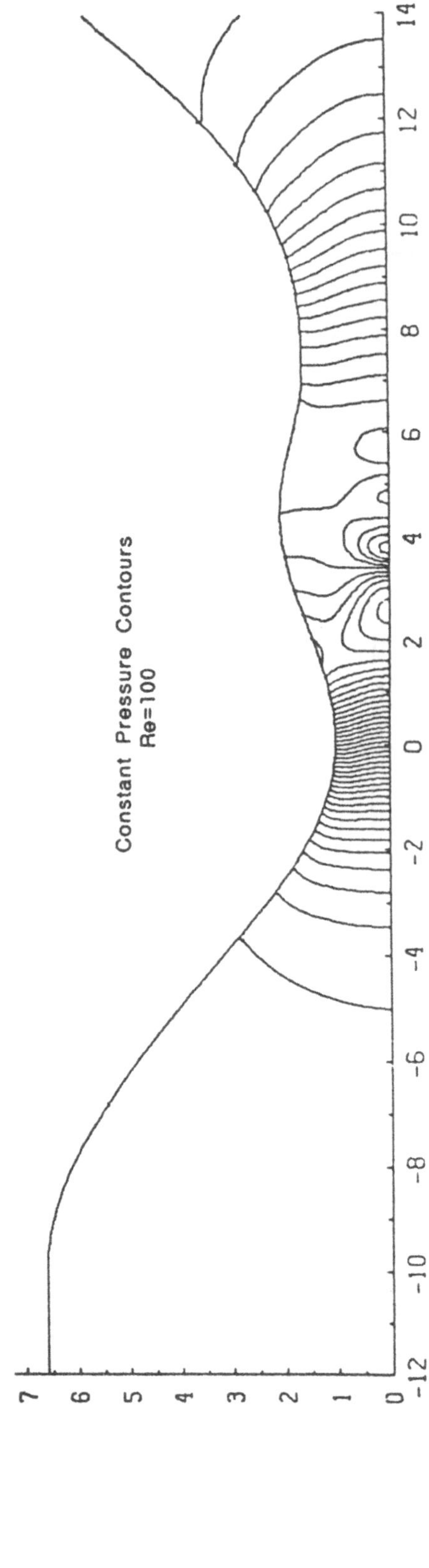
Constant Pressure Contours
Re=100
-12 -10 -8 -6 -4 -2 0 2 4 6 8 10 12 14
0 1 2 3 4 5 6 7

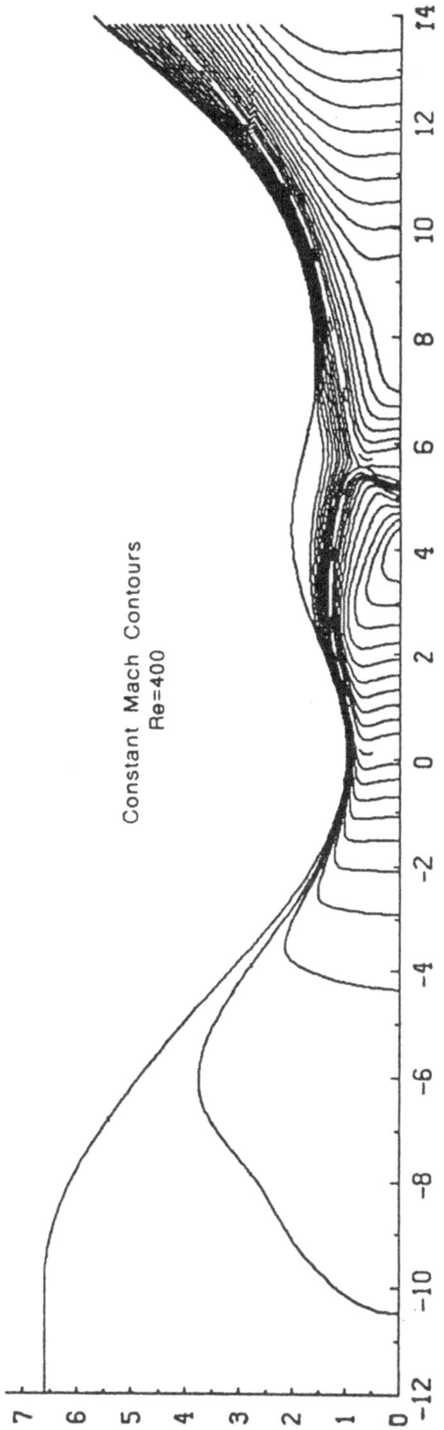
Constant Mach Contours
Re=400
-12 -10 -8 -6 -4 -2 0 2 4 6 8 10 12 14
0 1 2 3 4 5 6 7

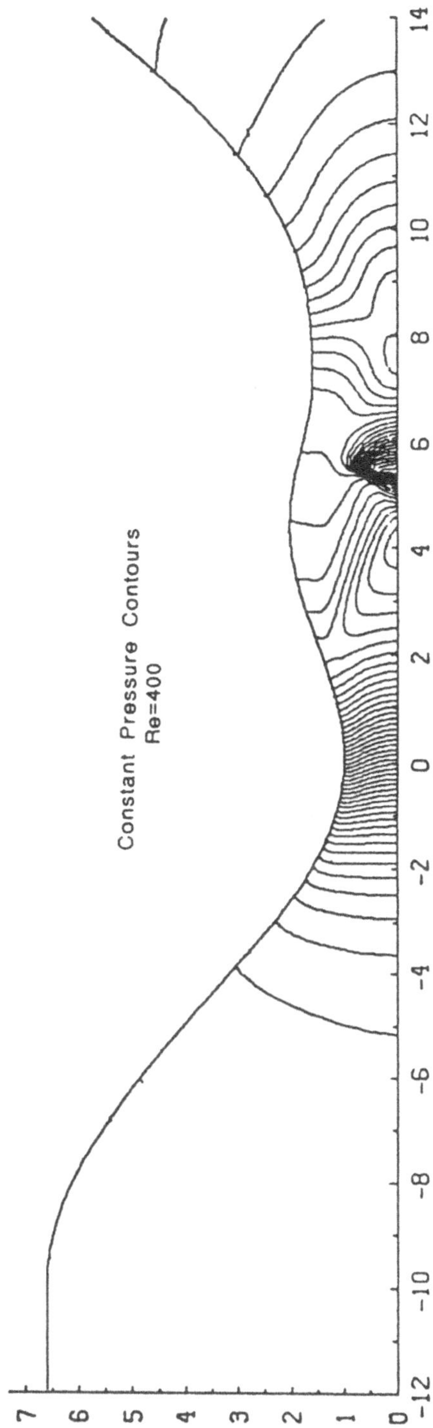
Constant Pressure Contours
Re=400
-12 -10 -8 -6 -4 -2 0 2 4 6 8 10 12 14
0 1 2 3 4 5 6 7

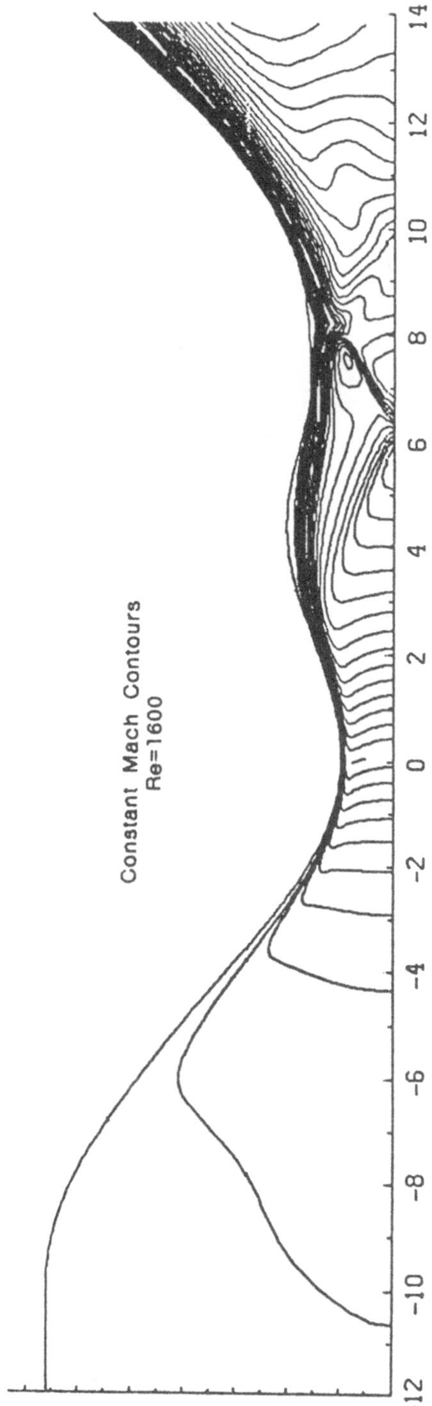
Constant Mach Contours
Re=1600
-12 -10 -8 -6 -4 -2 0 2 4 6 8 10 12 14
0 1 2 3 4 5 6 7

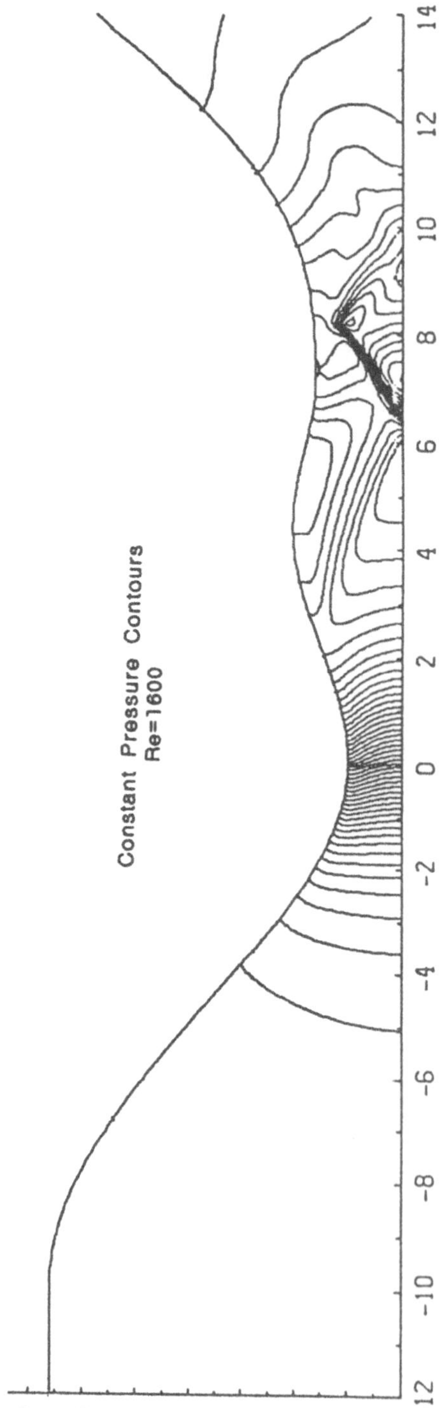
Constant Pressure Contours
Re=1600
-12 -10 -8 -6 -4 -2 0 2 4 6 8 10 12 14
0 1 2 3 4 5 6 7

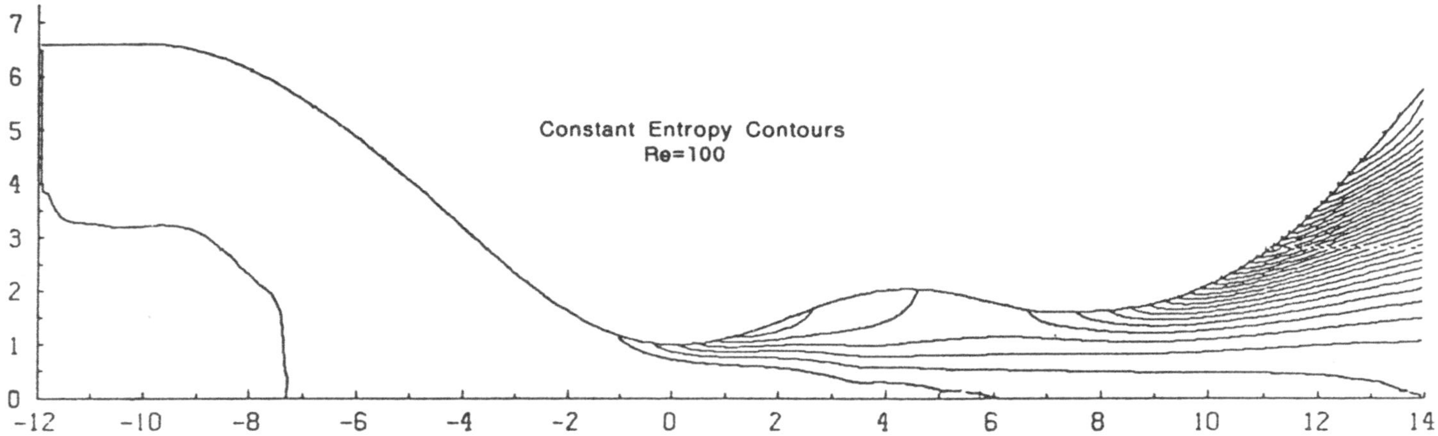

Constant Entropy Contours
Re=100
7
6
5
4
3
2
1
0
-12
-10
-8
-6
-4
-2
0
2
4
6
8
10
12
14

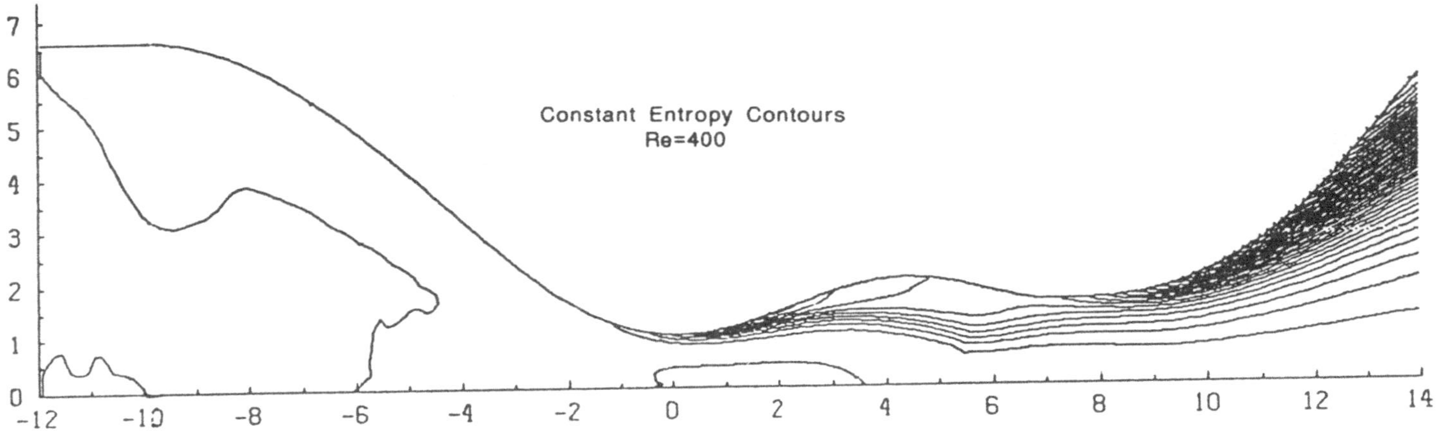

Constant Entropy Contours
Re=400
7
6
5
4
3
2
1
0
-12
-10
-8
-6
-4
-2
0
2
4
6
8
10
12
14

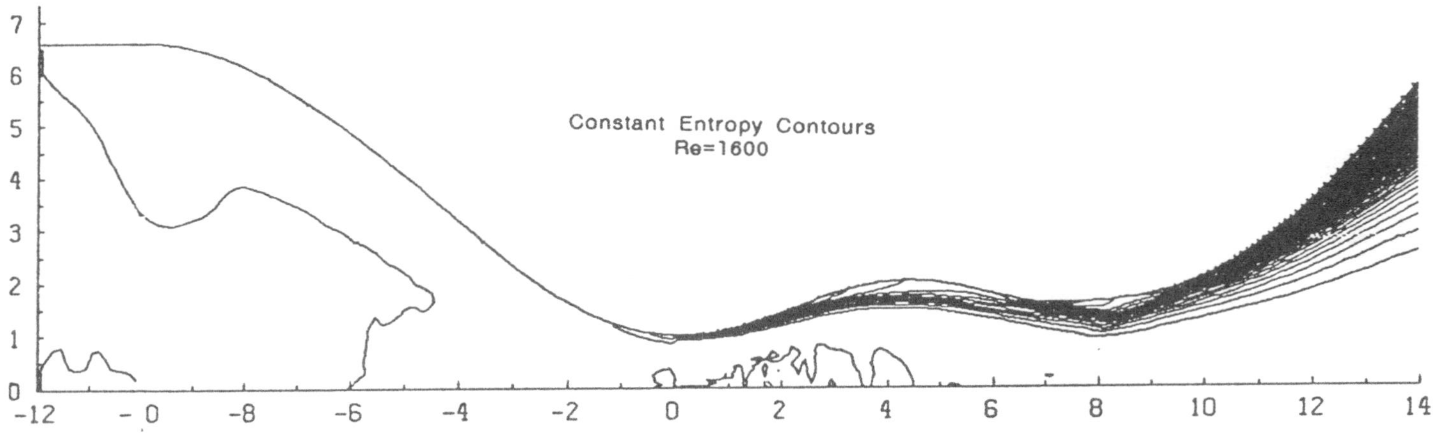

Constant Entropy Contours
Re=1600
7
6
5
4
3
2
1
0
-12
- 0
-8
-6
-4
-2
0
2
4
6
8
10
12
14

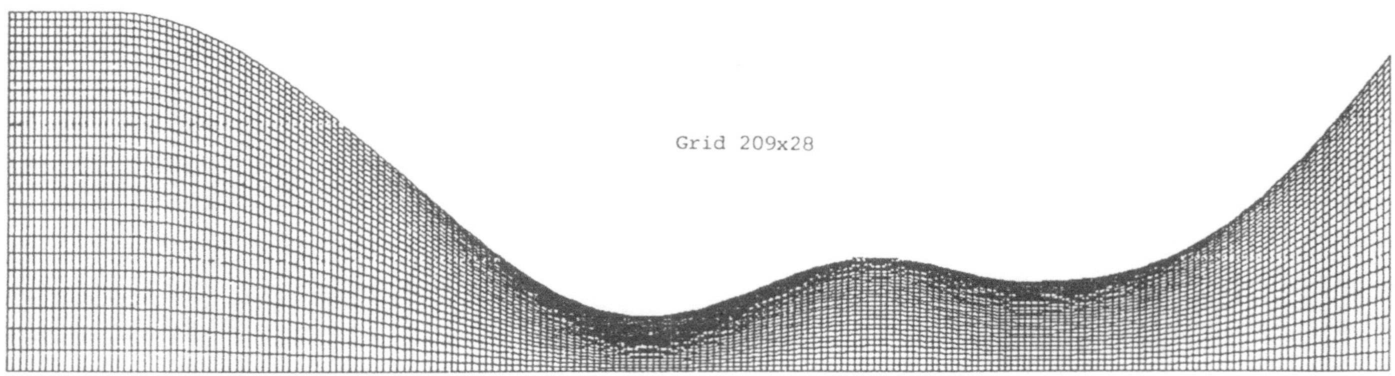

Grid 209x28

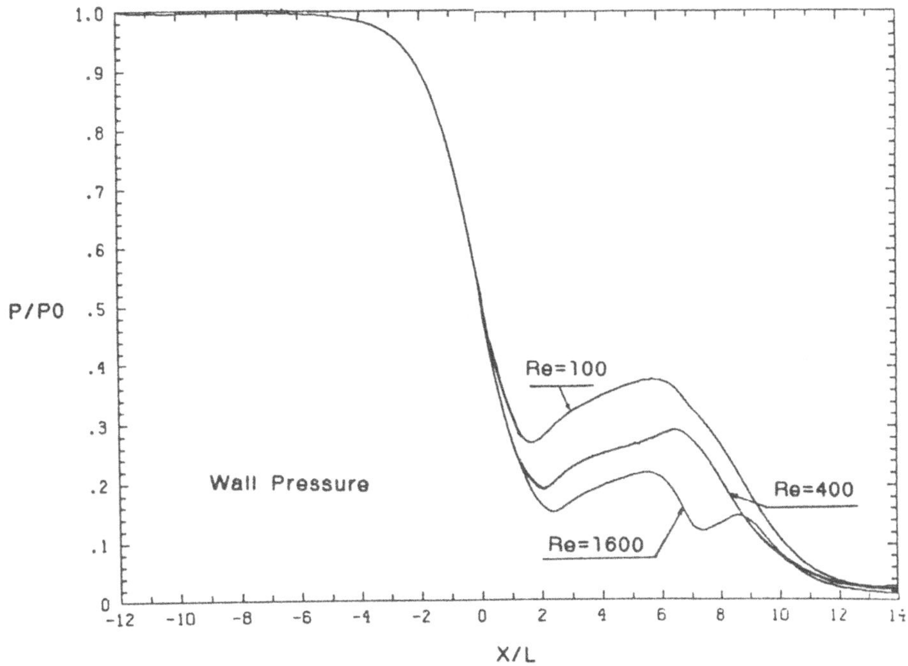

1.0
.9
.8
.7
.6
P/P0
.5
.4
.3
.2
.1
0
Re=100
Wall Pressure
Re=400
Re=1600
-12
-10
-8
-6
-4
-2
0
2
4
6
8
10
12
X/L

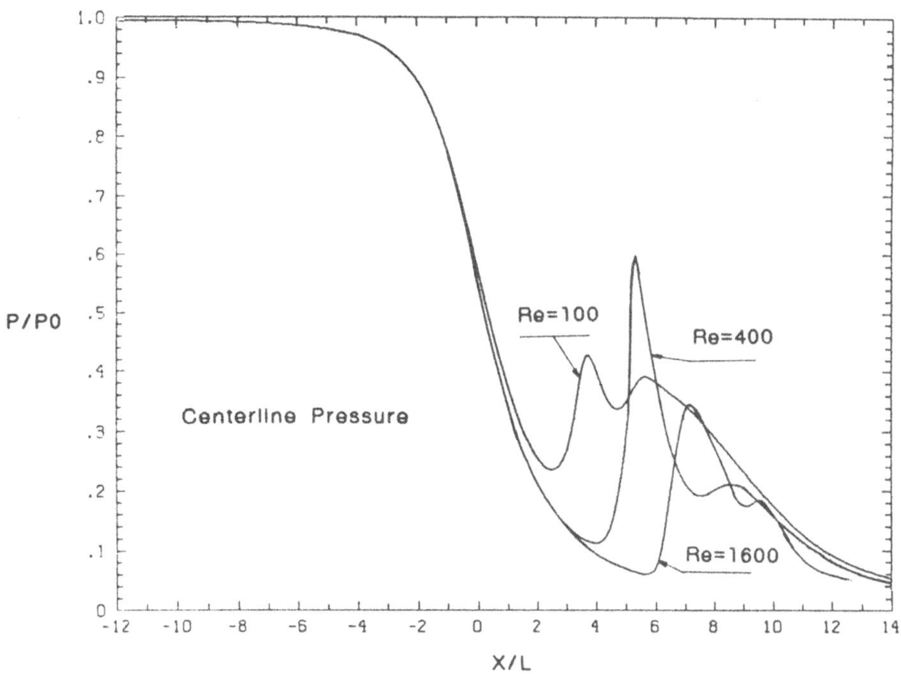

1.0
.9
.8
.7
.6
P/P0
.5
.4
.3
.2
.1
0
Re=100
Re=400
Centerline Pressure
Re=1600
-12
-10
-8
-6
-4
-2
0
2
4
6
8
10
12
14
X/L

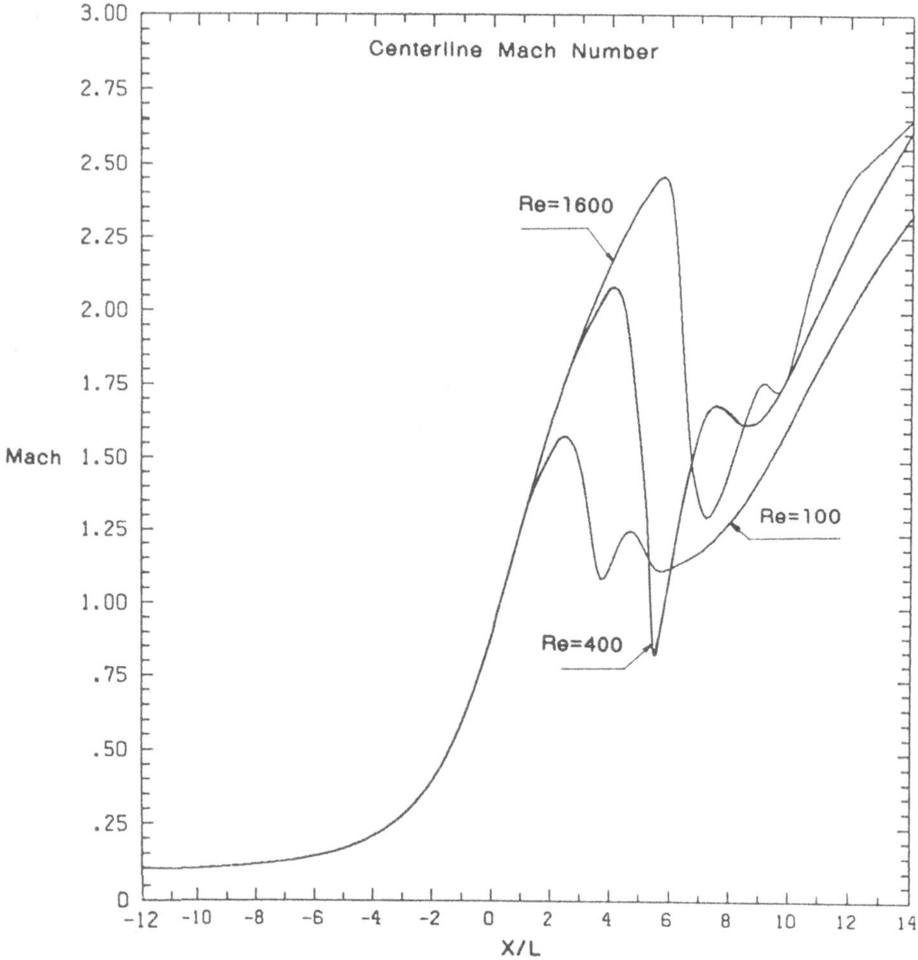

Centerline Mach Number
Mach
3.00
2.75
2.50
2.25
2.00
1.75
1.50
1.25
1.00
.75
.50
.25
0
Re=1600
Re=100
Re=400
-12 -10 -8 -6 -4 -2 0 2 4 6 8 10 12 14
X/L

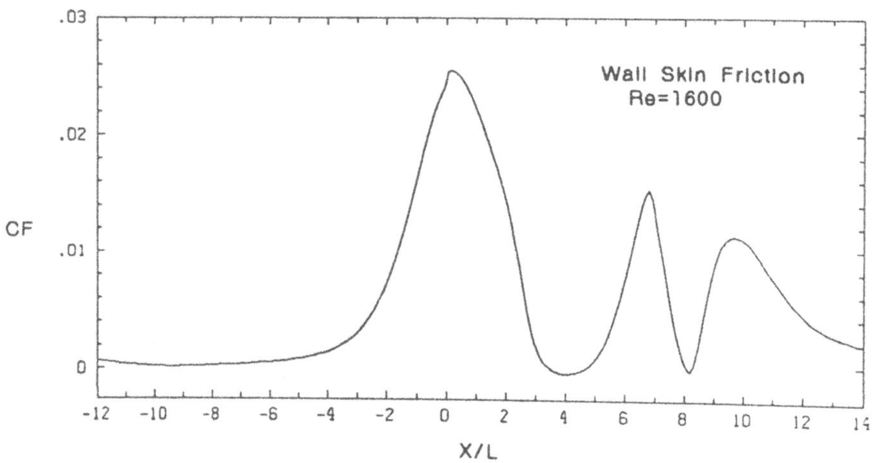

Wall Skin Friction
Re=1600
CF
.03
.02
.01
0
-12 -10 -8 -6 -4 -2 0 2 4 6 8 10 12 14
X/L

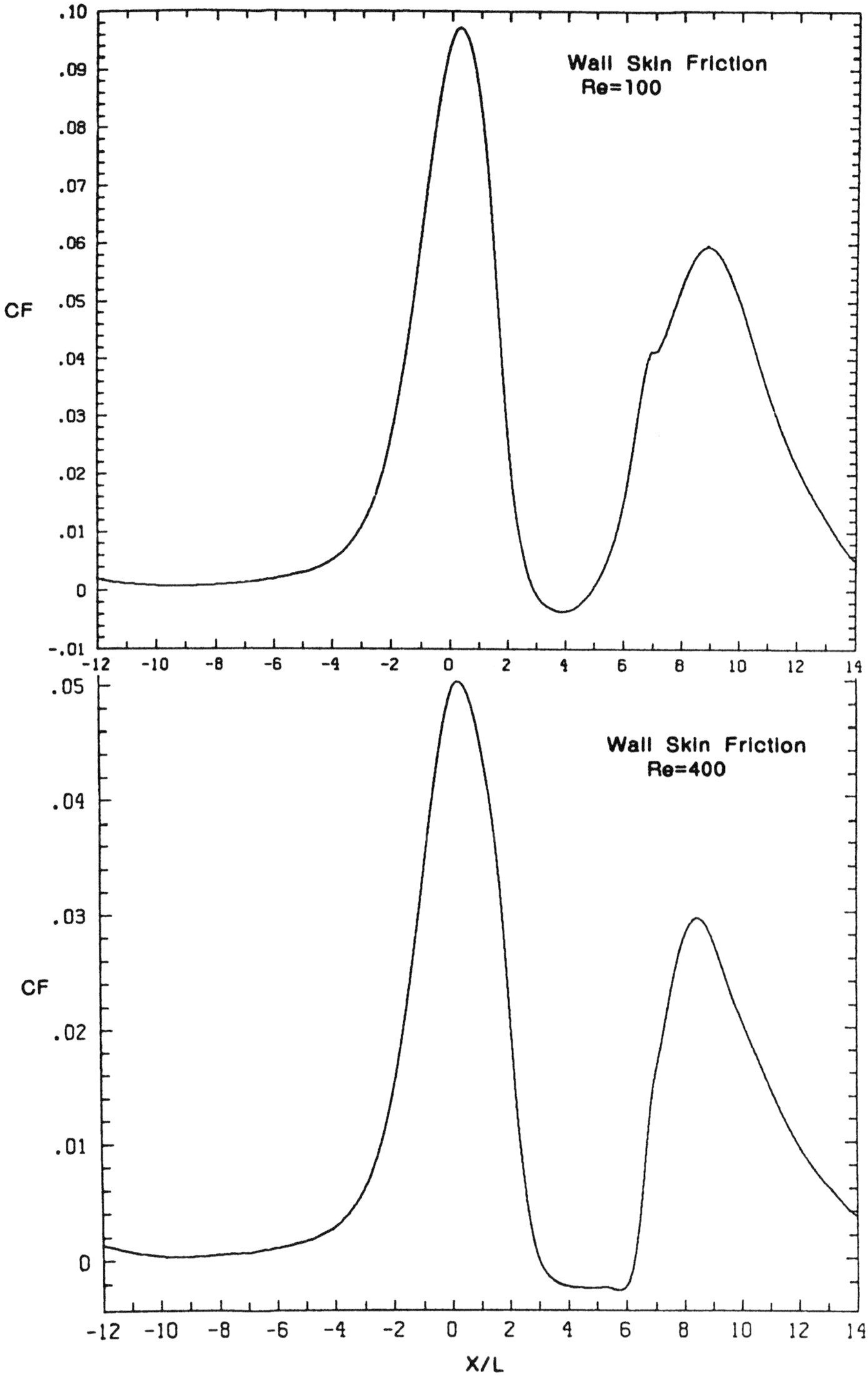

Wall Skin Friction
Re=100
CF
.10
.09
.08
.07
.06
.05
.04
.03
.02
.01
0
-.01
-12 -10 -8 -6 -4 -2 0 2 4 6 8 10 12 14
Wall Skin Friction
Re=400
CF
.05
.04
.03
.02
.01
0
-12 -10 -8 -6 -4 -2 0 2 4 6 8 10 12 14
X/L

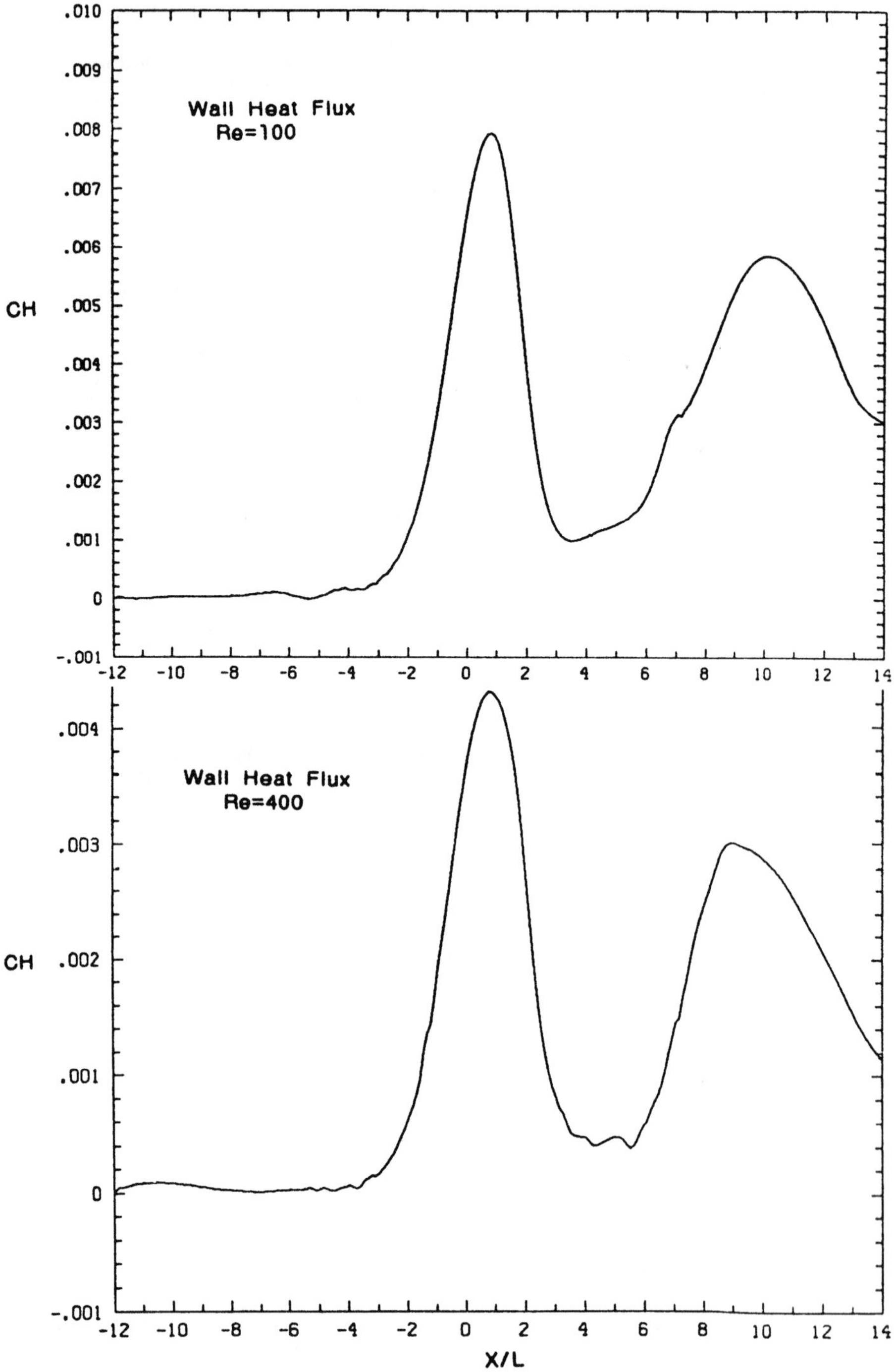
Wall Heat Flux
Re=100
CH
.010
.009
.008
.007
.006
.005
.004
.003
.002
.001
0
-.001
-12
-10
-8
-6
-4
-2
0
2
4
6
8
10
12
14
Wall Heat Flux
Re=400
CH
.004
.003
.002
.001
0
-.001
X/L

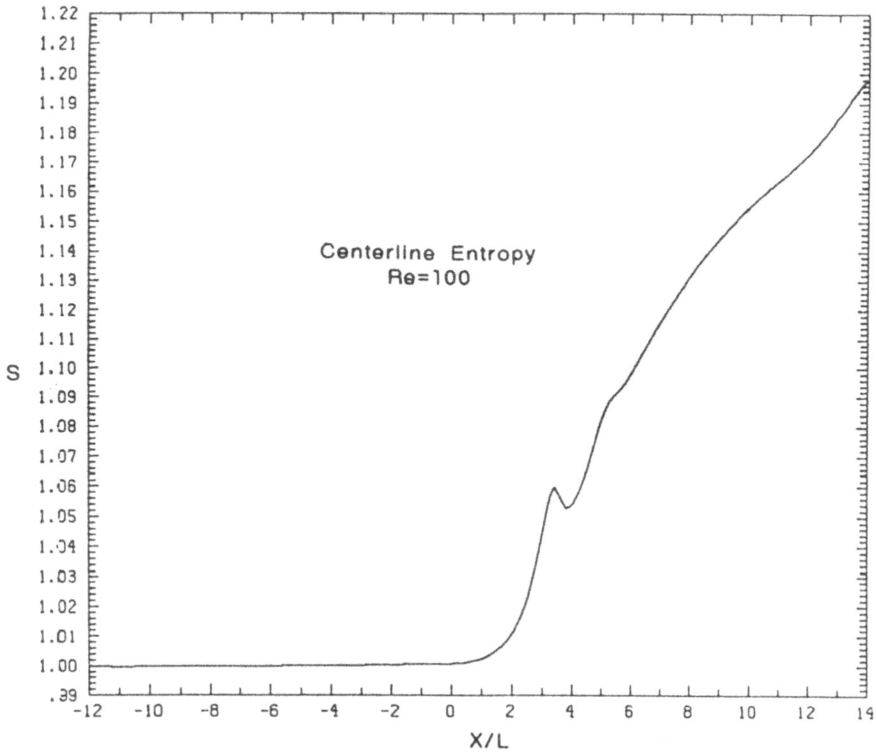
Centerline Entropy
Re=100
S
X/L

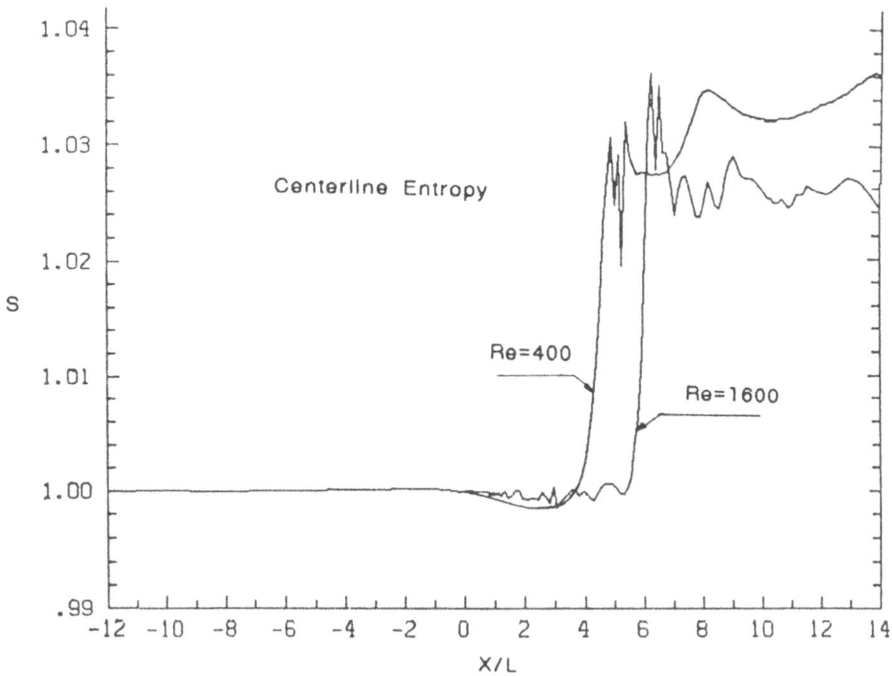
Centerline Entropy
Re=400
Re=1600
S
X/L

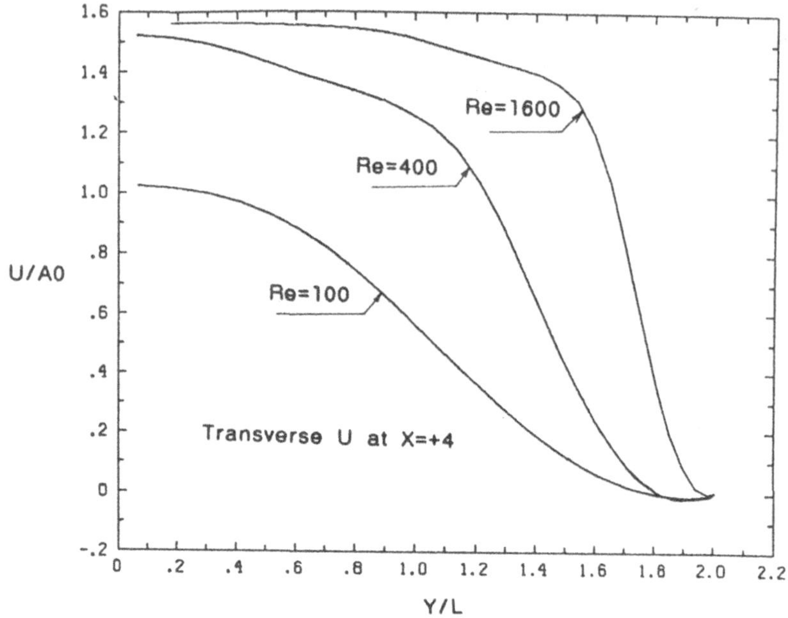
U/A0
Re=1600
Re=400
Re=100
Transverse U at X=+4
Y/L

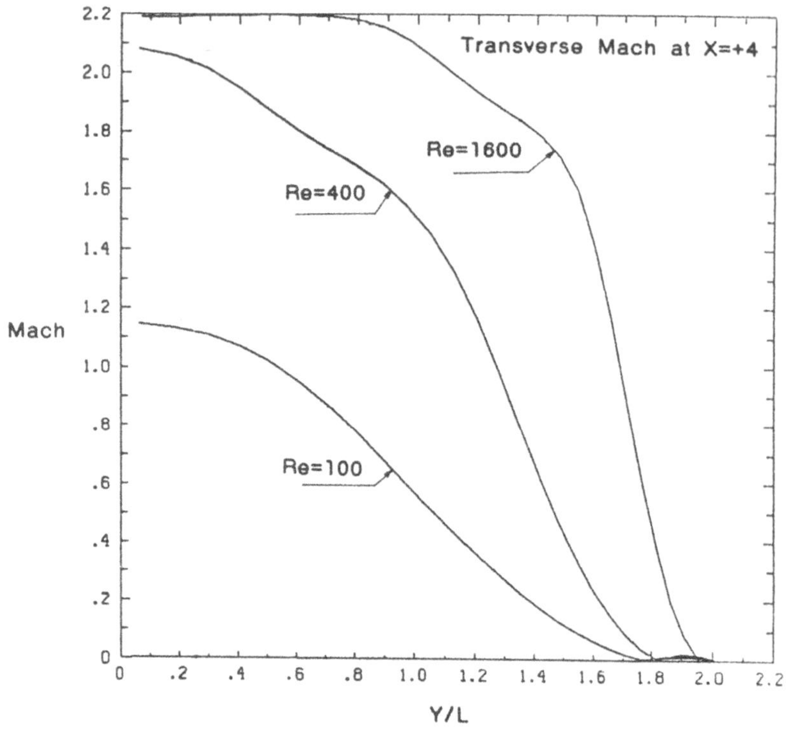
Transverse Mach at X=+4
Mach
Re=1600
Re=400
Re=100
Y/L

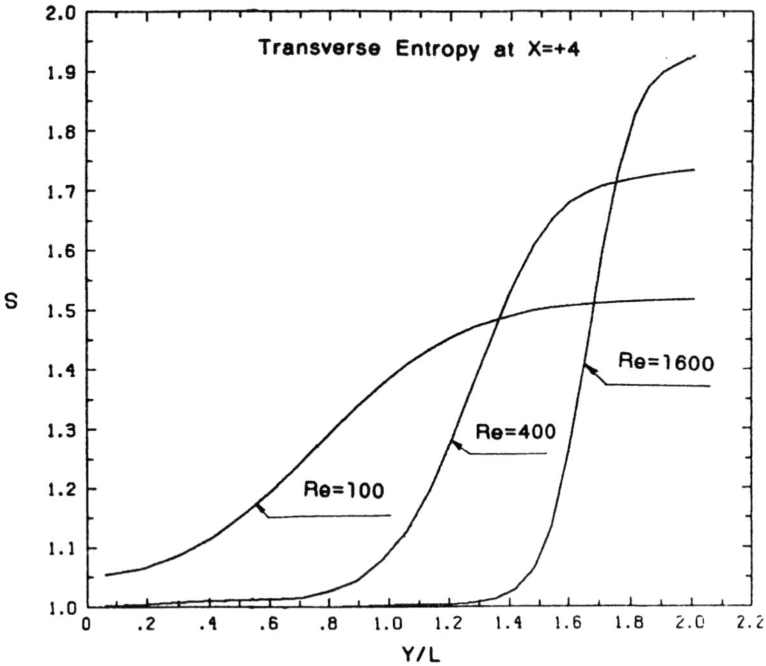
Transverse Entropy at X=+4
S
2.0
1.9
1.8
1.7
1.6
1.5
1.4
1.3
1.2
1.1
1.0
0
.2
.4
.6
.8
1.0
1.2
1.4
1.6
1.8
2.0
2.2
Y/L
Re=1600
Re=400
Re=100

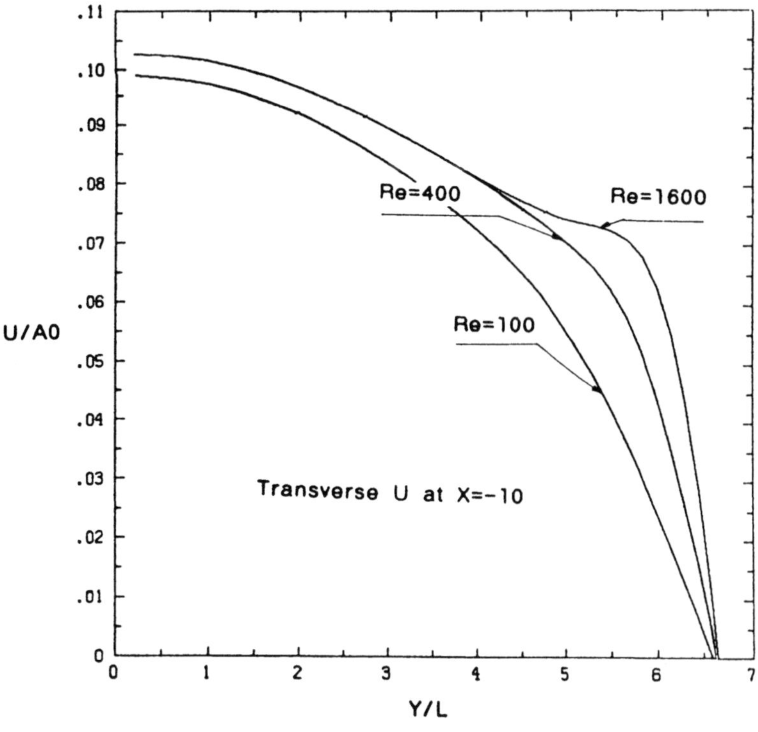
Transverse U at X=-10
U/A0
.11
.10
.09
.08
.07
.06
.05
.04
.03
.02
.01
0
1
2
3
4
5
6
7
Y/L
Re=400
Re=1600
Re=100

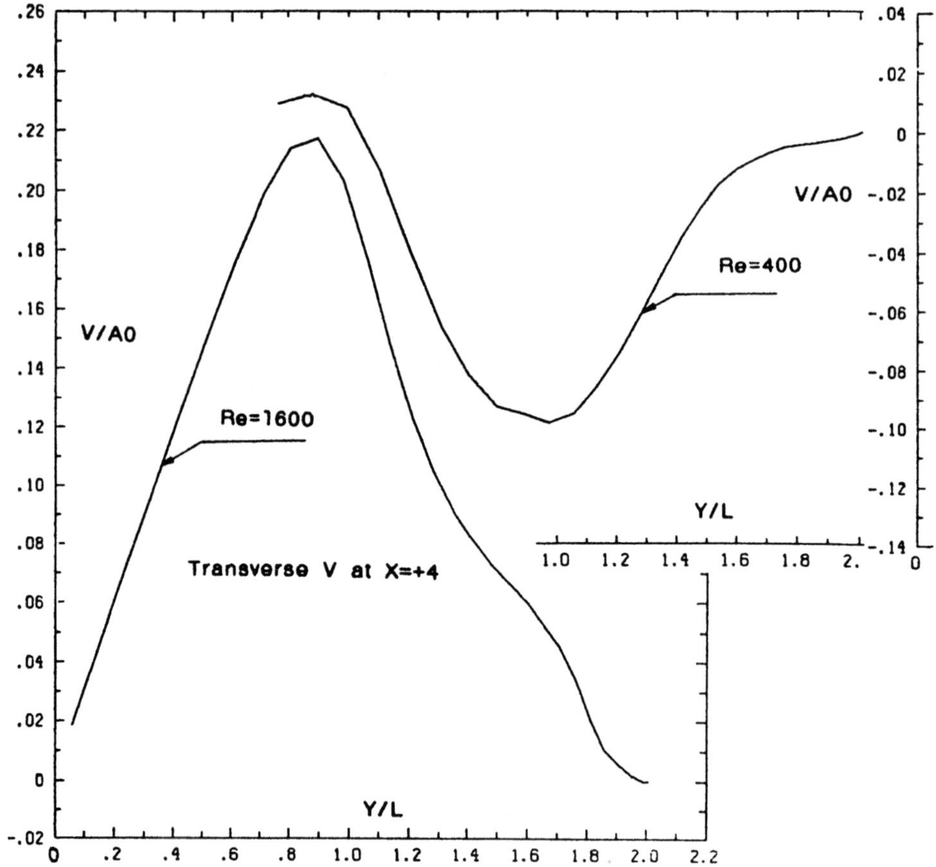

V/A0
Re=1600
Transverse V at X=+4
Y/L
V/A0
Re=400
Y/L

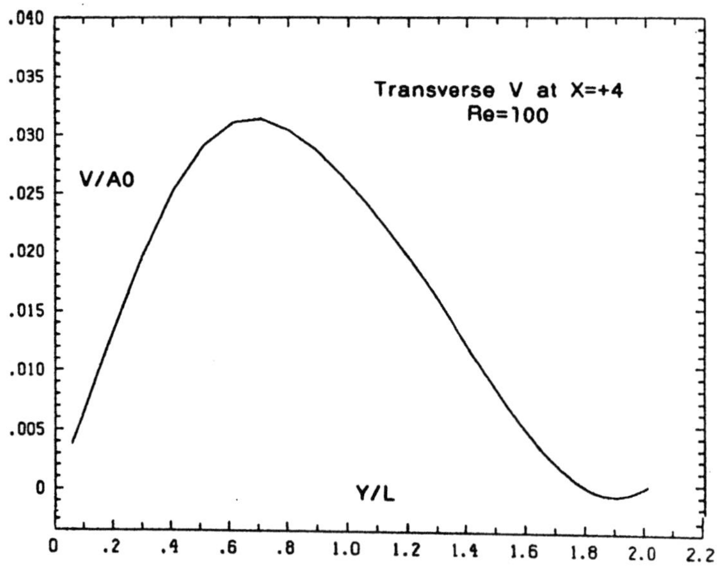

Transverse V at X=+4
Re=100
V/A0
Y/L

Numerical Simulation of Compressible Laminar Nozzle Flow

D. SCHWAMBORN
DFVLR, Institute for Theoretical Fluid Mechanics
Bunsenstrasse 10, D-3400 Goettingen, Germany

Summary

The full Navier-Stokes equations are solved by a finite-volume Runge-Kutta time-stepping approach. The artificial dissipation needed to stabilize the solution is modified to avoid too high damping near walls. The flow through a double throat nozzle is computed and analysed for different Reynolds number.

Numerical Method

The integral formulation of the full Navier-Stokes equations is given by

$$\frac{\partial}{\partial t} \int_V U dv + \int_S (F dy - G dx) = 0 \qquad (1)$$

with

$$U = \begin{pmatrix} \rho \\ \rho u \\ \rho v \\ \rho e \end{pmatrix} \quad F = \begin{pmatrix} \rho u \\ \rho u^2 + \sigma_x \\ \rho uv + \tau_{xy} \\ (\rho e + \sigma_x) u + \tau_{xy} v + q_x \end{pmatrix} \quad G = \begin{pmatrix} \rho v \\ \rho uv + \tau_{yx} \\ \rho v^2 + \sigma_y \\ (\rho e + \sigma_y) v + \tau_{yx} u + q_y \end{pmatrix}$$

where ρ, u, v, p, e, T and k denote density, cartesian velocity components, pressure, total internal energy, temperature and heat conduction coefficient, while σ , σ and $\tau = \tau$ are the components of the stress tensor. The finite volume method approximates equation (1) for each cell of the computational grid assuming U to be constant within each cell and located at the cell center. The fluxes through the cell boundaries are obtained as an average from both adjacent cells and the resulting set of ordinary differential equations in time is solved by an explicit Runge Kutta method. This technique was used by JAMESON et al. [1] to solve the Euler equations. In contrast to the case of the Euler equations the system now retains derivatives of the velocity components and the temperature which can be obtained by a local mesh transformation [2].

Due to the explicitness of this unsteady approach the time step of the method is limited. To accelerate the convergence, when only the steady state is of interest, we use local time stepping thus allowing for maximum signal progagation. A further acceleration is obtained by implicit residual averaging which allows for larger CFL numbers. Whereas residual averaging for the solution of the Euler equations works very well when applied at alternate stages of a Runge Kutta step only, it is found here that for the Navier Stokes equations it is needed at each stage.

Solving the Navier Stokes equations for a flow that is dominated by viscous effects in most of the computational domain does not require artificial dissipation if no or only very weak shocks are in the flow field. Otherwise, if there are strong shocks or large regions where the flow behaves almost inviscid artificial dissipation is needed to stabilize the solution. For the dissipation operator added to the Navier Stokes equations a combination of second and fourth differences is used as suggested in [1] for the Euler equations. Numerical experiments showed, however, that the corresponding artificial viscosity falsifies the solution in the vicinity of walls. Due to the large gradients in the boundary layer the artificial dissipation terms can be of the same magnitude as the physical ones. Therefore we modify the artificial dissipation terms by multiplying them with the square of the local velocity (as long as this is less then one), i.e.

$$U_t + PU + mDU = 0$$

with

$$m = \min\ (1.,\ u^2 + v^2)$$

where D and P are spacial operators representing the artificial viscosity and the physics, respectively. This modification is very simple and easy to implement, and it works very well for test cases where we compare the Navier-Stokes solution for flat plate flows with e.g. the well known similarity solution of Chapman and Rubesin [3] at different Reynolds and Mach numbers. We find that the wall shear stress and other properties compare very well with the theoretical values since the modified artificial dissipation no longer dominates the physical one in the vicinity of the wall. Although the dissipation terms are added at each stage of a Runge Kutta step, we compute these terms only at the first stage of each odd time-step. This 'freezing' of the dissipation saves more than one third of the computation time without influencing the convergence markedly. The same could be done with the physical dissipation terms resulting again in a 20 - 30 percent gain, whenever we are interested in steady-state solutions.

The boundary conditions used in the solution procedure are generally as follows: At the wall the no-slip condition is imposed, the pressure gradient normal to the wall equal is set to zero, and either a prescribed temperature or an adiabatic wall is assumed. For subsonic flow we obtain the farfield boundary data via characteristic relations at inflow (e.g. [4]) and by extrapolation plus a nonreflecting boundary condition for the pressure [5] at outflow. In the supersonic case four variables are prescribed at inflow and extrapolated at outflow.
The initial conditions for the workshop case are given below.

Discussion of the numerical results.

Mesh

The calculations were made on two different meshes, one with 100 x 30 cells and one with 201 x 30 cells (fig.1). Both meshes consist of straight lines x=const. which are distributed along the x-axis in a relatively smooth way. The distribution in the y-direction is obtained by the sum of two tanh functions, one for the distribution at the throat (x=0) and the other in order to allow for an almost constant smallest stepsize near the wall elsewhere. Since the distribution in x-direction takes into account that the interesting part of the flow-field is between the throats, the results on both mesh are qualitatively equal. The resolution, however, is better on the finer mesh and results are only given for this one.

Boundary and initial conditions

For the nozzle flow the boundary conditions are as prescribed in the specifications,i.e.

wall conditions as given above with wall temperature equal to reservoir temperature,
all conservative variables extrapolated at outflow,
symmetry conditions at the symmetry line and
at inflow as prescribed
    parallel flow (no vertical velocity),
    total enthalpy equal reservoir enthalpy and
    entropy equal to reservoir entropy.
Since an additional condition is needed numerically at inflow, we extrapolate the horizontal velocity component thus imposing no-slip at the wall at this point indirectly.

As initial conditions for the calculation on the coarse grid the solution of one-dimensional isentropic flow was used for Reynolds numbers 100 and 400. For Re = 1600 we started from the coarse grid solution of Re =400. The Reynolds number is here defined by the reservoir values of density, viscosity and sound speed and the half width of the first throat is taken as the reference length. The solution on the finer grid always started from a zeroth order interpolation of the corresponding coarse grid solution. An interpolation of first or second order could avoid the steplike initial distribution and would of course result in a somewhat faster convergence. In the following only the results on the finer mesh are given.

Results

The flow field at Re = 100 is clearly dominated by the viscous effects, as can be seen by the isomach-lines in figure 2. In this figure and also in the other isomach plots subsonic flow is shown by full lines and supersonic flow by short dashed lines; the longer dashed line is the sonic-line. Since a separation bubble is situated between the two throats (see table 1) and because of the thickness of the boundary layer we find a a smooth recompression of the flow (fig.3). Furthermore the low Reynolds number results in a subsonic part of the flow at the outlet which covers about one third of the outlet width (fig.2)

At Re = 400 the boundary layer and the separated region become thinner and we find an almost normal shock between x = 5 and x = 5.5 with a subsonic flow region behind (fig.4,5) The separation bubble is now longer then before due to the pressure rise at the shock. Behind the shock the flow is again accelerated resulting in an effective position of the second "throat" at about x = 5.5.

At Re = 1600 the separation bubble is much smaller and the shock is now oblique (fig.6,7). Keeping the symmetry in mind, we find that the oblique compression zone between x = 3. and 6.5 is continued by the oblique shock in the lower half plane, i.e. we have a system of two intersecting shocks where the shock strength in front of the intersection is smaller then behind it. These shocks (one in the lower and one in the upper half plane) interact with the bounday layer, resulting in another separation (see table 1).
It is well known, that a shock is reflected by the boundary layer on a flat plate as a combination of a compression zone, an expansion fan and another compression zone. Here we find a similar situation where the expansion is stronger than in the flat plate case and the second compression does not exist, due to the divergence of the nozzle behind the throat. The first compression, however, can be found as well in the mach number (fig.8) and pressure distribution (fig.9) along the symmetry line as in the iso-line plots of figures 6 and 7, although it is not very well resolved there.
Figures 9 and 10 show a comparison of the pressure distribution along the symmetry line and along the wall for all three Reynolds numbers. With increasing Reynolds number the beginning of the first recompression is shifted down-stream near the wall as well as at the line of symmetry. The pressure variation between the throats is smaller at the wall than along the symmetry line, because of the viscous effects and the resulting upstream travelling information in the boundary layer.

In figures 11 to 13 the skin-friction and heat-conduction coefficients are compared for the three Reynolds numbers. In all cases one recognizes the similarity between the two coefficients which is well known from the flat plate boundary layer. For the lowest Reynolds number we find small kinks in both curves at the position x=7 (fig.11), which result from the discontinuity in the wall curvature at this position. At Re = 400 these kinks are less distinct and at Re = 1600 we can not recognize them, since there are local maxima in both curves just at this position.

Figures 14 to 16 show the mass flow rates at the different Reynolds number. The error in the massflow rates is less then 1 percent for all cases and the wiggles at entrance and exit of the nozzle indicate that the mesh should be a little smoother and denser in this regions. The wiggles at the exit may also be caused by the extrapolation boundary condition there, which is not valid in the subsonic zone near the wall.

Finally we present in tables 2 to 4 the numerical data demanded for the workshop.

Computer

The calculations were made on a Cray-1s computer with formally one megaword of storage. Since our method is an explicit one vectorisation is simple and straightforward. It leads,however, to a higher amount of storage, because arrays are needed in DO-loops, where one has simple variables on a serial computer.
Our calculations needed about 3.3exp(-5) sec./(cell•time-step) and the storage for the fine mesh was 410 kilowords, however, the code is not yet optimized regarding the storage. We used the standard four-stage Runge-Kutta scheme which needs more storage than the linearized five or six stage versions and is less efficient regarding convergence. Furthermore we did not try to optimize the combination of the parameter in the residual averaging and CFL number for these calculations.
On the coarse mesh (101x31) it takes about 500 to 600 seconds - depending on the Reynolds number to - to obtain a sufficiently converged solution, i.e. a maximum residual of less than exp(-3) and a L_2-norm of the residual of less than exp(-5).
For the 202x31 mesh we obtained the same accuracy after about 1000 seconds (for Re = 1600), starting from the above mentioned zeroth order interpolation.

REFERENCES

[1] Jameson, A., Schmidt, W., Turkel, E., "Numerical Solutions of the Euler Equations by Finite Volume Methods Using Runge-Kutta Time-Stepping Schemes", AIAA-81-1259 (1981).

[2] Deiwert, G.S., "Numerical Simulation of High Reynolds Number Transonic Flows", AIAA Journal Vol.13, No.10 (Oct.75).

[3] Chapman, D.R., Rubesin, M.W., "Temperature and Velocity Profiles in the Compressible Laminar Boundary Layer with Arbitrary Distribution of Surface Temperature", J. Aeronautical Sciences, Vol.16 (1949), S.547-565.

[4] Schmidt, W., Jameson, A., "Recent Developments in Finite-Volume Time-Dependent Techniques for Two and Three Dimensional Transonic Flows", VKI Lecture Series 1982-04, 1982.

[5] Rudy, D.H., Strikwerda, J.C., "A Nonreflecting Outflow Boundary Condition for Subsonic Navier-Stokes Calculations", Jour. Computational Physics 36, pp. 55-70 (1980).

Table 1: Separation and reattachment points

	x - Position for		
	Re = 100	Re = 400	Re = 1600
Separation	2.9696	2.9616	3.4240
reattachment	4.8479	6.2304	4.7391
2. separation	-	-	8.0058
2. reattachment	-	-	8.2795

Table 2: Data for Re = 100

Re = 100	x	- 10	- 4	0	4	7	13
WALL VALUES	p/p_o	.997	.971	.549	.353	.335	2.88E-2
	c_f	1.12E-3	7.1E-3	9.5E-2	-2.6E-3	4.2E-2	1.63E-2
	c_h	1.24E-5	7.2E-3	7.05E-3	.98E-3	3.4E-3	2.93E-3
SYMMETRY AXIS VALUES	p/p_o	.994	.971	.593	.414	.344	.07
	M	.09	.208	.90	1.12	1.19	2.15
	s_1	1.00	1.00	1.00	1.06	1.12	1.19

Table 3: Data for Re = 400

Re = 400	x	- 10	- 4	0	4	7	13
WALL VALUES	p/p_o	.997	.981	.512	.256	.279	.019
	c_f	4.2E-4	3.9E-3	8.8E-3	-1.9E-3	1.68E-2	8.9E-3
	c_h	2.8E-6	4.3E-5	1.6E-4	5.0E-4	1.63E-3	1.48E-3
SYMMETRY AXIS VALUES	p/p_o	.994	.968	.575	.116	.216	.061
	M	.096	.216	.927	2.05	1.50	2.41
	s_1	1.00	1.00	1.00	1.004	1.030	1.038

Table 4: Data for Re = 1600

Re = 1600	x	- 10	- 4	0	4	7	13
WALL VALUES	p/p_o	.997	.981	.491	.196	.126	.014
	c_f	1.36E-4	2.04E-3	2.51E-2	-5.39E-4	1.36E-2	4.05E-3
	c_h	9.9E-7	2.27E-5	1.90E-3	3.38E-4	1.45E-3	6.40E-4
SYMMETRY AXIS VALUES	p/p_o	.993	.967	.566	.095	.345	.047
	M	.099	.221	.941	2.20	1.31	2.61
	s_1	1.00	1.00	1.00	1.00	1.021	1.023

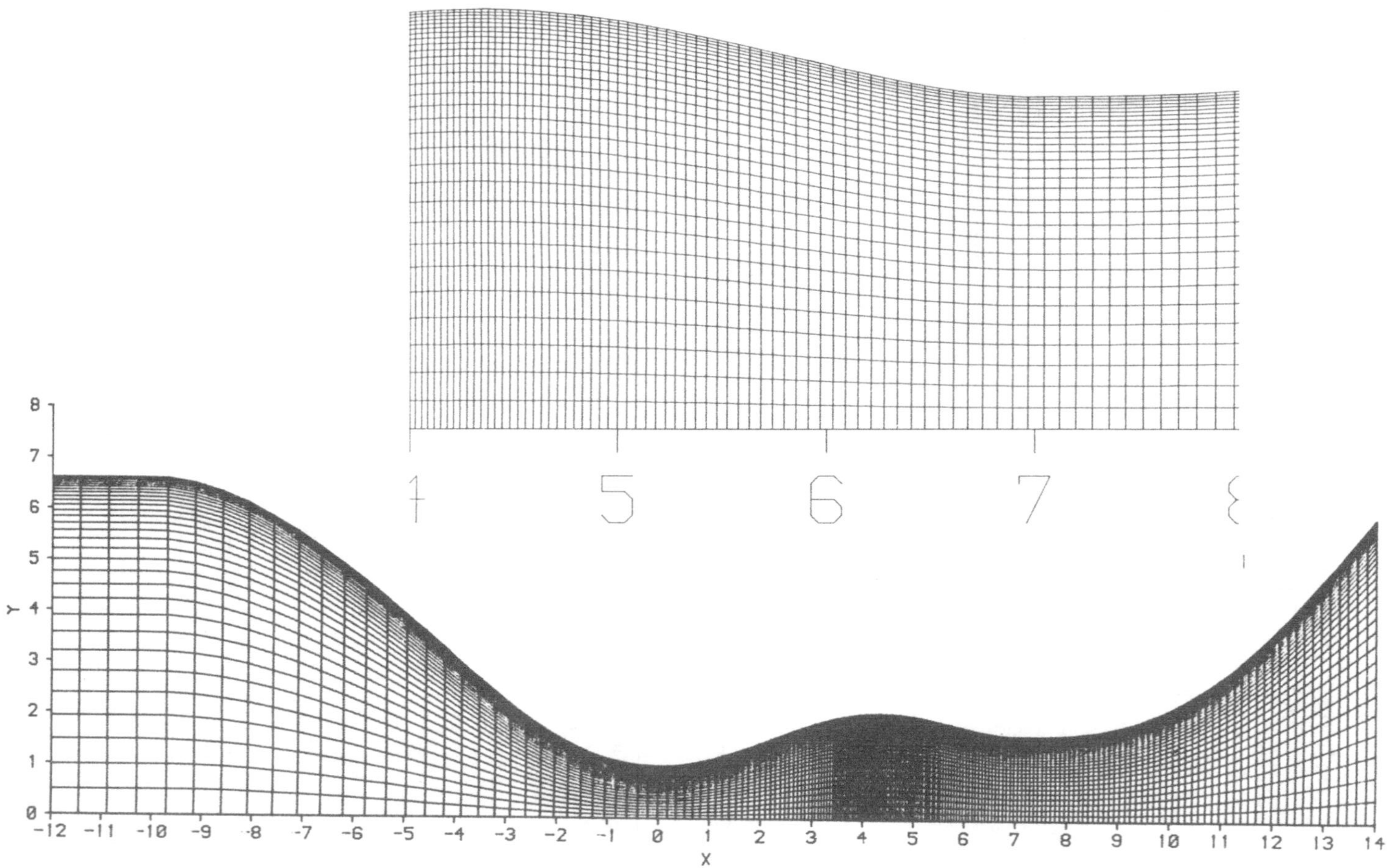

Figure 1: 202x31 mesh with section x = 4 to 8 enlarged

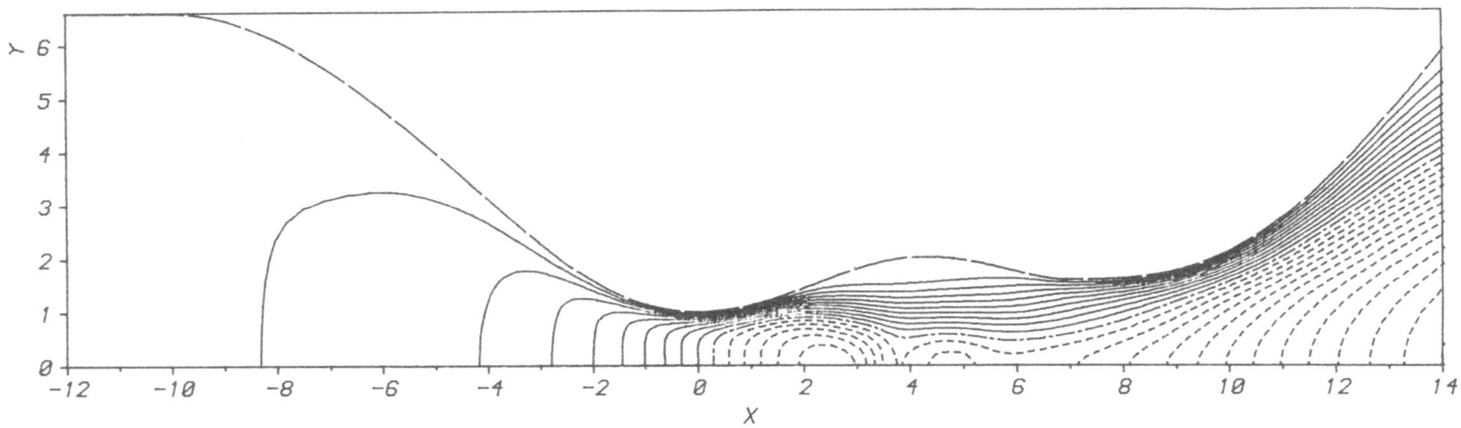

Figure 2: Isomach lines at Re = 100 (Δ= 0.1)

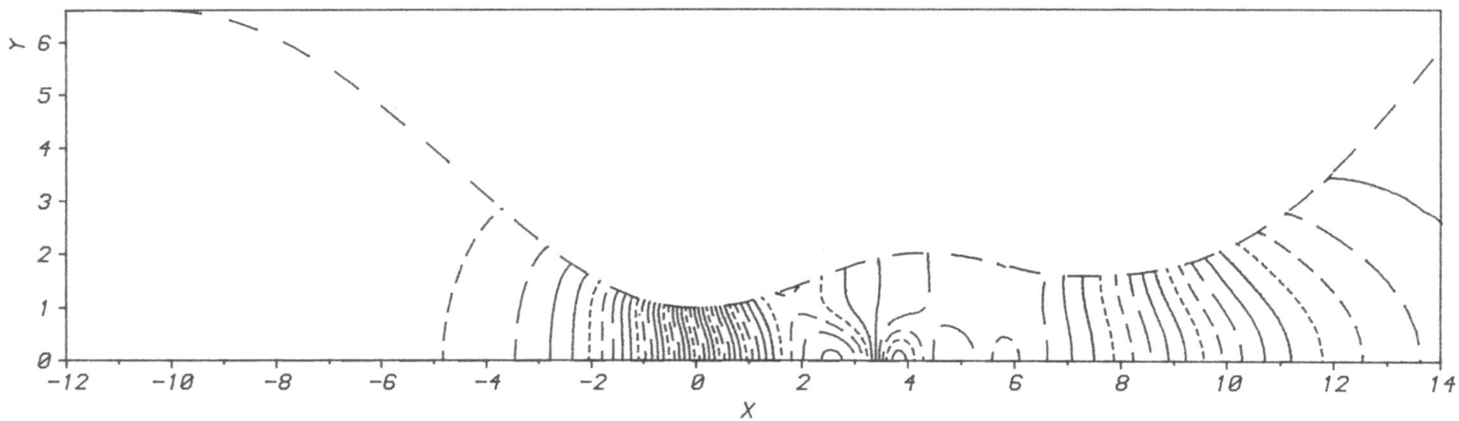

Figure 3: Isobars at Re = 100 (p/p_0 = 1.-n*0.02)

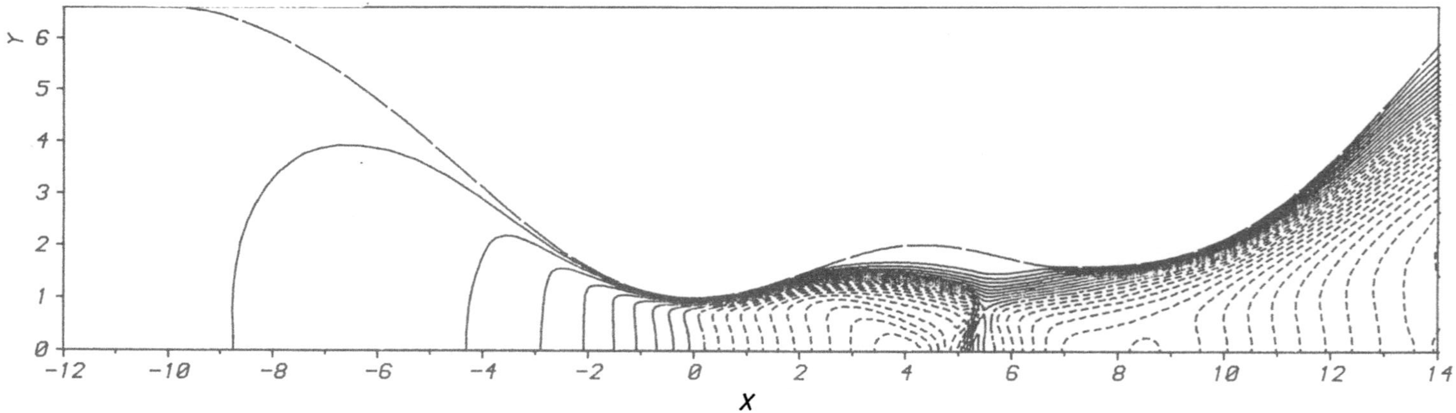

Figure 4: Isomach lines at Re = 400 (Δ= 0.1)

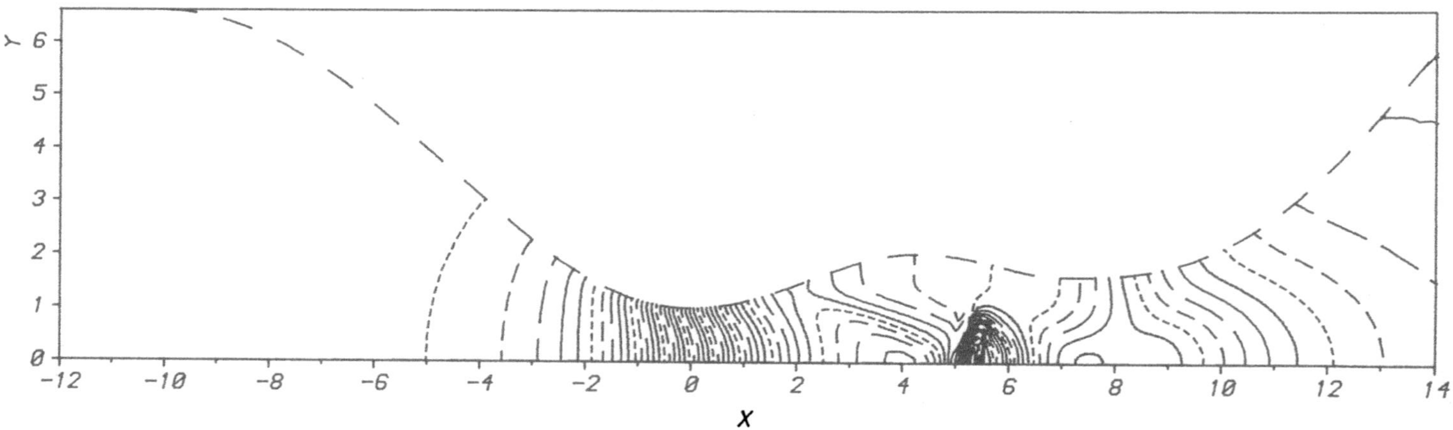

Figure 5: Isobars at Re = 400 (p/p_0 = 1.-n*0.02)

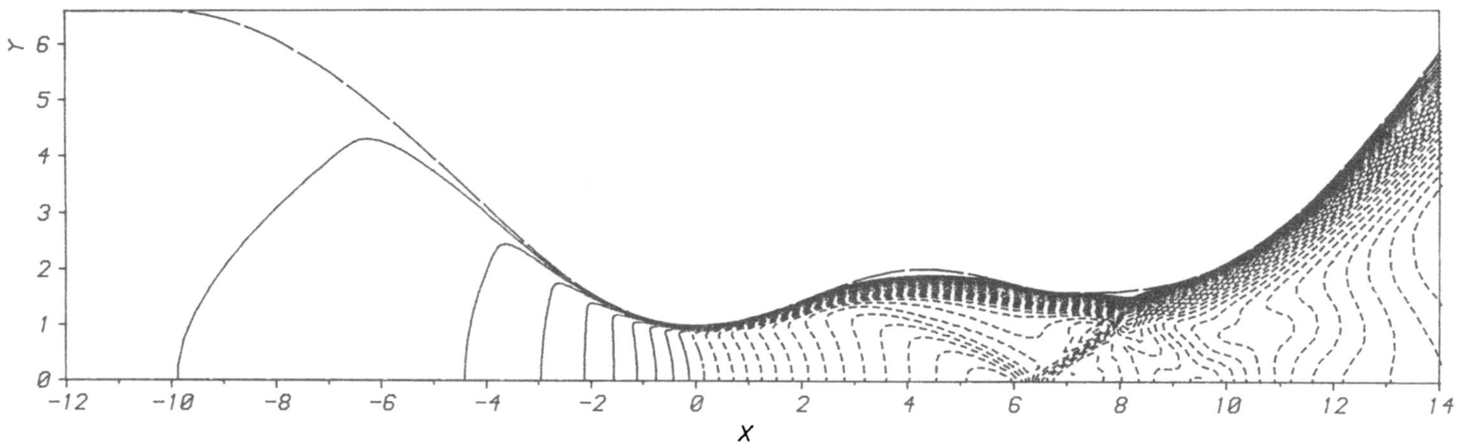

Figure 6: Isomach lines at Re = 1600 (Δ= 0.1)

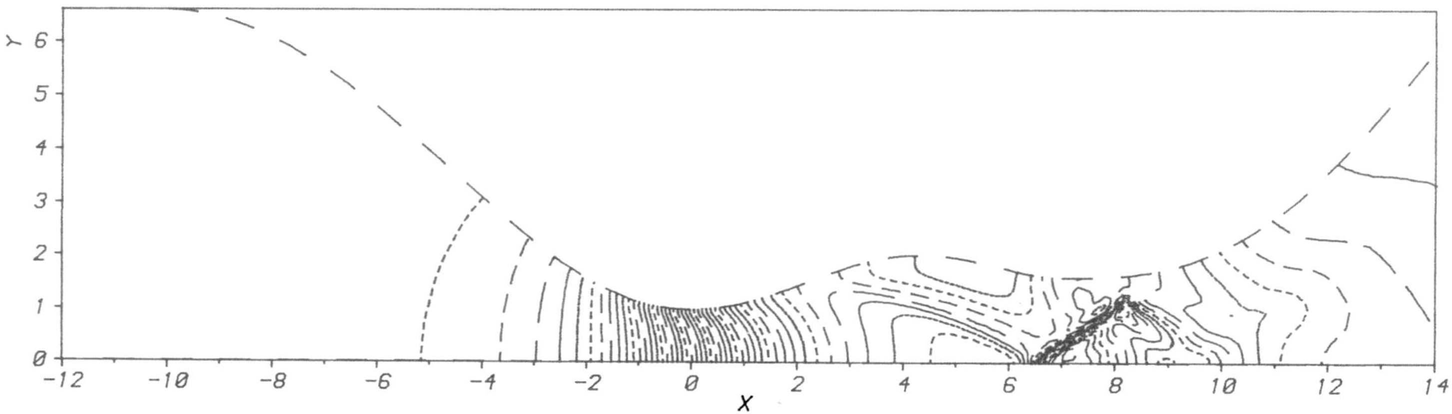

Figure 7: Isobars at Re = 1600 (p/p_0 = 1.-n*0.02)

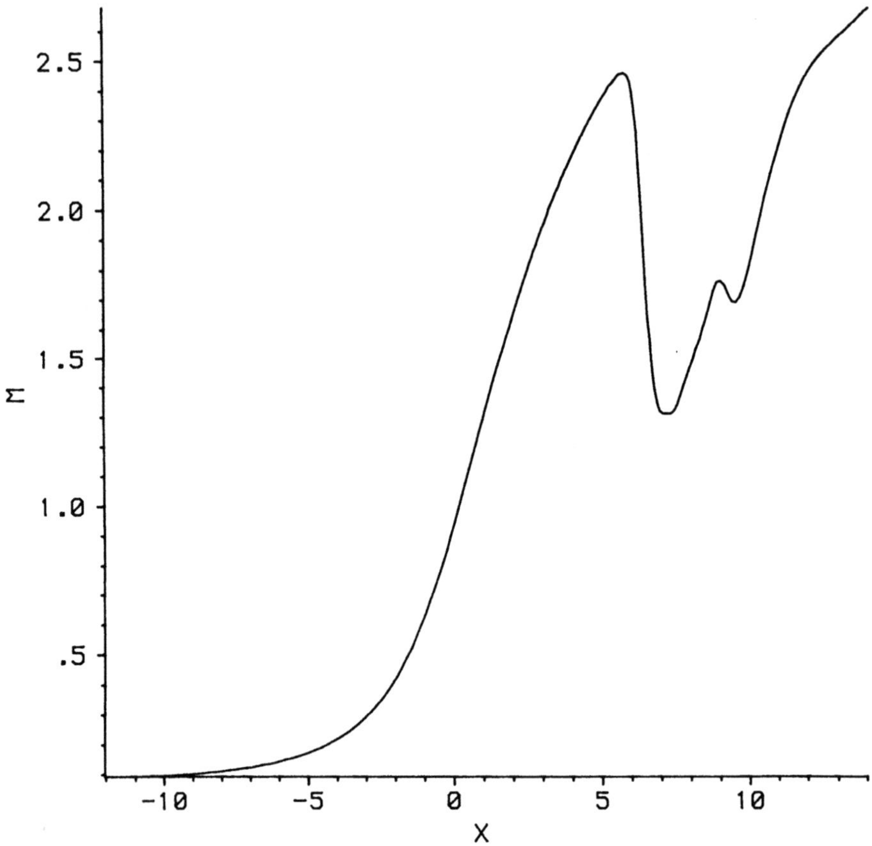

Figure 8: Local mach number at symmetry-axis for Re = 1600

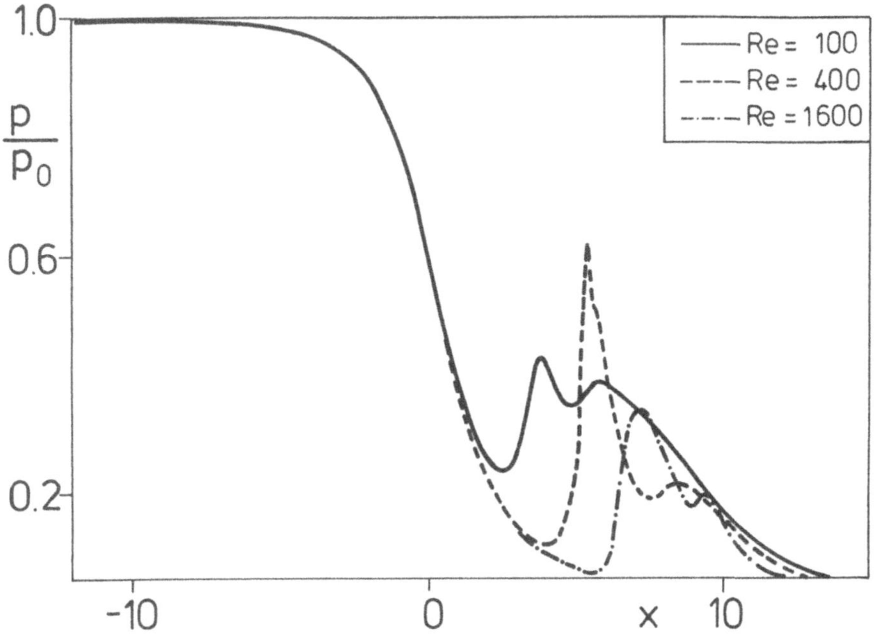

Figure 9: Pressure distribution at symmetry-axis

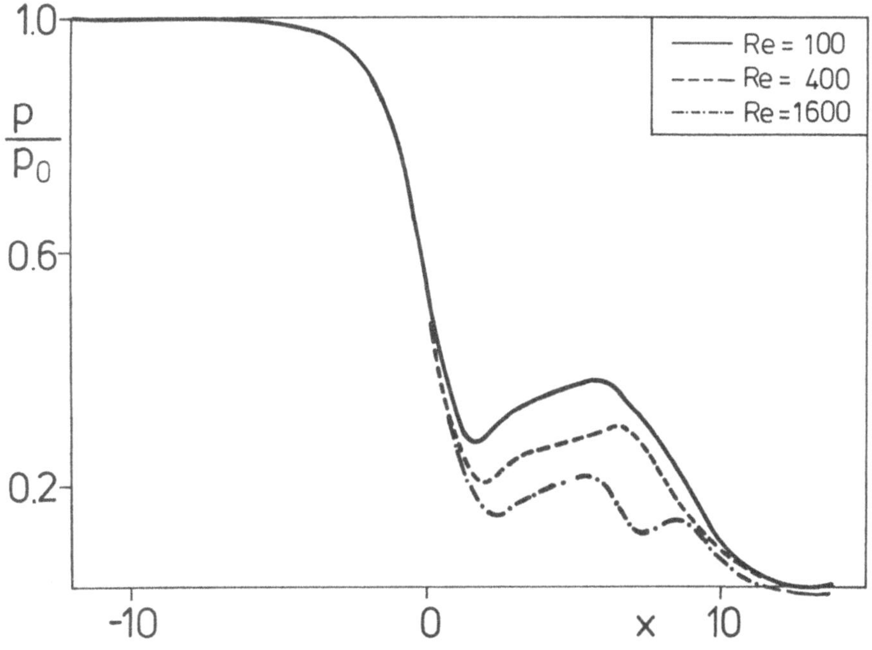

Figure 10: Pressure distribution along the wall

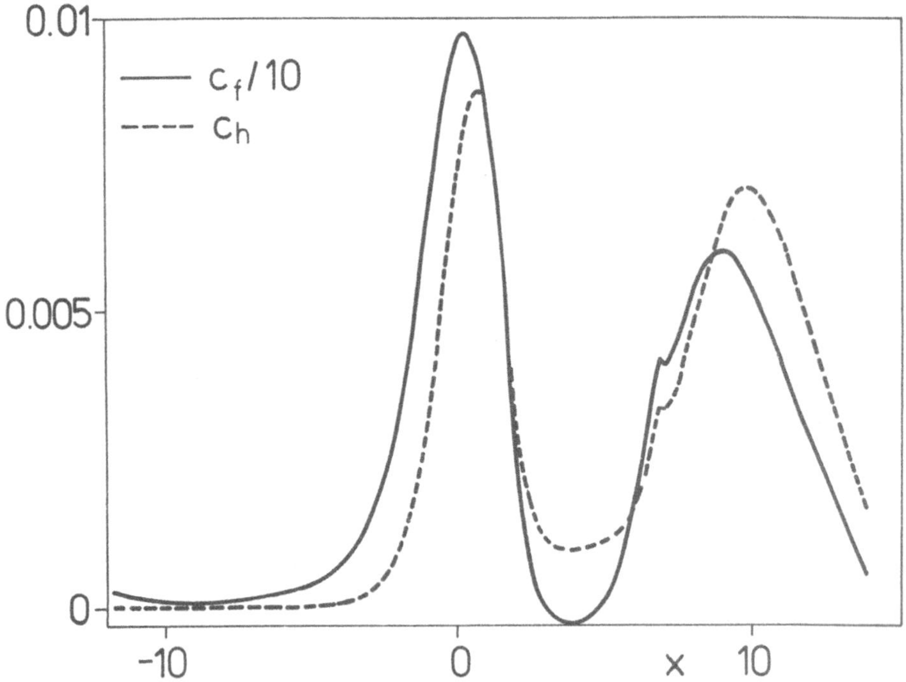

Figure 11: Distribution of skin-friction and heat-conduction coefficients along the wall at Re = 100

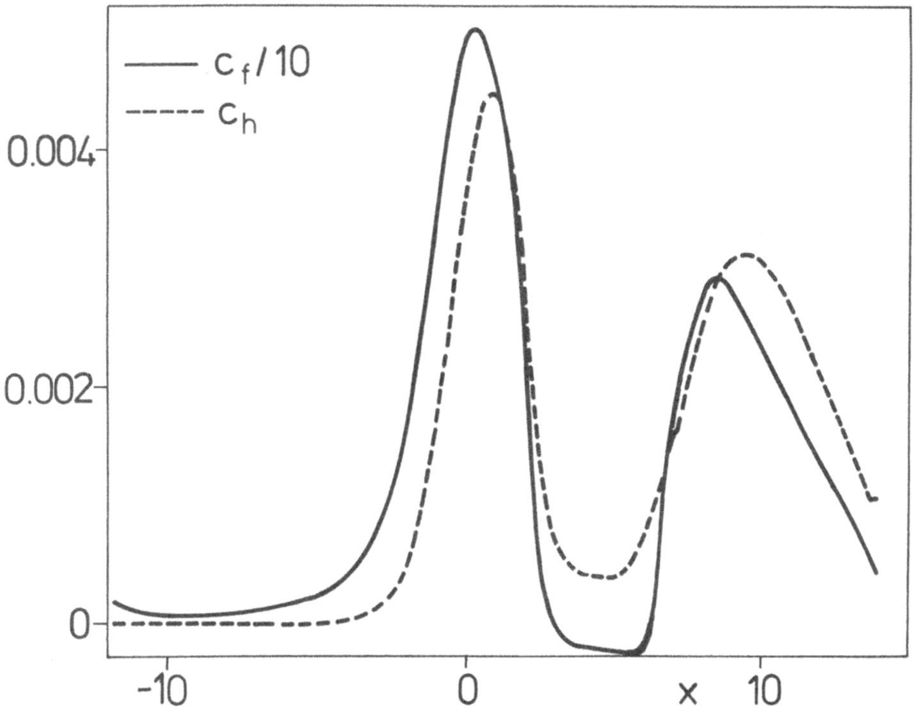

Figure 12: Distribution of skin-friction and heat-conduction coefficients along the wall at Re = 400

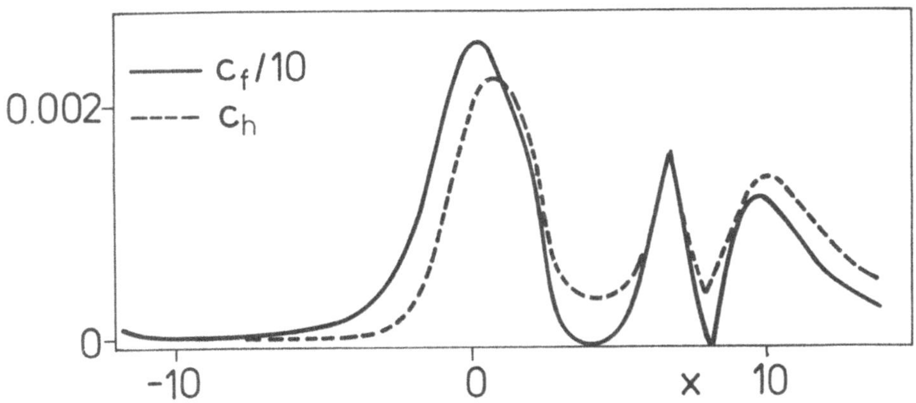

Figure 13: Distribution of skin-friction and heat-conduction coefficients along the wall at Re = 1600

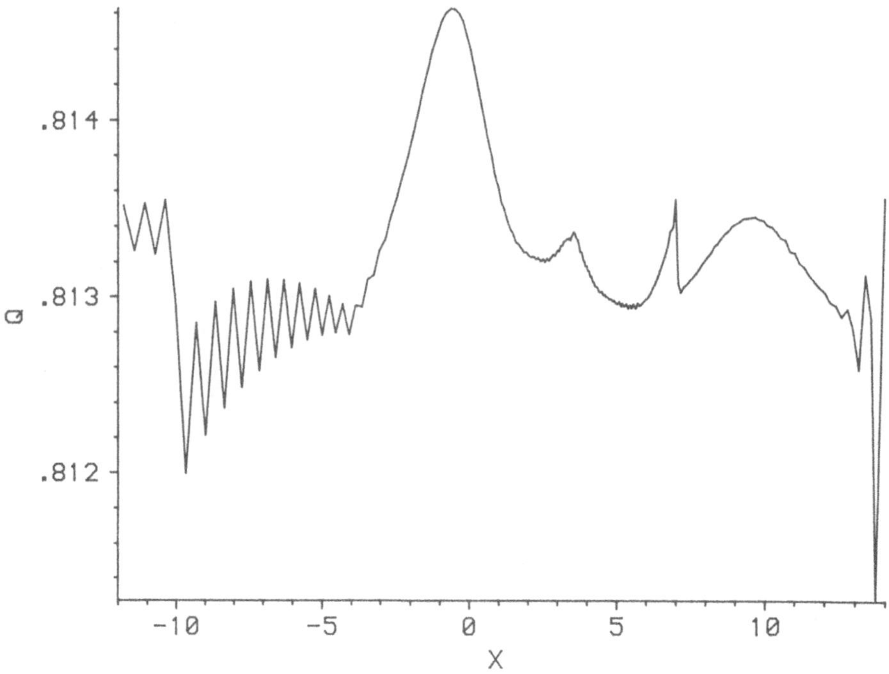

Figure 14: Mass flow rate for Re = 100

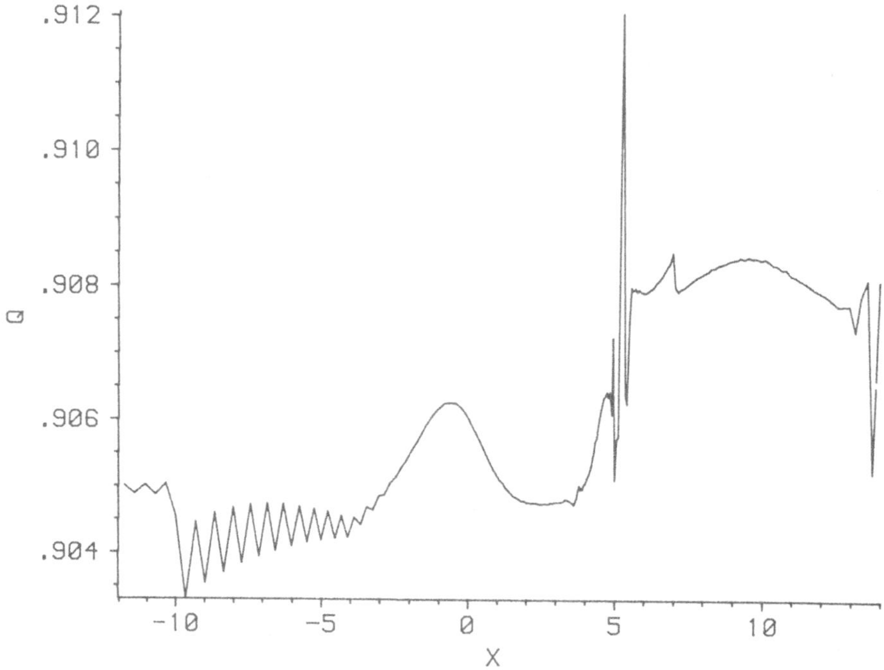

Figure 15: Mass flow rate for Re = 400

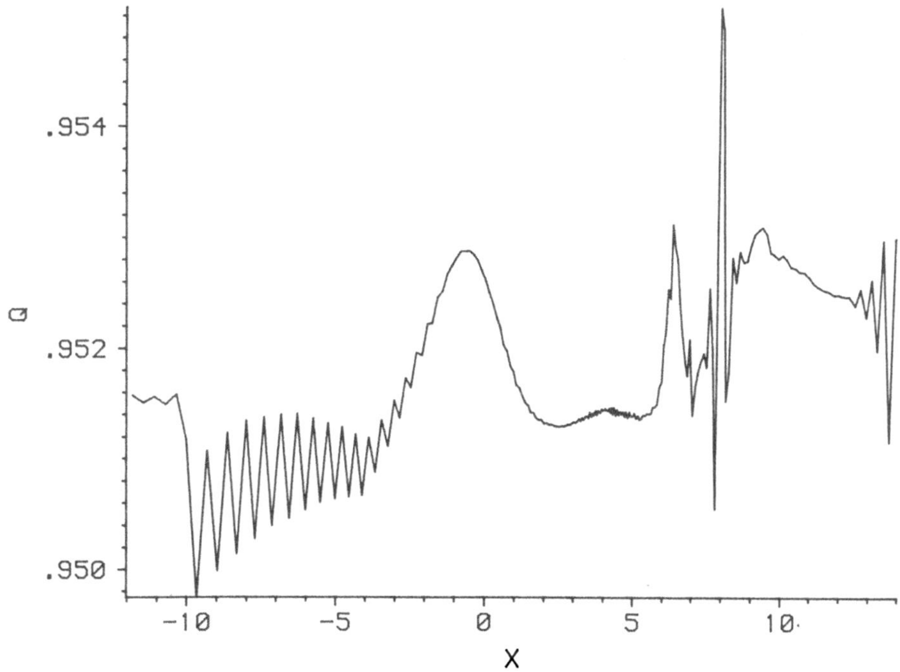

Figure 16: Mass flow rate for Re = 1600

NUMERICAL SIMULATION OF COMPRESSIBLE NAVIER-STOKES FLOW IN A DOUBLE THROAT NOZZLE

James N. Scott
University of Dayton Research Institute
300 College Park, Dayton, Ohio 45469

and

Miguel R. Visbal
Flight Dynamics Laboratory
Air Force Wright Aeronautical Laboratories
Wright-Patterson Air Force Base, Ohio 45433

INTRODUCTION

The flow through a double throat nozzle is computed using the complete time-dependent compressible Navier-Stokes equations. The computations were performed by using an existing working code, which uses MacCormack's explicit finite difference scheme. There were no special modifications made in the code for this particular application. The computations were performed on a Cyber 845 computer and a CRAY XMP-48 computer using three different grid sizes.

GOVERNING EQUATIONS

The complete time-dependent compressible Navier-Stokes equations are solved to simulate the flow through a double throat nozzle described as problem B for the GAMM Workshop on Numerical Simulation of Navier-Stokes Flows [1]. The equations are cast in what is known as chain rule conservation law form, which has been discussed extensively in the literature and is widely used for both internal and external flow analysis [2-5].

COMPUTATIONAL GRID

Body-fitted finite-difference grids were generated using the elliptic grid generation procedure described in Reference 6. Three different grids were employed in the present nozzle calculations. Details for these grids (which are referred to subsequently as coarse, medium, and fine) can be found in Table 1. The coarse (117 x 31) mesh, shown in Figure 1, displays near-orthogonality and mesh clustering in both coordinates directions. This coarse mesh was obtained using the technique of Reference 6, while the medium and fine grids were constructed from the coarse grid by successively halving the grid spacing in both coordinate directions.

Table 1 Grid details.

GRID	IL×JL	ΔS MIN.	ζ/L MAX.	ΔS MIN.	η/L MAX.
Coarse	117×31	0.1	0.5	0.025	0.24
Medium	227×61	0.05	0.25	0.0125	0.12
Fine	453×121	0.025	0.125	0.00625	0.06

Legend :

IL : No. of grid points in streamwise direction (ζ)

JL : No. of grid points in transverse direction (η)

ΔS_ζ : physical grid spacing in ζ-direction

ΔS_η : physical grid spacing in η-direction

NUMERICAL PROCEDURE

MacCormack's explicit finite difference scheme [7] is used for the present study. This algorithm has been used successfully for a wide range of internal and external flows [3,4,5]. The computer code used in this program is a modified version of a code developed by Shang [3,4] for operation on a vector processor. This algorithm requires a combination of alternating forward and backward differences for the predictor and corrector sweeps.

The efficiency of the code is maximized utilizing a local CFL condition on the allowable time increment for generalized coordinates. The range of CFL numbers used in the present investigation vary from 0.3 to as high as 0.95. This is governed by the Reynolds number and the grid spacing.

The numerical damping is also used in the present analysis in order to suppress numerical oscillations. Again the form of the damping terms is reported by Shang [4]. These damping or artificial viscosity terms are implemented in each sweep direction.

Solutions for this problem were also obtained using the Beam and Warming approximate factorization implicit finite difference algorithm. The details of this formulation are described in Reference 8. A similar version of the implicit Navier-Stokes code has been used successfully for a variety of flow configurations [9,10]. Selected results using this algorithm are presented in Figure 2 for purposes of comparison with the results from the MacCormack algorithm.

BOUNDARY CONDITIONS

The boundary conditions for these computations are formulated in accordance with the guidelines provided in the problem statement.

Upstream Boundary

$$P_0,\ T_0,\ v = 0 \quad \text{specified}$$

$$\frac{\partial u}{\partial x} = 0$$

Downstream Boundary

$$\frac{\partial f}{\partial \zeta} = 0 \quad , \quad \text{where } f = \rho,\ \rho u,\ \rho v,\ \rho e$$

Symmetry Axis

$$v = 0$$

$$\frac{\partial g}{\partial \eta} = 0 \quad , \quad \text{where } g = \rho,\ \rho u,\ \rho e$$

Nozzle Wall

$$u = v = 0 \quad , \ T_w = T_0$$

$$\frac{\partial p}{\partial \eta} = 0.$$

COMPUTING DETAILS

In performing the computations with the MacCormack algorithm, the boundary conditions were implemented and verified using a coarse grid (117 x 31) on the Aeronautical Systems Division's CDC Cyber 845 computer system. Computations were performed using this computer for all three Reynolds numbers until converged solutions were achieved based on the variation of mass flow rate at three different axial locations.

Upon completing these preliminary computations the grid was refined to 227 x 61 and transferred to the CRAY XMP-48 at CRAY Research, Inc., Mendota Heights, Minnesota, where CRAY Research graciously provided computer time and personnel support to complete the computations using the refined grid for all three Reynolds numbers. In addition, the computer capability and support from CRAY personnel made it possible to further refine the grid to 453 x 121 for a total of 54,813 grid points and run the computation for a Reynolds number of 400. The convergence history for the 110 x 31 mesh at a Reynolds number of 400 is shown in Figure 3. The computing rate on the XMP-48 is 9.7×10^{-6} sec/iteration/grid point and this rate is essentially the same for each Reynolds number and grid size on this computer while it is 6.04×10^{-4} sec/iteration/grid point on the Cyber 845 computer. It should be noted that the code had been optimized for operation on vector processor machines such as CRAY machines. Keeping this in mind the code ran at a rate in excess of 100M flops. The acceleration achieved in running on the XMP-48 as opposed to the Cyber is a factor of 62.

For the coarse grid the memory required on the CRAY was 235,008 words while the memory requirement for the Cyber for the same grid was 311,000 words.

Table 2 Summary of computations for double throat nozzle.

GRID / ALGORITHM	COARSE 117 x 31	MEDIUM 227 x 61	FINE 453 x 121
MacCormack	Re = 100^A, $400^{A,B}$, 1600^A	100^B, 400^B, 1600^B	400^B
Beam-Warming	Re = 100^A, 400^B	400^B	

A - computed on U.S. Air Force ASD Cyber 845 computer, Wright-Patterson Air Force Base, Ohio

B - computed on CRAY XMP-48 computer, CRAY Research, Mendota Heights, Minnesota

The results of the computations are presented in the standardized form established by the GAMM Committee and are shown in Figures 4-18, and Table 3 shows the extent of axial separation for each Reynolds number.

Table 3 Axial extent of separation (medium grid, 227 x 61).

Re		x Separation	x Reattachment
100		2.951	4.811
400		3.005	6.241
1600	a	3.432	4.728
	b	8.043	8.234

ACKNOWLEDGEMENT

This work was supported by AFOSR Grant AFOSR-85-0032 for which Dr. James Wilson is the project monitor. The authors would like to thank the Computational Aerodynamics Group of the AFWAL Flight Dynamics Laboratory for providing computer support on the ASD Cyber 845 computer system and graphics support on the Prime 750 computer. In particular, we would like to thank Mr. Jerry Trummer for his assistance with tape logistics and data handling. In addition, we would like to express our deepest gratitude to CRAY Research for providing the opportunity to perform the bulk of the computations in this study on their XMP-48 at Mendota Heights, Minnesota. We would especially like to thank Mr. Kent Misegades for making this arrangement possible and for the many hours of his time and that of Mr. Jeff Dawson in assisting with the computations.

REFERENCES

1. Announcement of the GAMM-Workshop on Numerical Simulation of Compressible Navier-Stokes Flows.

2. Shang, J. S., "An Assessment of Numerical Solutions of the Compressible Navier-Stokes Equations," AIAA Paper 84-1549, June 1984.

3. Shang, J. S., "Oscillatory Compressible Flow Around a Cylinder," AIAA 82-0098, AIAA 20th Aerospace Sciences Meeting, Orlando, Florida, 11-14 January 1982.

4. Shang, J. S., "Numerical Simulation of Wing-Fuselage Aerodynamic Interaction," AIAA 83-0225, AIAA 21st Aerospace Sciences Meeting, Reno, Nevada, 10-13 January 1983.

5. Scott, J. N. and Hankey, W. L. Jr., "Numerical Solution of Cold Flow in an Axisymmetric Centerbody Combustor," AIAA 83-1741, AIAA 16th Fluid and Plasma Dynamics Conference, Danvers, Massachusetts, 12-14 July 1983.

6. Visbal, M. and Knight, D., "Generation of Orthogonal and Nearly Orthogonal Coordinates with Grid Control Near Boundaries," AIAA Journal, Vol. 20 (3), 1982.

7. MacCormack, R. W., "Numerical Solution of the Interaction of a Shock Wave with a Laminar Boundary Layer," Lecture Notes in Physics, Vol. 59, Springer-Verlag, 1971.

8. Beam, R. and Warming, R., "Am Implicit Factored Scheme for the Compressible Navier-Stokes Equations," AIAA Journal, Vol. 16, 1978.

9. Visbal, M. and Knight, D., "The Baldwin-Lomax Turbulence Model for Two-Dimensional Shock-Wave Boundary Layer Interactions," AIAA Journal, Vol. 22 (7), 1984.

10. Visbal, M. and Shang, J., "A Comparative Study Between an Implicit and Explicit Algorithm for Transonic Airfoils," AIAA Paper 85-0480, 1985.

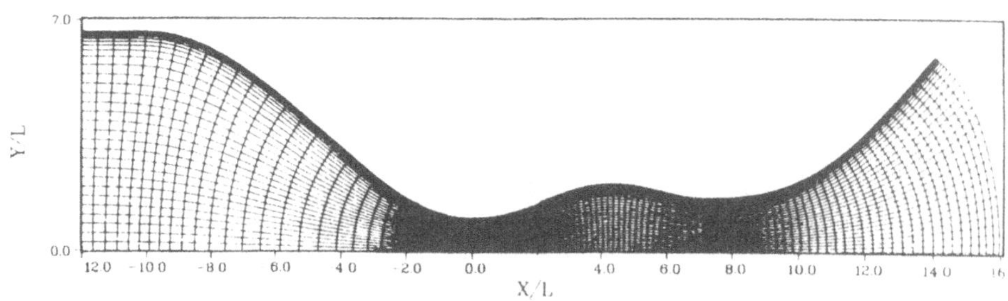

Fig. 1 Coarse-mesh grid (117 x 31).

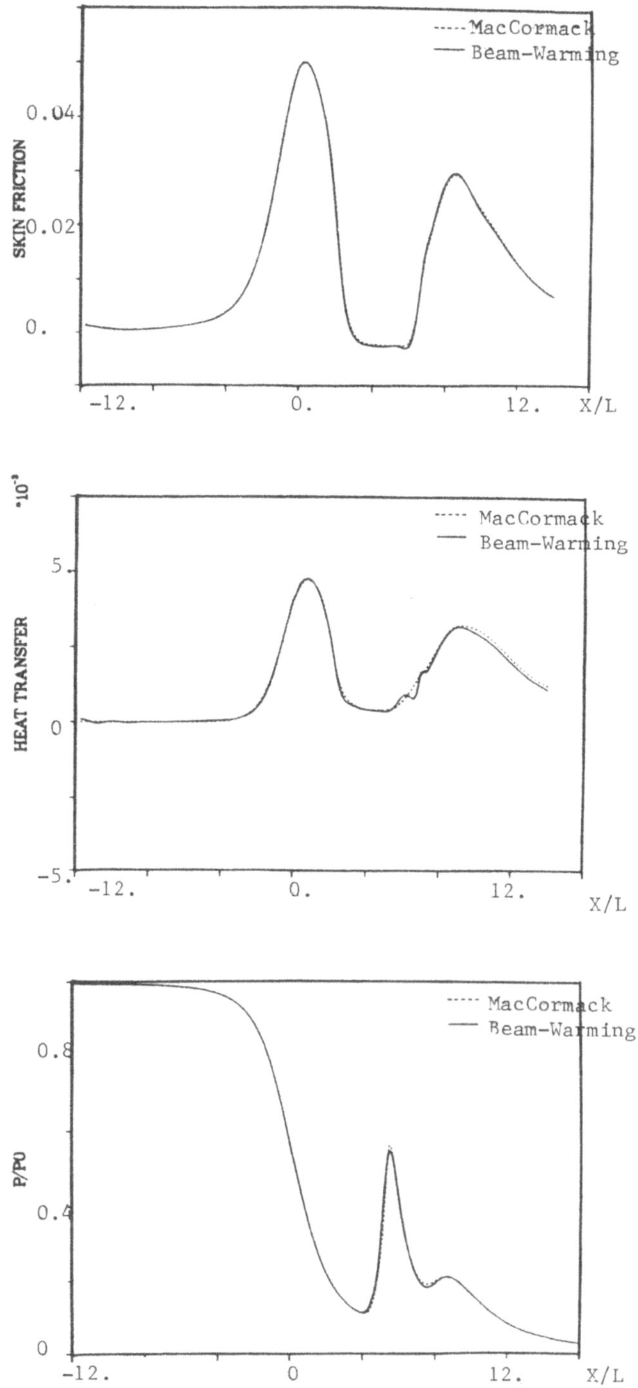

Fig. 2 MacCormack - Beam-Warming comparison.

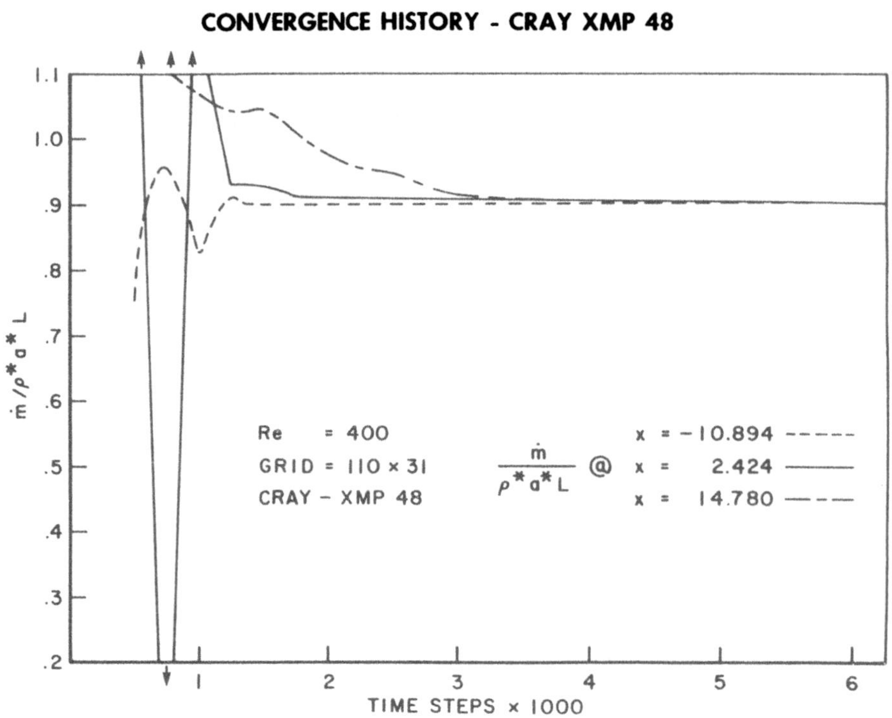

Fig. 3 Convergence history.

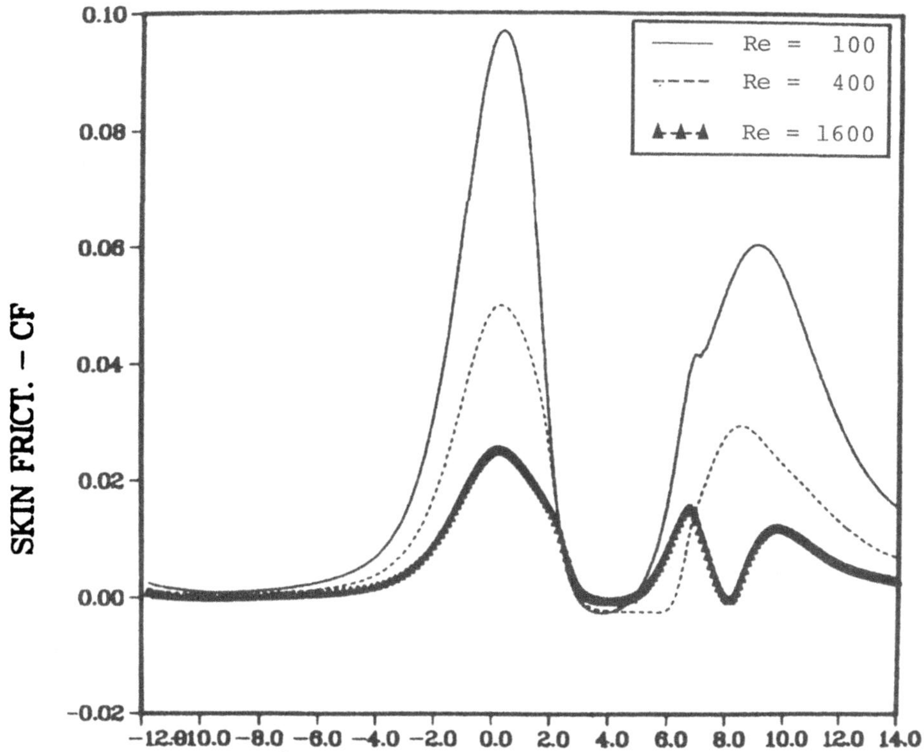

Fig. 4 Skin friction coefficient.

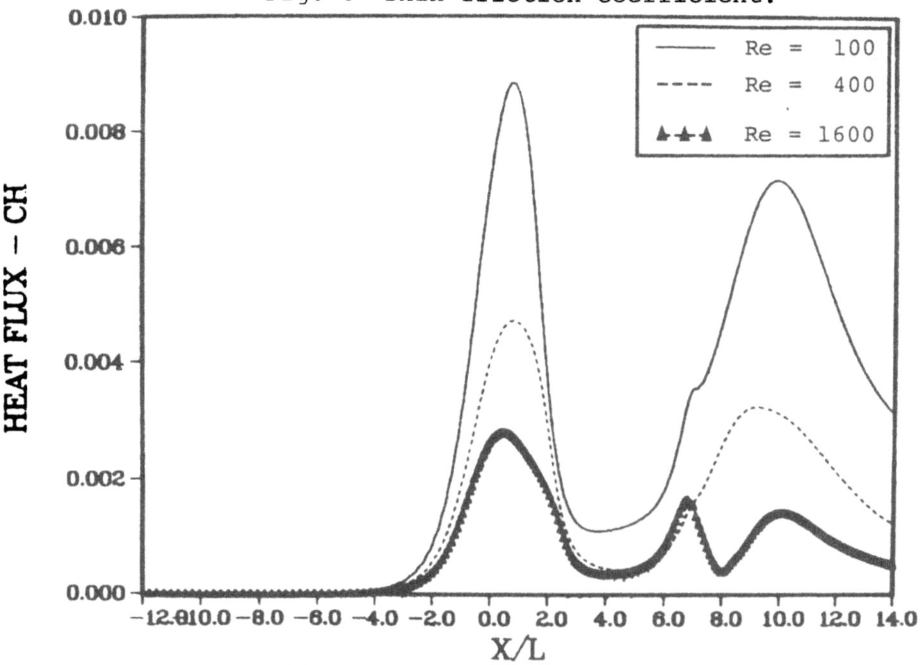

Fig. 5 Heat flux coefficient.

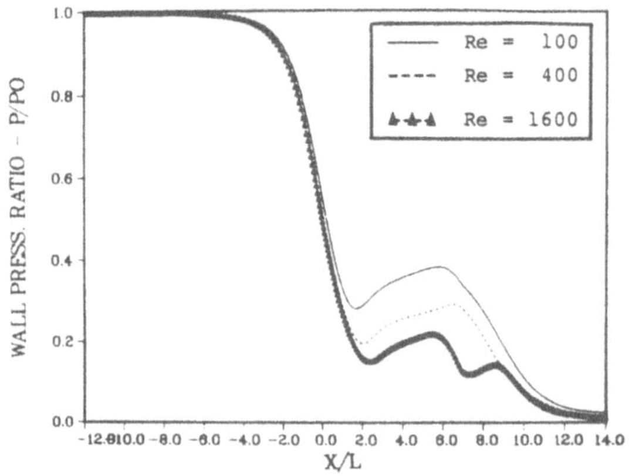

Fig. 6 Wall pressure distribution.

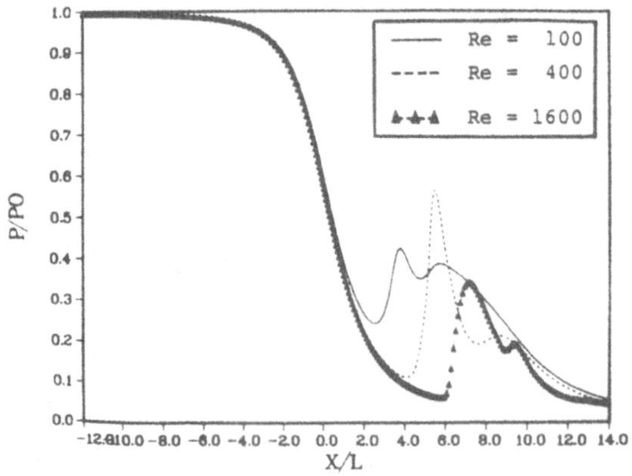

Fig. 7 Pressure distribution on symmetry axis.

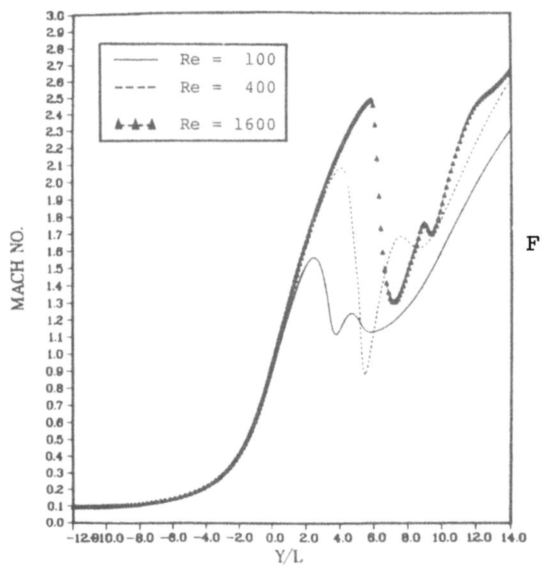

Fig. 8 Mach number variation on symmetry axis.

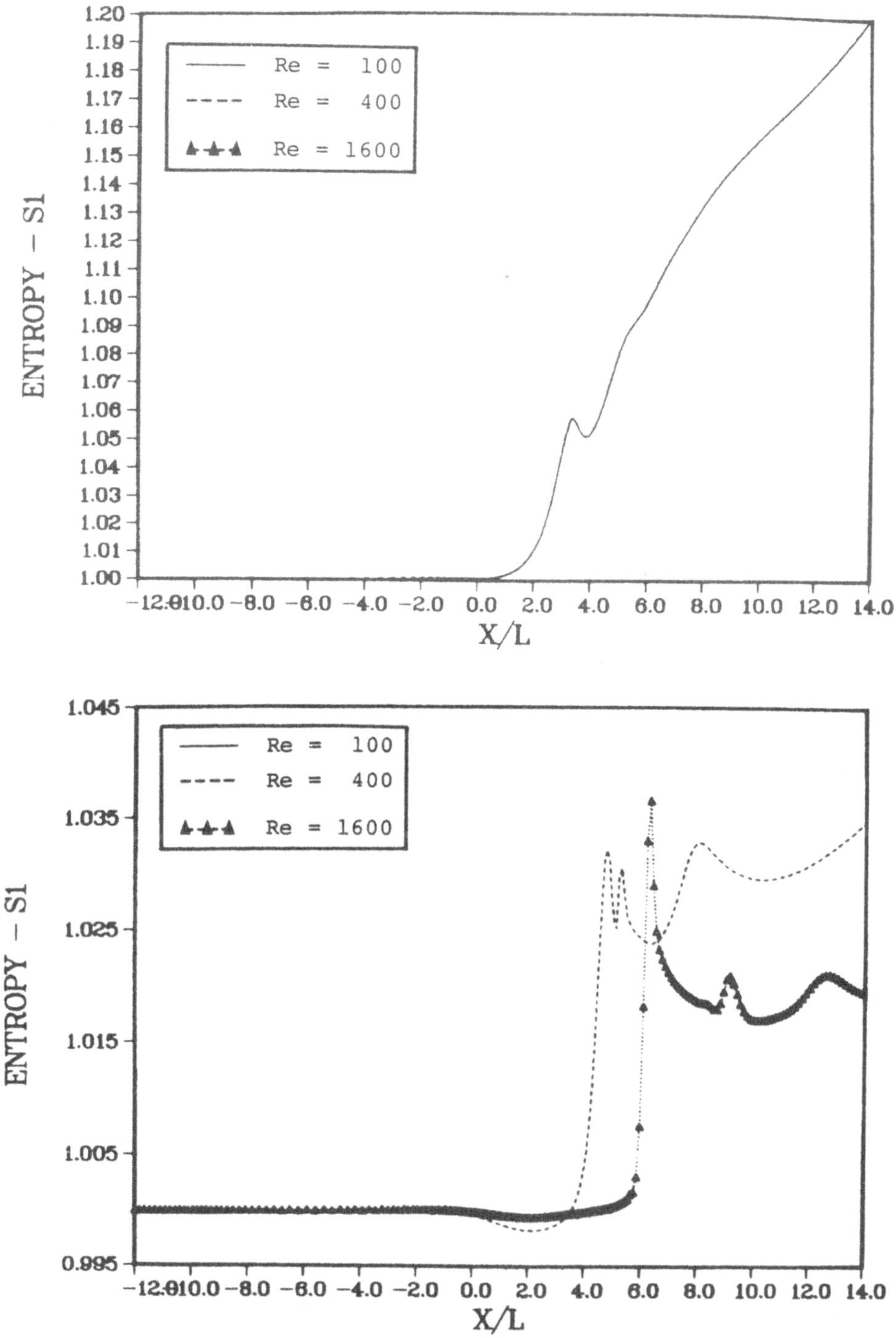

Fig. 9 Entropy variation on symmetry axis.

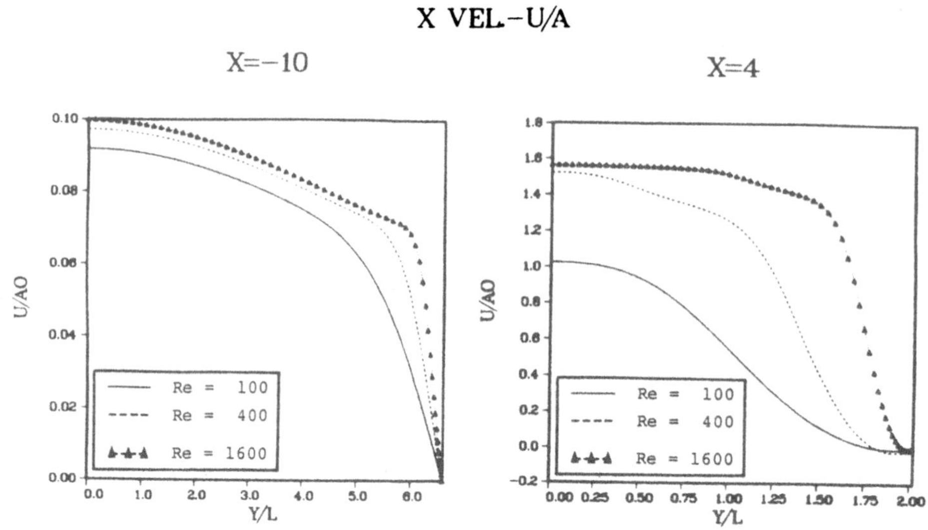

Fig. 10 x velocity profiles at x = -10 and x = 4.

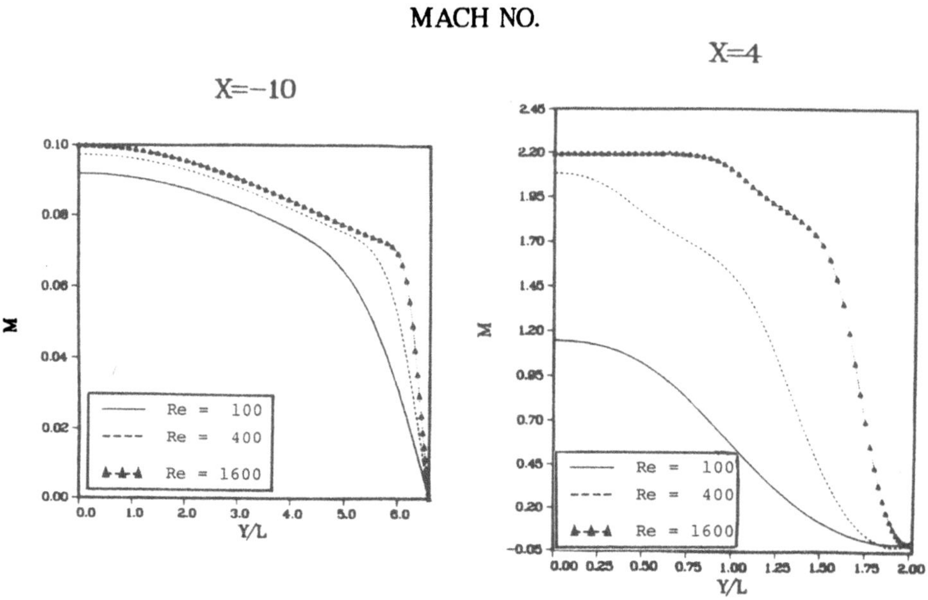

Fig. 11 Mach number profiles at x = -10 and x = 4.

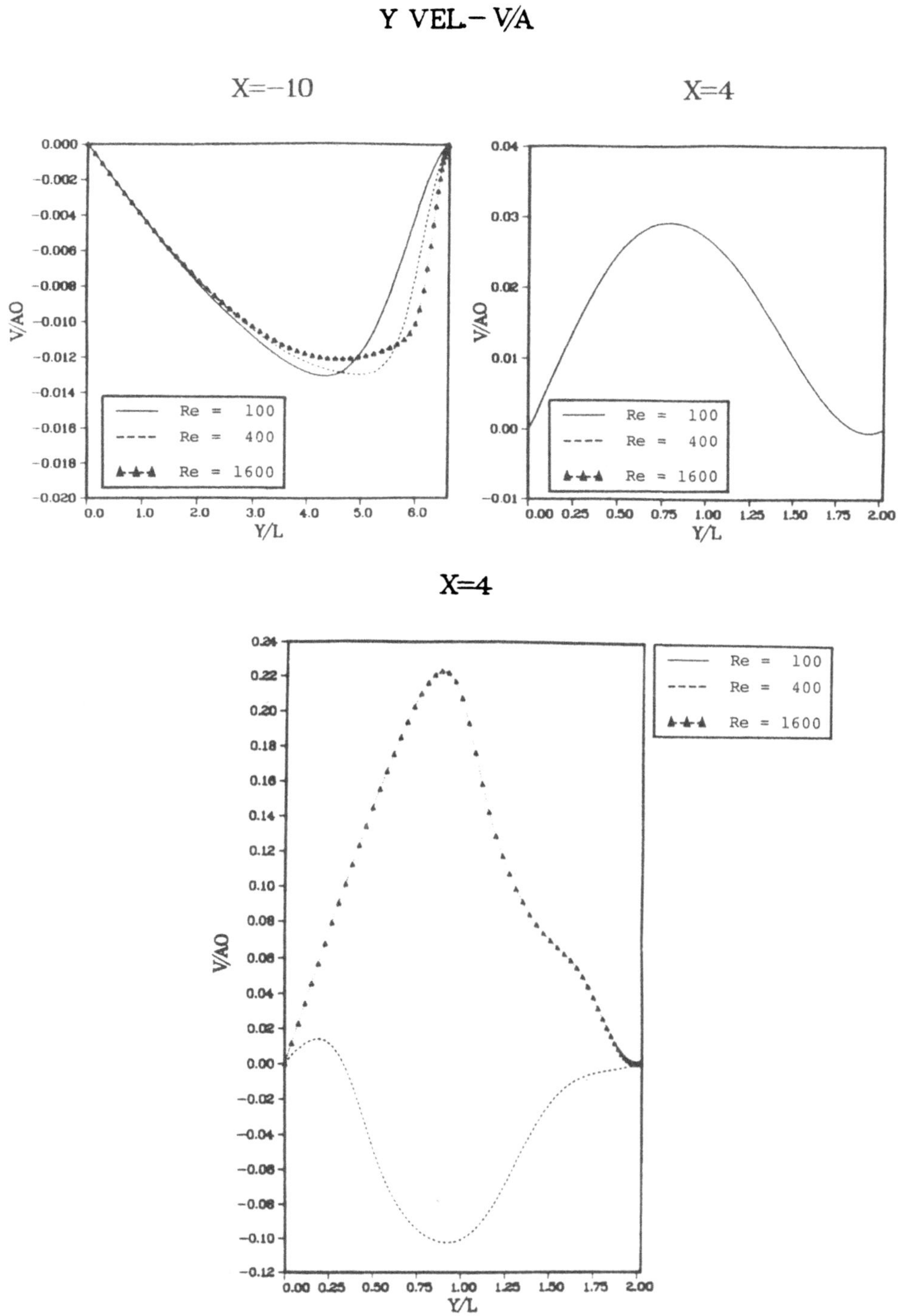

Fig. 12 y velocity profiles at x = -10 and x = 4.

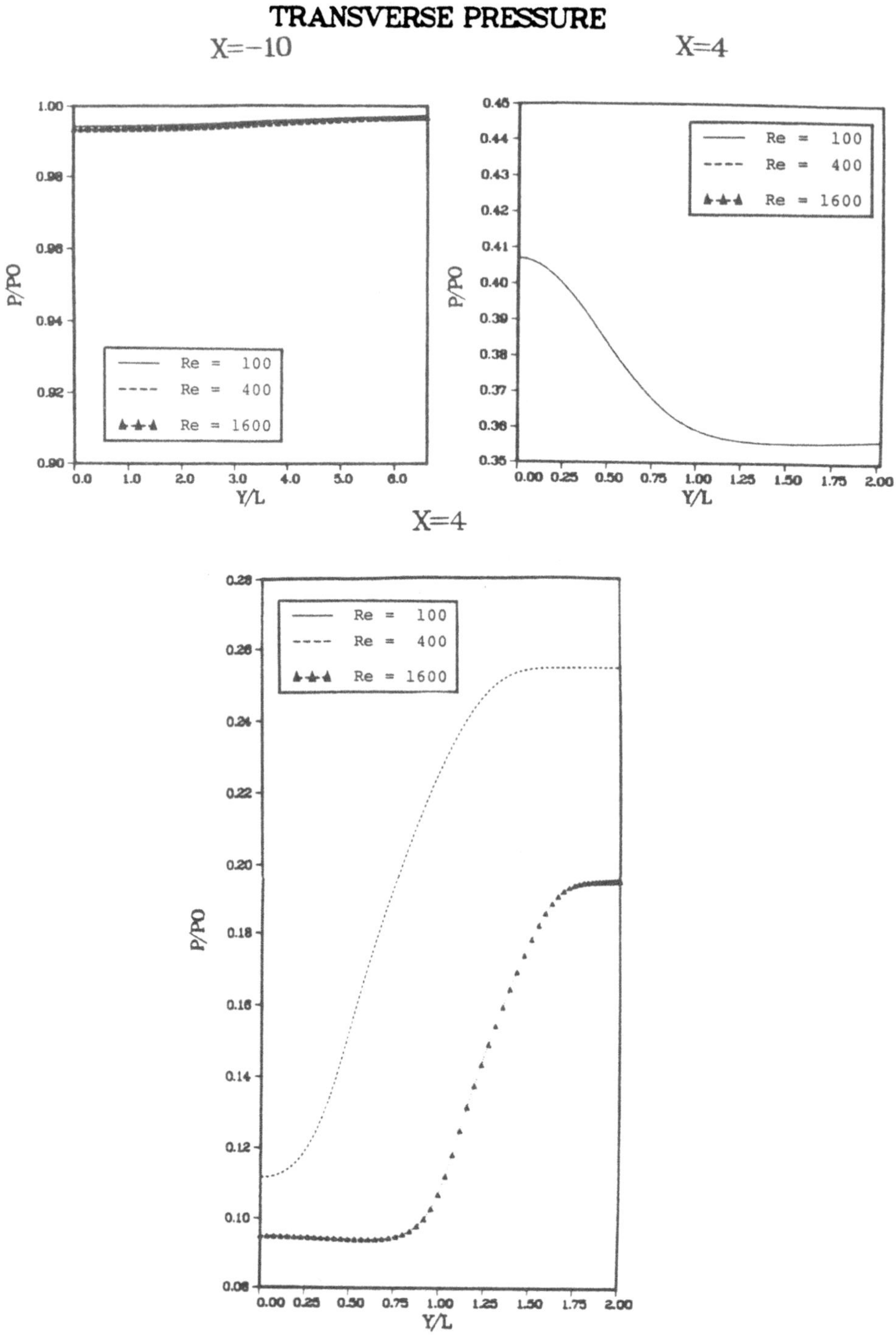

Fig. 13 Pressure profiles at x = -10 and x = 4.

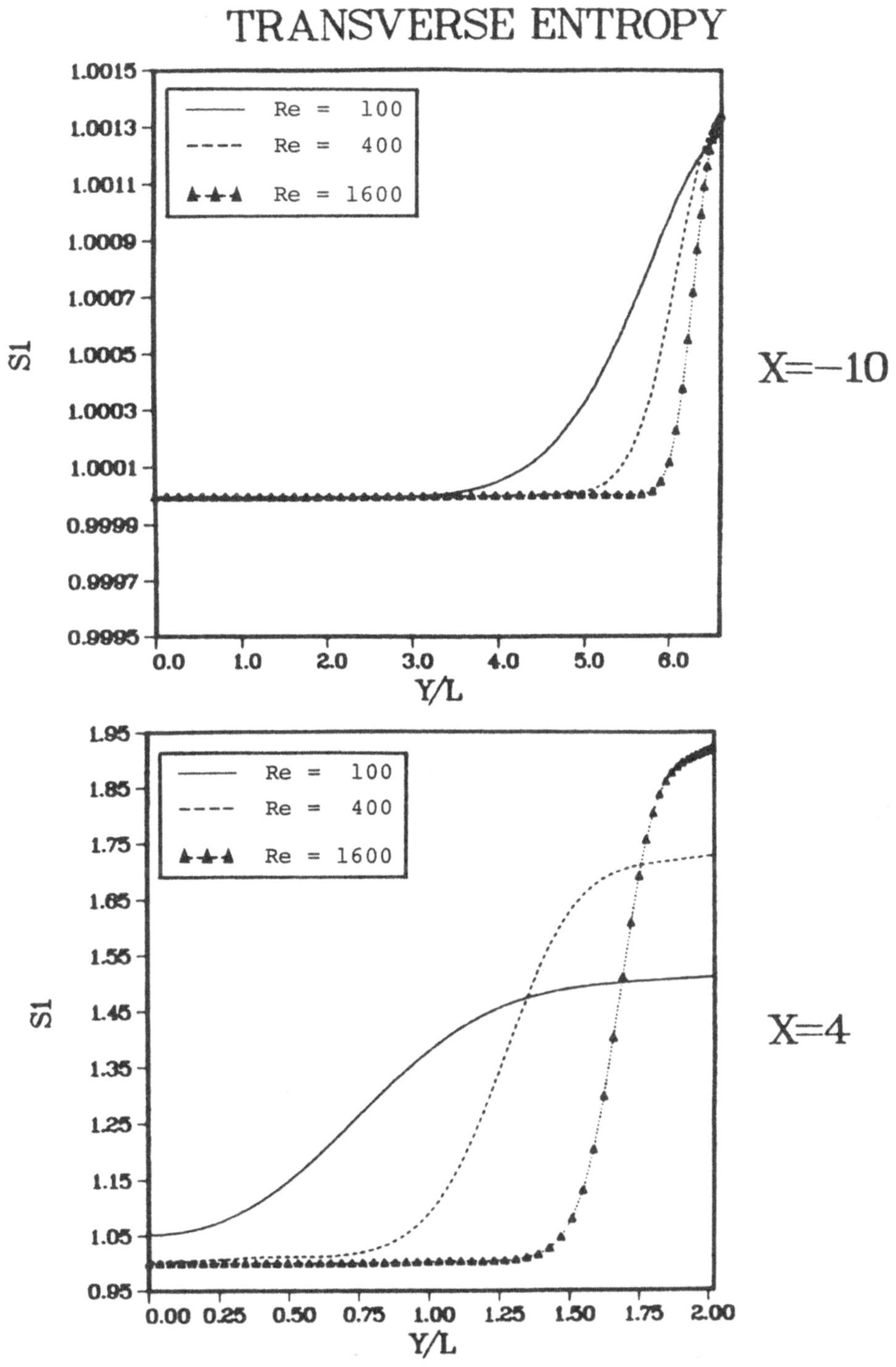

Fig. 14 Entropy profiles at x = -10 and x = 4.

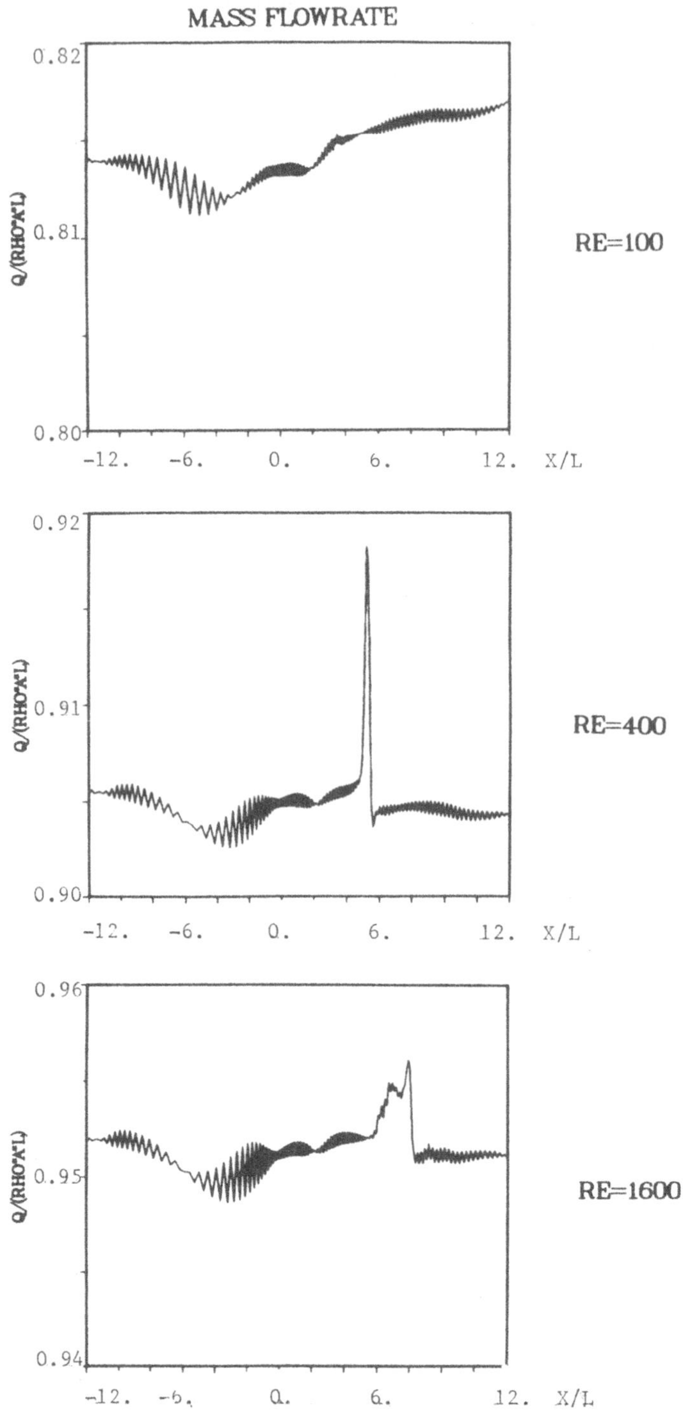

Fig. 15 Mass flow rate.

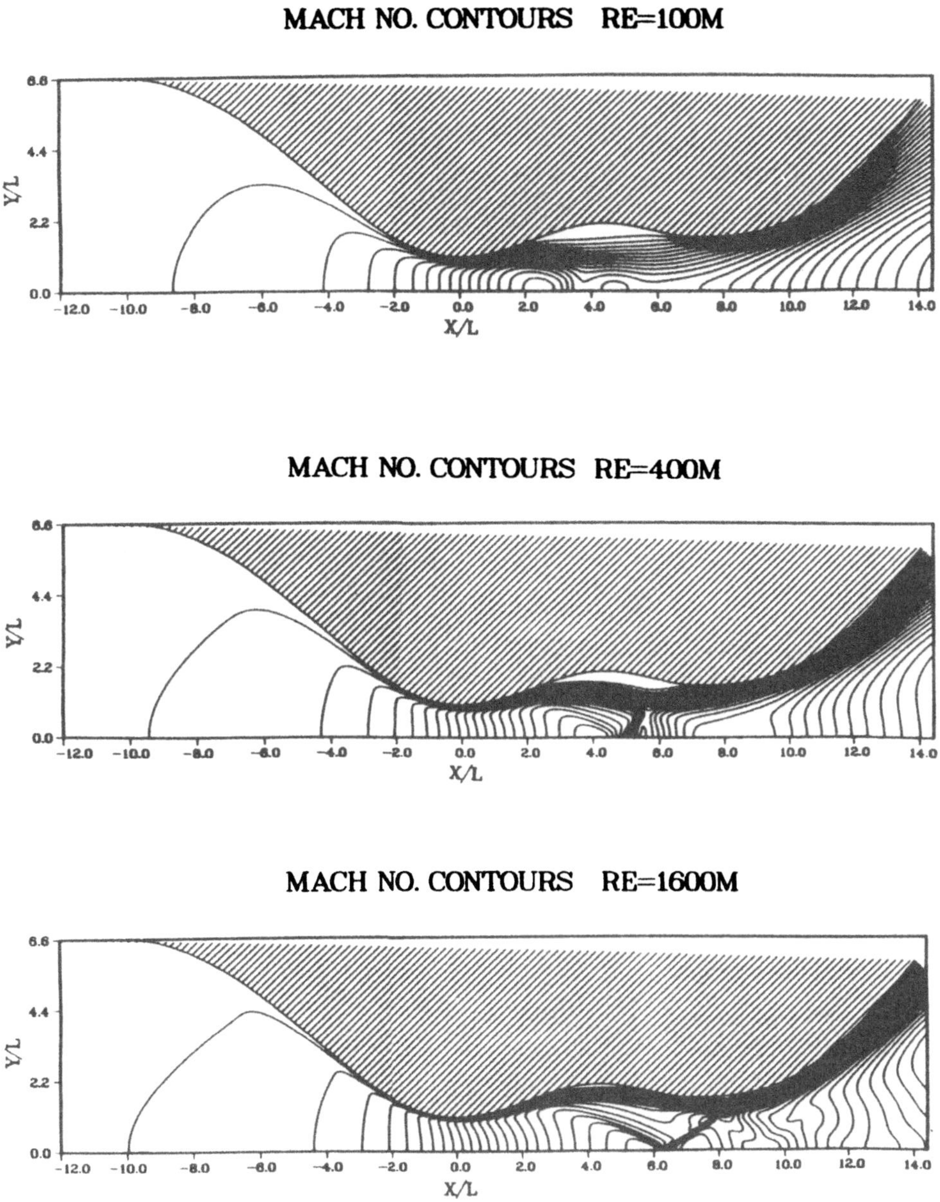

Fig. 16 Mach number contours.

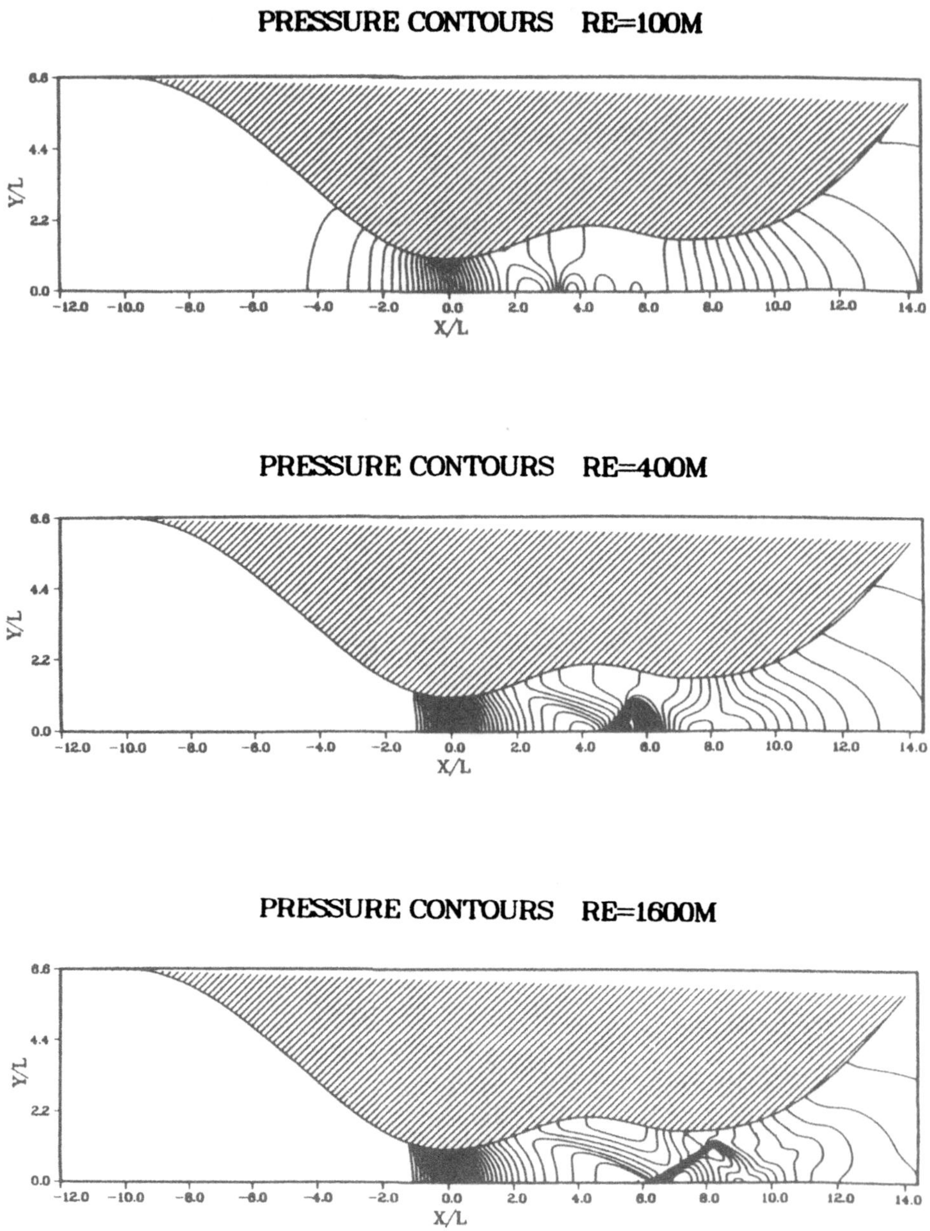

Fig. 17 Pressure contours.

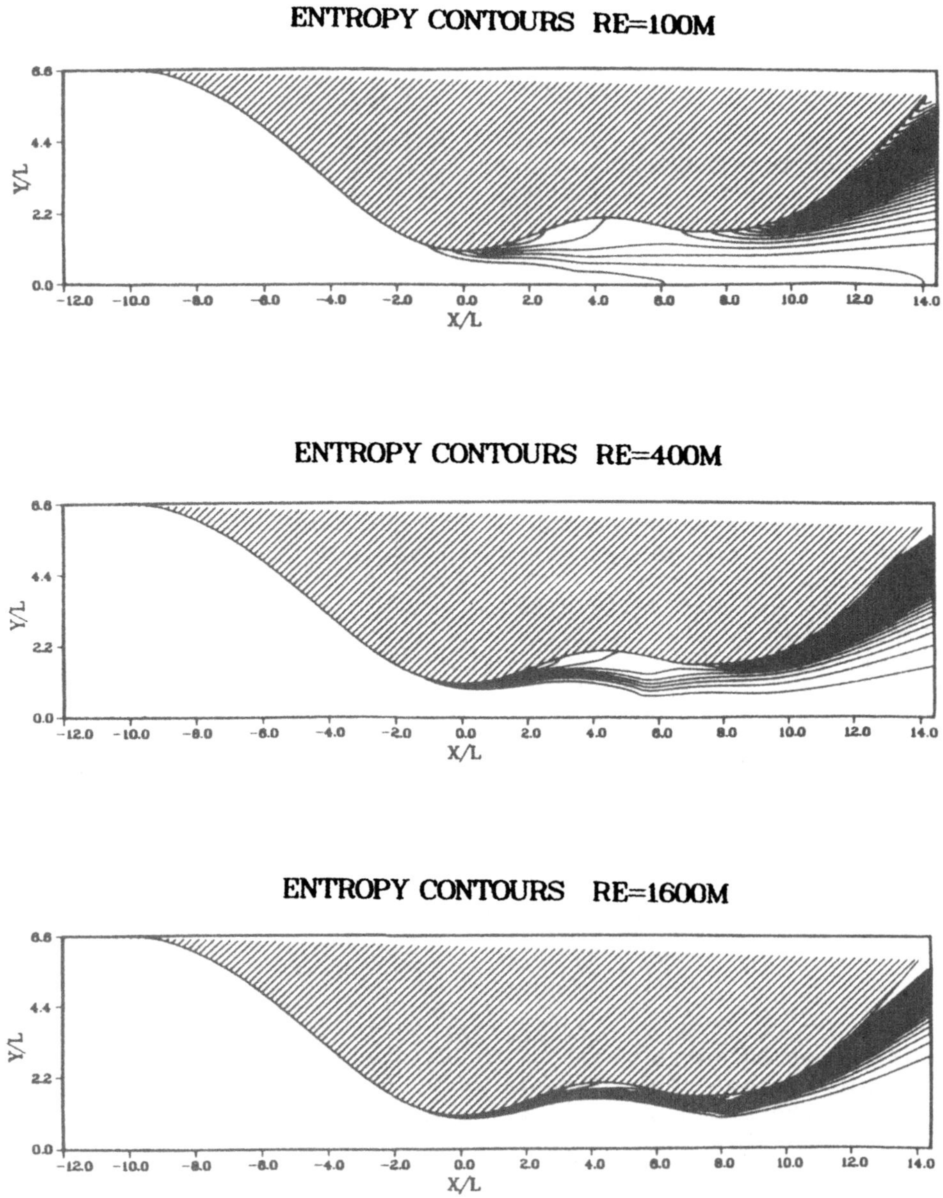

Fig. 18 Entropy contours.

AN IMPLICIT FLUX-SPLIT ALGORITHM FOR THE COMPRESSIBLE NAVIER-STOKES EQUATIONS

James L. Thomas
NASA Langley Research Center
Hampton, VA U.S.A.

Robert W. Walters
Virginia Polytechnic Institute and State University
Blacksburg, VA U.S.A.

Bram van Leer
Delft University of Technology
Delft, The Netherlands

Christopher L. Rumsey
NASA Langley Research Center
Hampton, VA U.S.A.

SUMMARY

An implicit upwind scheme for the compressible Navier-Stokes equations is described and applied to the internal flow in a dual throat nozzle. The method is second-order accurate spatially and naturally dissipative. Upwind differencing is implemented by flux splitting the convective and pressure terms; the elements of the shear stress tensor and heat flux vector are centrally differenced. A spatially-split approximate factorization method is used to obtain efficient steady-state solutions on the NASA Langley VPS-32 (CYBER 205) supercomputer.

GOVERNING EQUATIONS

The vector form of the complete time-dependent Navier-Stokes equations can be written in generalized coordinates as

$$\frac{\partial Q}{\partial t} + \frac{\partial F}{\partial \xi} + \frac{\partial G}{\partial \eta} = \frac{1}{Re} \left(\frac{\partial R}{\partial \xi} + \frac{\partial S}{\partial \eta}\right) \tag{1}$$

where $Q = [\rho, \rho u, \rho v, e]^T/J$ represents the standard form of the conserved variables and J is the Jacobian of the transformation. The elements of F, G, R, and S have appeared in the literature numerous times and will not be repeated here. The thin-layer form of the Navier-Stokes equations are obtained by neglecting the viscous terms in the streamwise (ξ) direction. The Euler equations are obtained with R and S set to zero. Though the governing equations are written in generalized coordinates, a finite volume formulation is employed so that the discrete equations are exactly satisfied for freestream flow on arbitrary grids. A brief synopsis of the algorithm follows; for a more complete description see [1] and [2].

SPATIAL DIFFERENCING

The spatial differencing of the Navier-Stokes equations is determined by considering the inviscid (convective and pressure) and viscous terms separately. The mixed hyperbolic/elliptic nature of the steady-state Euler equations has led to the advent of time-dependent upwind differencing schemes that closely model the propagation of waves in the solution domain. On the other hand, central differencing techniques are consistent with the diffusive nature of the shear stress and heat flux terms.

The inviscid fluxes, F and G, are split into forward and backward contribution ($F^{\pm}$ and $G^{\pm}$) based on the method developed by Van Leer [3] for Cartesian coordinates. The extension to generalized coordinates in multi-dimensions is developed in [4] and [5]. As an example of the differencing, the spatial gradient of the flux, F, is written as a flux difference across a cell, i.e.

$$\left(\frac{\partial F}{\partial \xi}\right)_j = \{[F^+(Q^-) + F^-(Q^+)]_{j+1/2} - [F^+(Q^-) + F^-(Q^+)]_{j-1/2}\}/\Delta\xi_j . \tag{2}$$

In the above, $F^{\pm}(Q^{\mp})$ denotes that the individual split flux contributions are evaluated with upwind-biased interpolations of Q to cell interfaces j+1/2. $Q^n_{j,k}$ denotes the average of Q at time t^n in the cell centered on (ξ_j, η_k); for simplicity, whenever the script notation is simply j, k, or n, it is most often dropped. The interpolation formula for the state quantities can be written as

$$Q^{\mp}_{j+1/2} = Q_j \pm \frac{s}{4}[(1 \mp \kappa s)\nabla_\xi + (1 \pm \kappa s)\Delta_\xi]Q_j . \tag{3}$$

The parameter κ determines the spatial accuracy of the difference approximation: $\kappa = -1$ corresponds to a fully-upwind second-order scheme and $\kappa = 1/3$ to an upwind-biased third-order scheme. All of the results presented below are obtained with $\kappa=1/3$. The operators Δ_ξ and ∇_ξ represent forward and backward difference operators, respectively, in the ξ-direction. The parameter s serves to limit higher-order terms in the interpolations to cell interfaces in order to avoid oscillations at discontinuities in the solutions, such as shock waves. The limiting is implemented by locally modifying the difference values in the interpolations to ensure monotone interpolation, as

$$s = \frac{2\Delta_\xi Q \; \nabla_\xi Q + \varepsilon}{(\Delta_\xi Q)^2 + (\nabla_\xi Q)^2 + \varepsilon} , \tag{4}$$

where ε is a small number ($\varepsilon = 10^{-6}$) preventing division by zero in regions of null gradients. The fully-upwind differencing ($\kappa = -1$) is the only technique that can capture weak transonic shocks in a monotone profile without limiting. Similar formulas as those above hold in the η-direction.

The viscous and heat flux terms are treated conservatively as differences across cell interfaces of first derivative terms, corresponding to second-order accurate central differencing. For example, the thin-layer approximations to the diffusion terms in Cartesian coordinates are differenced as

$$\frac{\partial \hat{S}}{\partial y} = \frac{1}{\Delta y} [\hat{S}_{k+1/2} - \hat{S}_{k-1/2}]$$

$$= \frac{1}{(\Delta y)^2} [\mu_{k+1/2}(T_{k+1} - T_k) - \mu_{k-1/2} (T_k - T_{k-1})], \tag{5}$$

where the vector T contains both velocity and temperature field data and μ is the molecular viscosity. In generalized coordinates, chain rule differentiation is applied to evaluate the first derivative terms required at cell interfaces.

An analysis of the preceeding differencing scheme when applied to the scalar model equation representing convection and diffusion

$$c \frac{\partial u}{\partial x} - \nu \frac{\partial^2 u}{\partial x^2} = 0, \quad c, \nu > 0 , \tag{6}$$

shows that the truncation error is

$$\frac{\Delta x^2}{12} [c(3\kappa-1) \frac{\partial^3 u}{\partial x^3} + \nu \frac{\partial^4 u}{\partial x^4}] + \frac{\Delta x^3}{8} c(1-\kappa) \frac{\partial^4 u}{\partial x^4} + O(\Delta x^4).$$

In addition to the dissipation arising from the diffusive term, a third-order dissipative term appears due to the upwind differencing of the convective term. Consequently, in the limit of $\nu \to 0$, the scheme remains naturally dissipative for any choice of $\kappa \neq 1$ (central differencing corresponds to $\kappa = 1$). Since the leading truncation error is $O(\Delta x)^2$, the scheme is second-order accurate in space, independent of the value of κ.

IMPLICIT TIME DIFFERENCING

The implicit, approximate factorization (AF) algorithm corresponds to a spatial factorization of the linearized, backward-time implicit integration scheme, as below

$$\left[\frac{I}{J\Delta t} + \partial_\xi\left(A - \frac{1}{Re}\frac{N}{J}\right)\right]\left[\frac{I}{J\Delta t}\right]^{-1}$$

$$\left[\frac{I}{J\Delta t} + \partial_\eta\left(B - \frac{1}{Re}\frac{M}{J}\right)\right] \Delta Q = -RHS \qquad (7)$$

where RHS denotes the discretized steady-state Navier-Stokes equations and $\Delta Q = Q^{n+1} - Q^n$. The Jacobian matrices A and B arise from linearization of the F and G fluxes, and M and N arise from the linearization of the viscous terms in the ξ- and η-directions, respectively. The implicit cross-derivative terms are neglected in the formulation and first order accurate derivatives of the implicit convective and pressure terms are used, so that the algorithm can be implemented as two block-tridiagonal sweeps through the mesh.

The computations were made using a spatially-varying time step, corresponding to a local Courant number of ten based on inviscid criterion. Since the algorithm is cast in delta form, the steady-state solution is independent of the time step. The algorithm is vectorizable on current supercomputers, with the minimum vector length corresponding to the number of rows or columns in the grid. By updating the implicit matrix terms only every twentieth iteration, roughly a factor of two reduction in the computational rate per iteration is effected at no loss in the observed convergence rate. The algorithm acquires 2.4 million words of 64-bit word requires storage for a grid of 11,000 points. The computational rate is 14 microseconds per grid point per iteration on the NASA Langley VPS-32 (CYBER 205) computer. Algorithms based on upwind relaxation [1],[6] have been developed which allow larger time steps to be taken, although the computational rate per iteration is less on the VPS-32 because of the scalar (recursive) operations required.

COMPUTATIONAL GRID AND BOUNDARY CONDITIONS

Computations for the dual throat nozzle have been made using the two grids shown in figure 1. Both grids contain 81 points in the transverse direction, clustered near the wall region. The 131 x 81 grid contains 131 equally spaced points in the streamwise direction, 10 of which are upstream of the specified nozzle inlet and are not shown. The 246 x 81 grid is clustered in the streamwise direction near the throat regions and does not extend upstream of the specified inlet. Computations were also made for a coarser mesh (124 x 41), obtained by deleting every other point from the 246 x 81 grid.

The boundary conditions for the dual throat nozzle problem are as follows. On the solid wall, no-slip conditions are applied and the wall temperature is held constant at the stagnation temperature. The additional thermodynamic

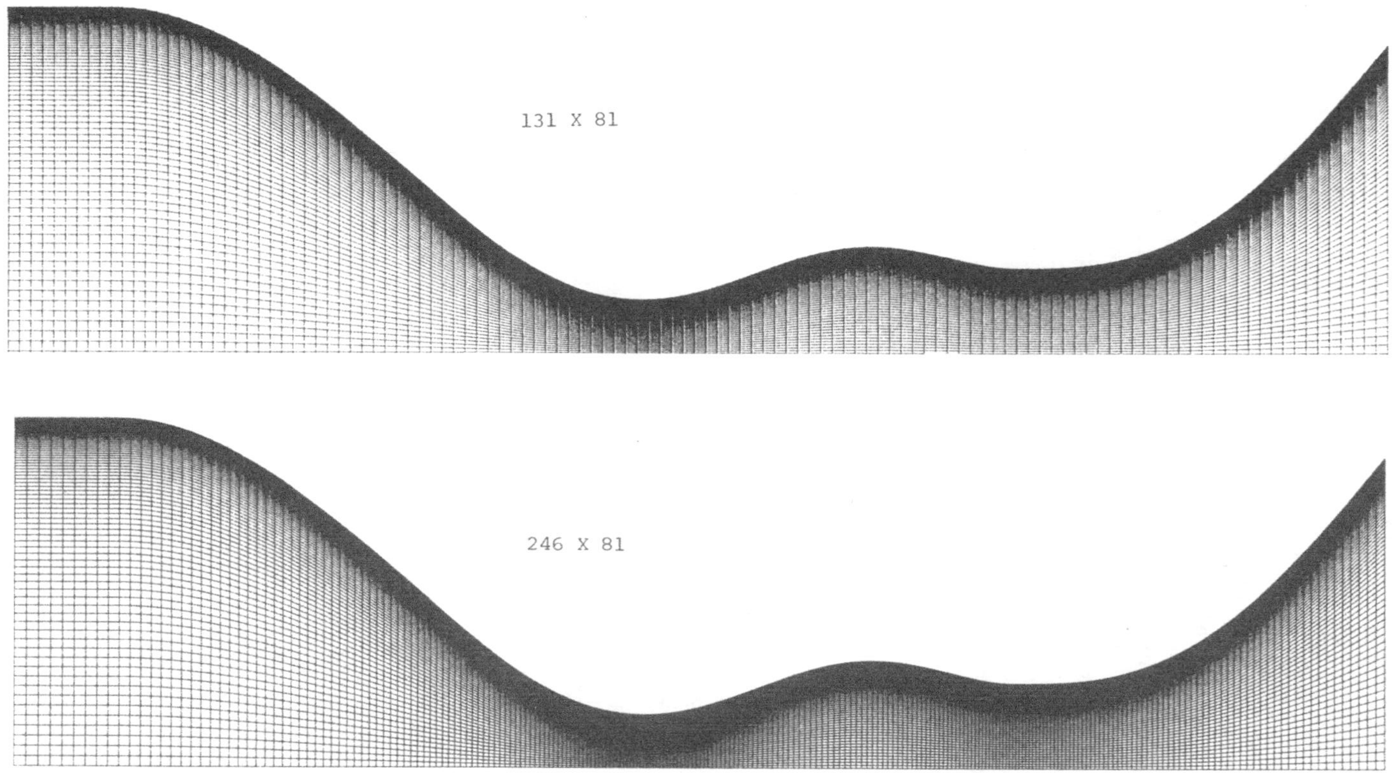

Figure 1. - Computational grids for the dual throat nozzle.

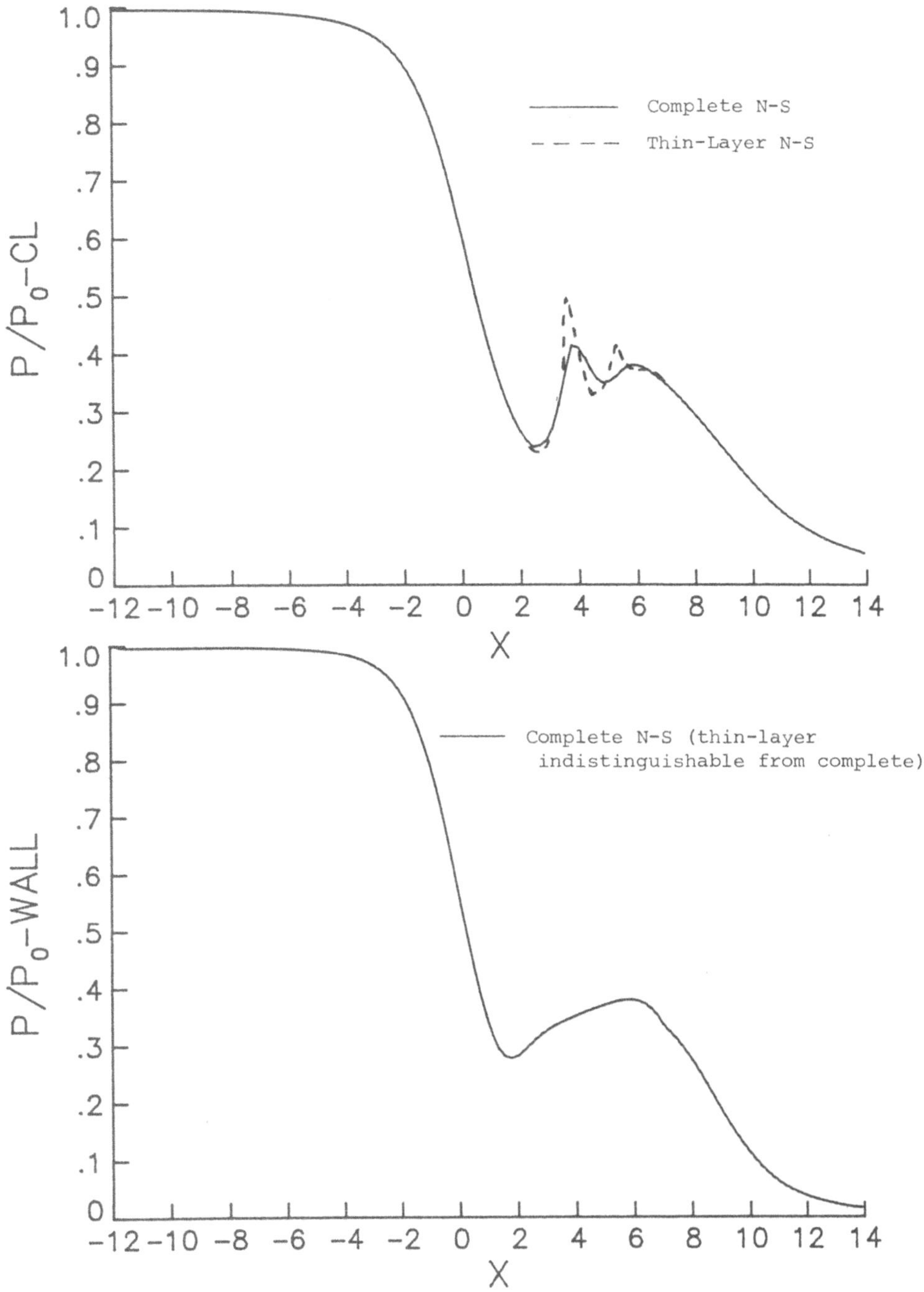

Figure 2. - Wall and centerline properties; Re = 100; 131 x 81 grid.

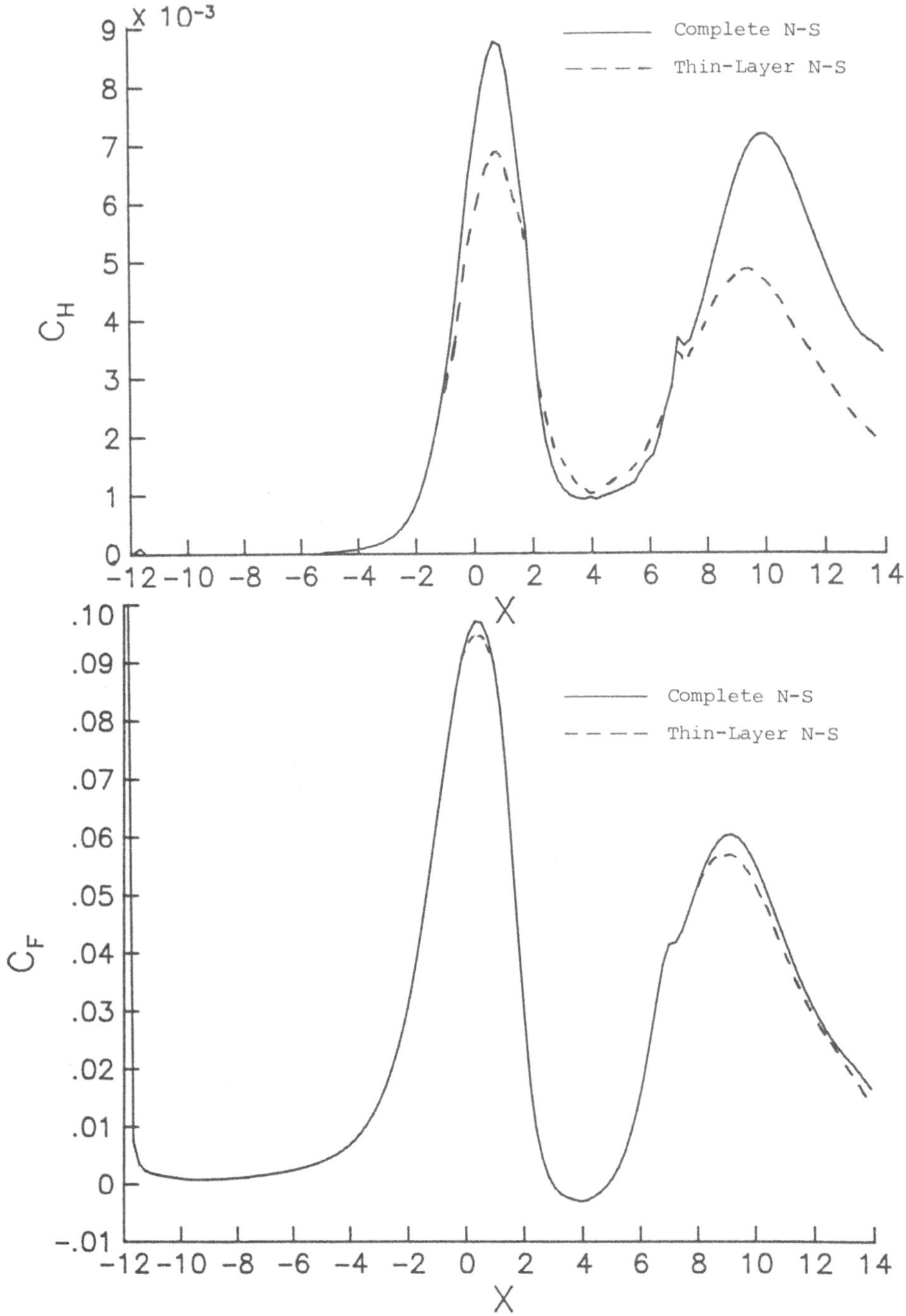

Figure 2. - Concluded.

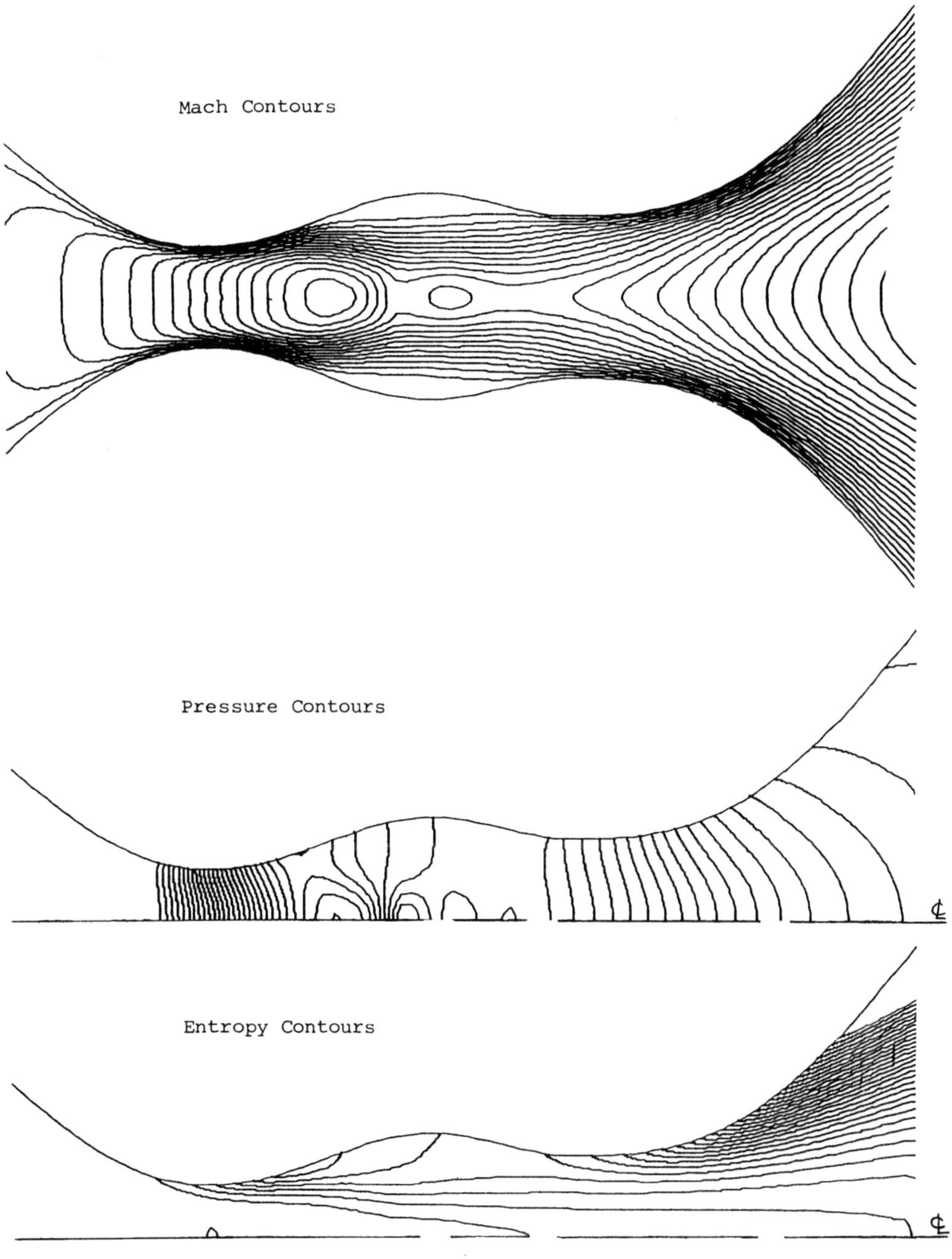

Figure 3. - Flowfield contours; Re = 100; 131 x 81 grid.

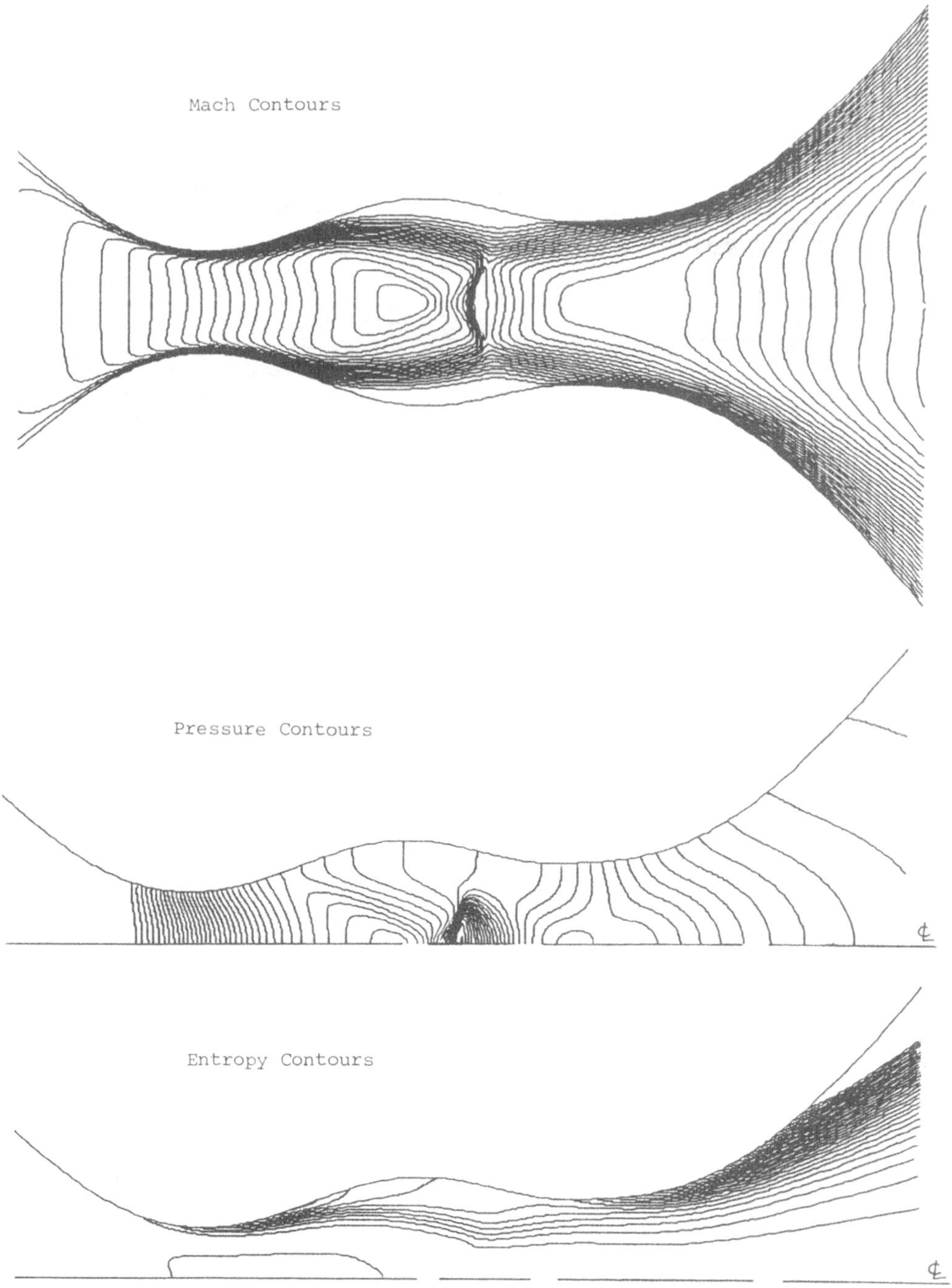

Figure 4. - Flowfield contours; Re = 400; 246 x 81 grid.

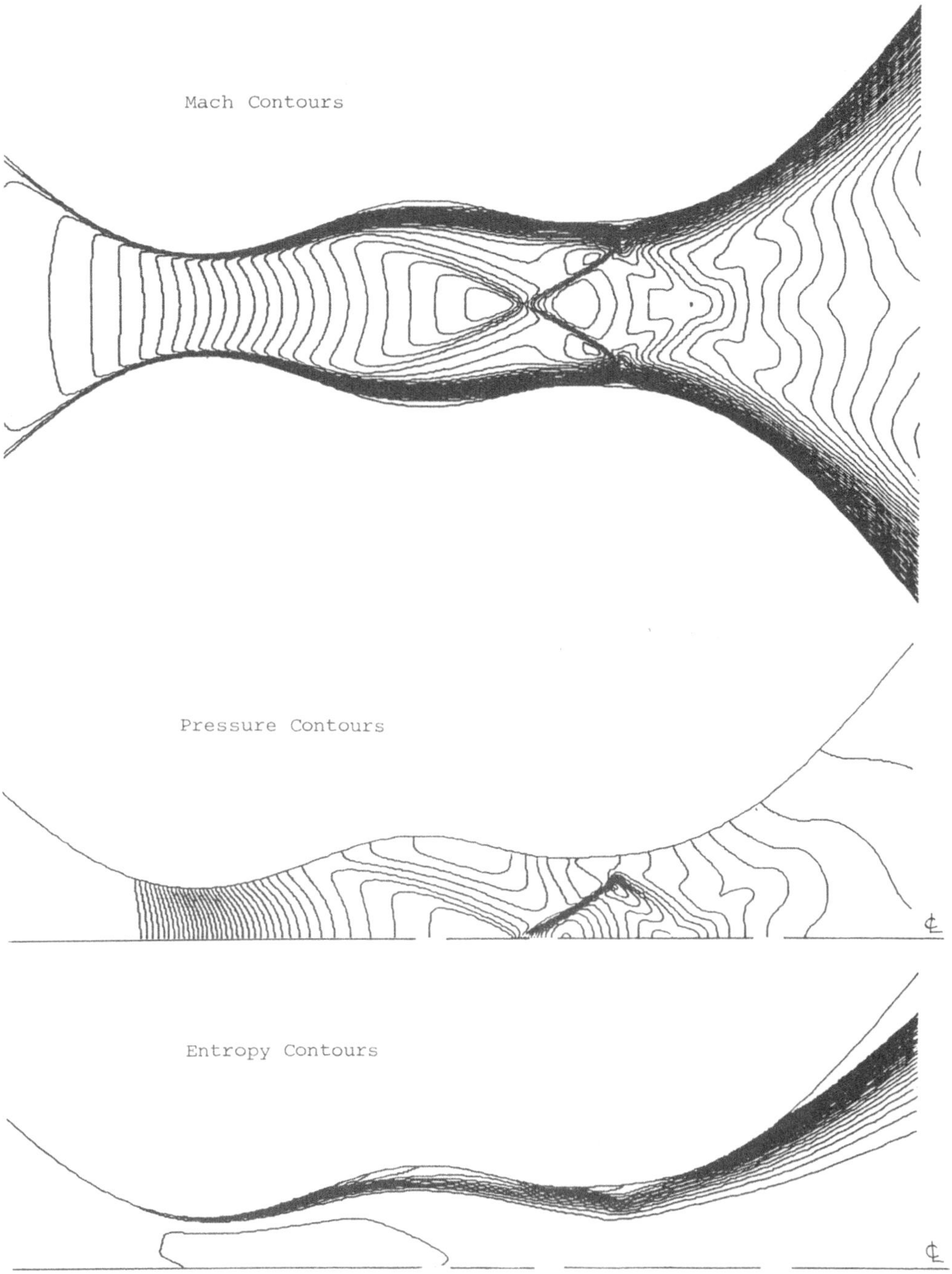

Figure 5. - Flowfield contours; Re = 1600; 246 x 81 grid.

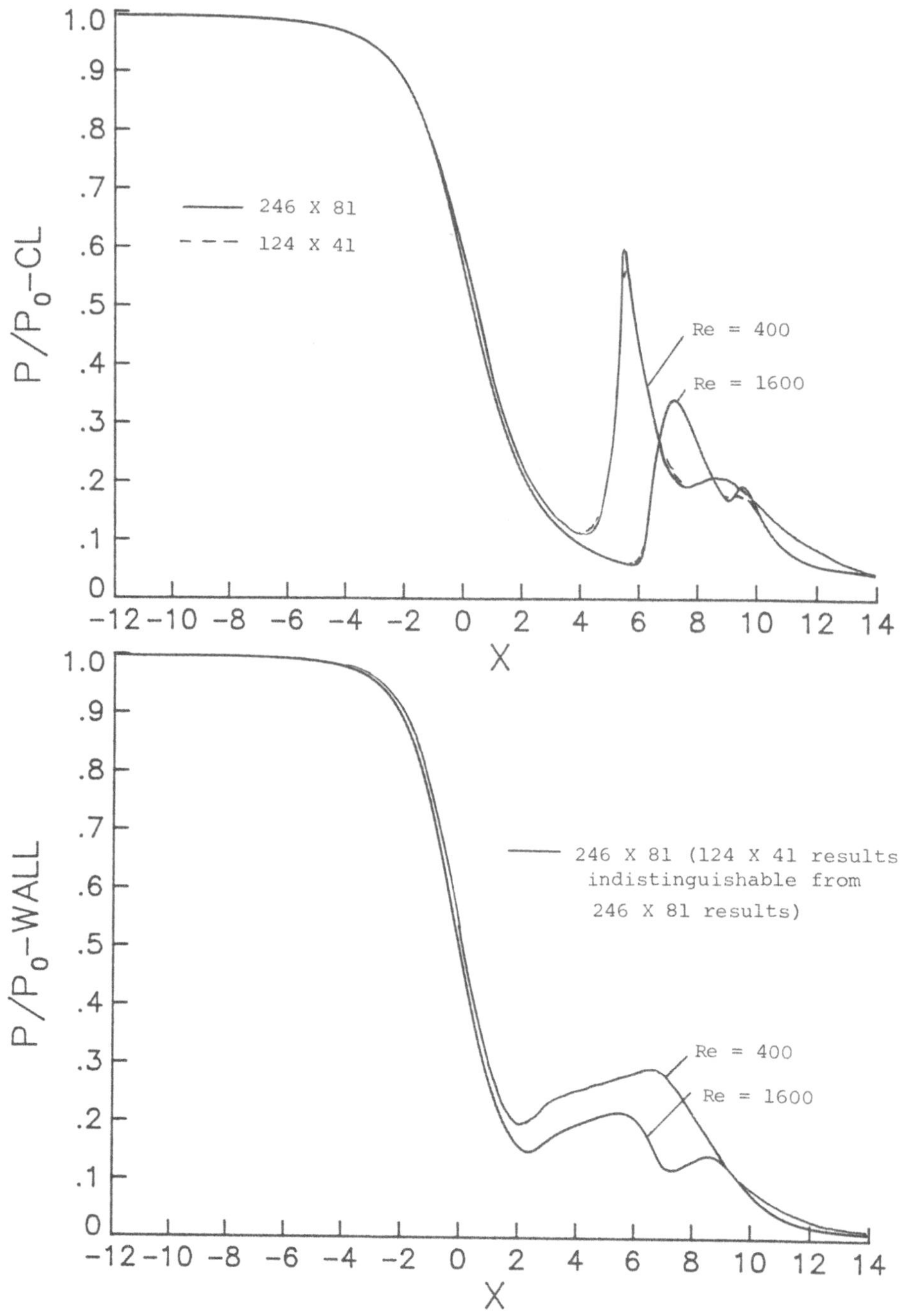

Figure 6. - Wall and centerline properties at Re = 400 and 1600.

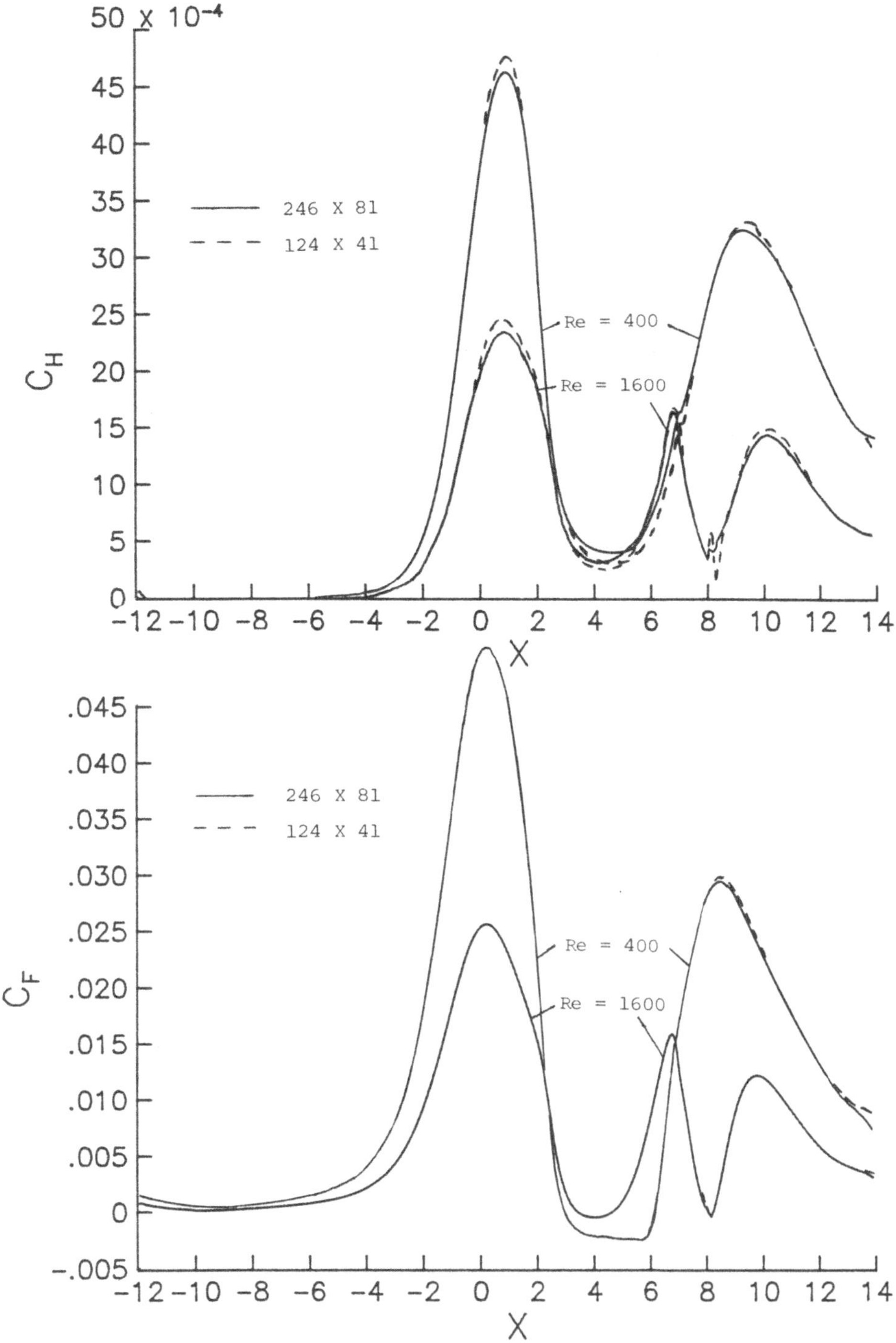

Figure 6. - Concluded.

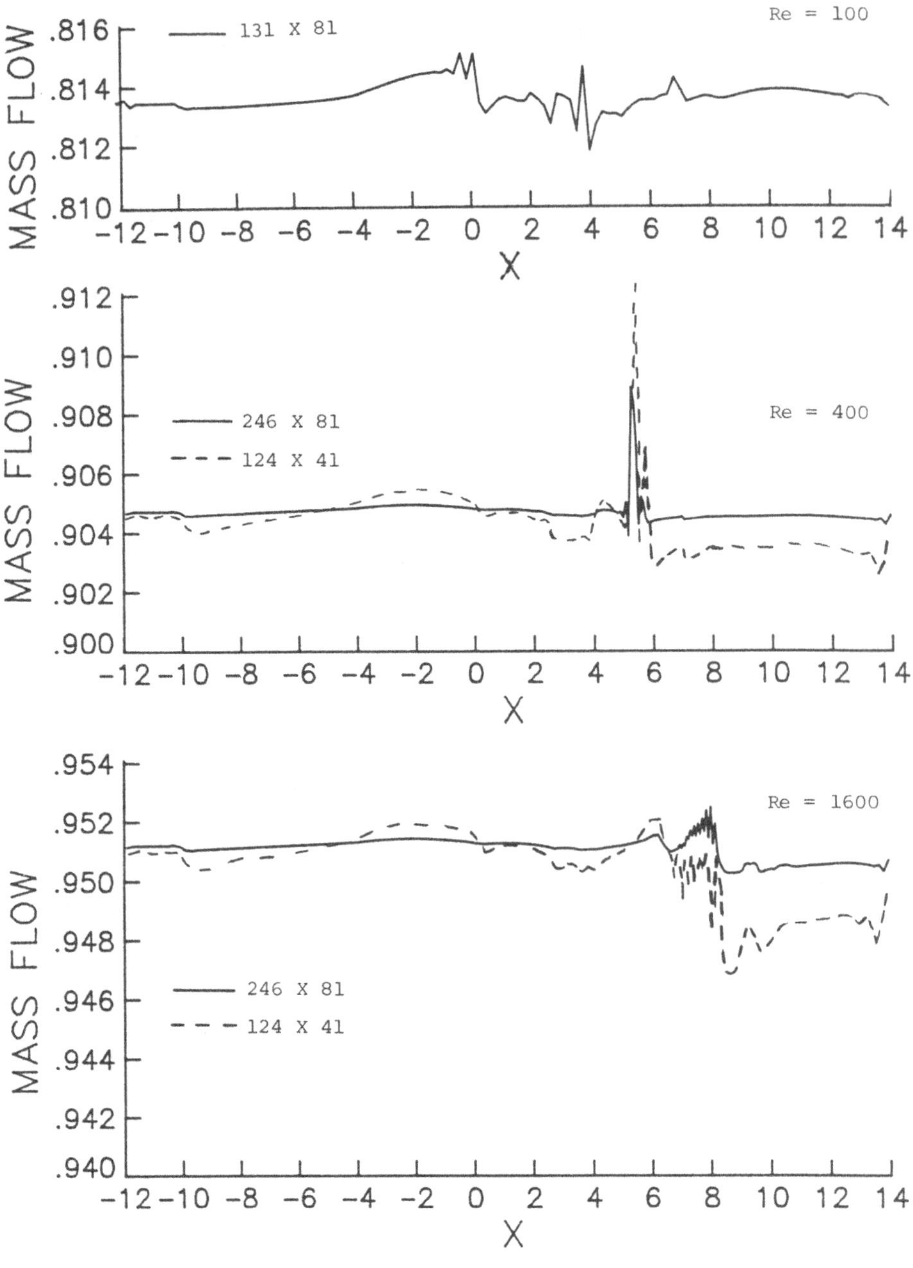

Figure 7. - Axial variation of mass flow rate.

constraint imposed on the solid surface is $\partial p/\partial \eta = 0$. At the downstream (outflow) boundary, the inviscid field is entirely supersonic and thus all state quantities, including the boundary-layer region near the wall, are extrapolated there.

Along the centerline, symmetry conditions were imposed, i.e., the normal gradients of ρ, u, and e were set to zero and $v = 0$. At the inflow boundary, the entropy and enthalpy were held constant at the stagnation conditions and the flow was assumed to be parallel ($v = 0$). Since the flow is subsonic at the inflow, one variation must be extrapolated from the interior. In this study, extrapolation of two different variables were tried, the static pressure, p, and the axial component of velocity, u. It was found that the solution was insensitive to this choice upon convergence, but extrapolating u did result in an improved rate of convergence over the static pressure extrapolation. A symmetry condition was applied in the region of the 131 x 81 grid which extended upstream of the specified inlet.

COMPUTATIONAL RESULTS

Computations with the 131 x 81 grid were made at all three Reynolds numbers using the complete form of the Navier-Stokes equations as well as the thin-layer form. The differences between the complete and thin-layer forms are evident primarily at the lowest Reynolds number, as could be expected, and are evidenced most noticeably in the centerline properties and heat transfer coefficients, as shown in figure 2. The standardized presentation of all results as set by the GAMM Committee is used throughout. The wall pressure and skin friction coefficients are insensitive to the thin-layer assumption. Flowfield contours at Re=100 using the complete Navier-Stokes equations are shown in figure 3. Flowfield contours using the thin-layer form are similar, except that a stronger shock is evident in the region between the throats. At the higher Reynolds numbers, the flowfield contours are insensitive to the thin-layer assumption; the only noticeable effect is a decrease in the maximum heat transfer coefficient.

At the two higher Reynolds numbers, computations using the complete Navier-Stokes equations were made with the grids clustered in the streamwise direction. Flowfield contours from the 246 x 81 grid at Re = 400 and Re = 1600 are shown in figures 4 and 5, respectively. The flow transitions from a normal shock pattern at Re = 400 to an oblique shock pattern at Re = 1600, the latter pattern more similar to the inviscid flow results for this nozzle shown in [6]. The results shown are very similar to results obtained on the mesh with a uniform axial distribution of points. The clustered mesh leads to improved resolution in the region of the second throat. Wall and centerline properties obtained on the 124 x 41 and 246 x 81 meshes are shown in figure 6; the small differences between results on the two meshes indicate the fine grid results are spatially converged.

The axial variation of the mass flow through the nozzle is shown in figure 7. Variations on the order of the truncation error are shown. The longitudinal extent of the separated flow regions at all three Reynolds numbers is shown in the table below.

Table 1 Axial extent of separation.

Re	Axial extent of separation	Grid
100	$3.004 < x < 4.820$	131 x 81
1400	$2.986 < x < 6.221$	246 x 81
1600	$3.432 < x < 4.709$ and $8.077 < x < 8.216$	246 x 81

The convergence of the algorithm to the solution of the steady-state equations is illustrated in figures 8 and 9. The convergence of the fine grid (246 x 81) starting from freestream conditions corresponding to a Mach number of 0.10 is shown in figure 8. The L_2 norm of the residual of all four equations remains nearly constant until the mass flow is established, after which the residual decreases rapidly. Solutions at other Reynolds numbers can be obtained within 1000 iterations (2-3 minutes) if restarted from a converged solution, as illustrated in figure 9 for the 131 x 81 mesh, since the solution is for practical purposes converged after a two order of magnitude decrease in the residual.

REFERENCES

[1] THOMAS, J. L., and WALTERS, R. W.: "Upwind Relaxation Algorithms for the Navier-Stokes Equations" in Proceedings of the AIAA 7th Computational Fluid Dynamics Conference, AIAA 85-1501 CP. July 1985.

[2] RUMSEY, C. L., THOMAS, J. L., WARREN, G. P., and LIU, G. C.: "Upwind Navier-Stokes Solutions for Separated Periodic Flows," AIAA 86-0247, January 1986.

[3] VAN LEER, B.: "Flux-Vector Splitting for the Euler Equations," ICASE Report No. 82-30, September 1982; also: Lecture Notes in Physics, Vol. 170, 1982, pp. 507-512.

[4] ANDERSON, W. K., THOMAS, J. L., and VAN LEER, B.: "A Comparison of Finite Volume Flux Vector Splittings for the Euler Equations," AIAA 85-0122, January 1985.

[5] THOMAS, J. L., VAN LEER, B., and WALTERS, R. W.: "Implicit Flux-Split Schemes for the Euler Equations." AIAA 85-1680, July 1985.

[6] WALTERS, R. W., and DWOYER, D. L.: "An Efficient Strategy Based on Upwind/Relaxation Schemes for the Euler Equations," Proceedings of the AIAA 7th Computational Fluid Dynamics Conference, AIAA 85-1501 CP, July 1985.

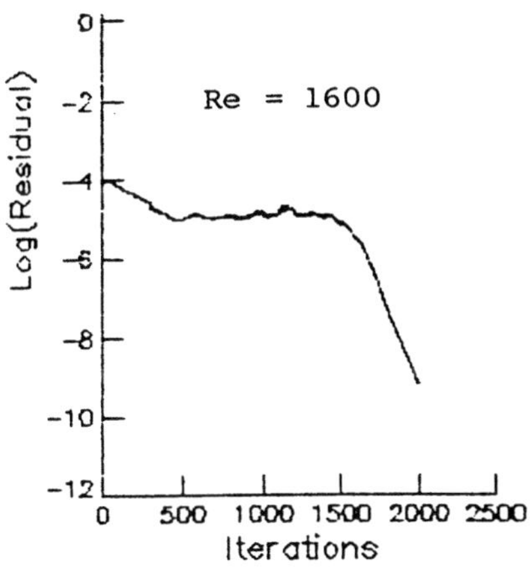

Figure 8. - Convergence of the fine (246 x 81) mesh starting from freestream initialization.

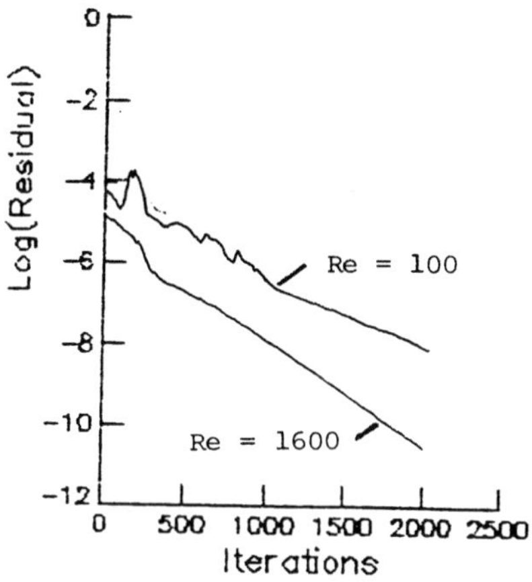

Figure 9. - Convergence of the 131 x 81 mesh starting from converged solution at Re = 400.

Addresses of the editors of the series „Notes on Numerical Fluid Mechanics":

Prof. Dr. *Ernst Heinrich Hirschel* (general editor)
Herzog-Heinrich-Weg 6
D-8011 Zorneding
FRG

Prof. Dr. *Keith William Morton*
Oxford University Computing Laboratory
Numerical Analysis Group
8-11 Keble Road
Oxford OX1 3QD
Great Britain

Prof. Dr. *Earll M. Murman*
Department of Aeronautics and Astronautics
Massachusetts Institut of Technology (M.I.T.)
Cambridge, MA 02139
U.S.A.

Prof. Dr. *Maurizio Pandolfi*
Dipartimento di Ingegneria Aeronautica e Spaziale
Polytecnico di Torino
Corso Duca Degli Abruzzi, 24
I-10129 Torino
Italy

Prof. Dr. *Arthur Rizzi*
FFA Stockholm
Box 11021
S-16111 Bromma 11
Sweden

Dr. *Bernard Roux*
Institut de Mécanique des Fluides
Laboratoire Associé au C.R.N.S. LA 03
1, Rue Honnorat
F-13003 Marseille
France

GPSR Compliance

The European Union's (EU) General Product Safety Regulation (GPSR) is a set of rules that requires consumer products to be safe and our obligations to ensure this.

If you have any concerns about our products, you can contact us on ProductSafety@springernature.com

In case Publisher is established outside the EU, the EU authorized representative is:

Springer Nature Customer Service Center GmbH
Europaplatz 3
69115 Heidelberg, Germany

Batch number: 10584556

Printed by Printforce, the Netherlands